EFFICIENCY AND EQUITY OF CLIMATE CHANGE POLICY

FONDAZIONE ENI ENRICO MATTEI (FEEM) SERIES ON ECONOMICS, ENERGY AND ENVIRONMENT

This series serves as an outlet for the main results of FEEM's research programmes in the areas of economics, energy and environment.

The titles published in this series are listed at the end of this volume.

Efficiency and Equity of Climate Change Policy

Edited by

CARLO CARRARO

University of Venice, Italy
and Fondazione Eni Enrico Mattei, Milan, Italy

KLUWER ACADEMIC PUBLISHERS
DORDRECHT / BOSTON / LONDON

Library of Congress Cataloging-in-Publication data is available.

ISBN 0-7923-6262-4

Published by Kluwer Academic Publishers
PO Box 17, 3300 AA Dordrecht, The Netherlands

Sold and distributed in North, Central and South America
by Kluwer Academic Publishers,
101 Philip Drive, Norwell, MA 02018, USA

In all other countries, sold and distributed
by Kluwer Academic Publishers,
PO Box 322, 3300 AH Dordrecht, The Netherlands

Printed on acid-free paper

Printed and bound in Great Britain by MPG Books Ltd, Bodmin, Cornwall

Table of Contents

Preface

The climate change problem has been framed by natural scientists concerned about CO_2 and other greenhouse gas emissions affecting the atmosphere and the global climate. The issue rose rapidly on the political agenda in the late 1980s and early 1990s, reaching a summit in the Kyoto Protocol agreed upon in 1997.

Implementation of the Climate Change Convention and the emission targets of the Kyoto Protocol is an unprecedented challenge in international policies, not least because the societal aspects, such as costing issues (efficiency) and distributional issues (equity), have not received enough attention.

Through this book, the European Forum on Integrated Environmental Assessment (EFIEA) wishes to contribute to a better understanding of these costing and distributional issues. During the EFIEA workshop, hosted by the Fondazione Eni Enrico Mattei in Milan, the most recent findings of the scientific community on efficiency and equity issues were elaborated. They are presented in a coherent way in this book.

The perspectives illustrated in this book will help policy makers, stakeholders and other interested parties to understand the complexities involved. The authors also indicate pathways and promising next steps to reach the goal of the Climate Change Convention: avoiding dangerous anthropogenic interference with the climate system.

I want to extend special thanks to the workshop organiser and editor of this book, Carlo Carraro. His work is exemplary of the goals EFIEA pursues: improving the science–policy interface for the benefit of the environment.

Pier Vellinga

Chairman of the European Forum on Integrated Environmental Assessment

CHAPTER 1

Efficiency and Equity of Climate Change Policy: an Introduction

Carlo Carraro

The Kyoto Protocol has raised almost as many questions as it has solved, and scientific and political arguments have tended to become intertwined in the resultant controversies. In particular, if the necessity of short-term climate policies is often debated, the optimal policy to be implemented is even more controversial. Both efficiency and equity of different policy proposals are currently being discussed, without achieving the sufficient consensus which would be necessary to adequately implement the Kyoto Protocol. One of the goals of the European Forum on Integrated Environmental Assessment (hereafter referred to as EFIEA) is to assess how integrated assessment approaches can help to identify the relevant policy questions on the one hand, and consensus answers to these questions on the other hand. The second EFIEA Policy Workshop[1] was specifically devoted to highlighting policy questions and answers related to climate issues by gathering representatives from science and economics, governments and NGOs, business and the EU Commission.

The main objective of the workshop was to analyse the costs and benefits of climate change policies from an integrated assessment perspective, addressing equity criteria in costs assessment, as well as identifying the mutual benefits and opportunities of international climate policies. The workshop focused on the analysis of climate change mitigation policies with the aim of providing an assessment of the decisions taken in Kyoto and Buenos Aires, and of identifying the critical issues to be addressed and tackled in the near future.

The two-and-a-half-day workshop was structured in three main sessions:

1. *Costs: Fairness, Equity, Burden-Sharing*, which addressed the international, intergenerational and social equity issues emerging from the distribution of costs and long-term benefits related to climate change mitigation and adaptation policies.
2. *Opportunities and Mutual Benefits*, which aimed at exploring the dynamics of the interaction of climate policies with different policy tools in a few relevant sectors, identifying the costs as well as the direct and secondary

C. Carraro (ed.), Efficiency and Equity of Climate Change Policy, 1–13.

benefits involved in mitigation and adaptation strategies. The session mainly focused on the energy and transport sectors, including urban air policies.
3. *Institutions and Policies after Buenos Aires*, which offered blue-prints for a climate policy, focusing on the integration of environmental policies with industrial and trade policies, as well as on the design of appropriate institutions and norms.

A worldwide call for papers was launched nine months before the workshop in order to collect the best and most recent research analyses on climate issues. In order to avoid a large number of scientific presentations to give time to policy discussions, scientific papers were circulated but not presented. The contents of the scientific papers were summarized by five surveys: 'International Equity in Climate Change Policy', presented by Bert Metz, 'Intergenerational Equity and Discounting', presented by Ferenc Toth, 'Energy and Climate Policies', by Alessandro Lanza, 'Transport, Urban Air Policies and Climate Change Policy', by Stef Proost, and 'Blue-Prints for a Climate Policy', by Frank Convery. All surveys are published in a special issue of *Integrated Assessment*.

The workshop gathered academics, policy makers, and representatives of the industry in various sectors, as well as Non-Governmental Organizations' and environmental associations' representatives. A lively discussion took place during the three days of the workshop. The outcomes of this discussion, as well as a synthesis of the main ideas and concerns expressed in the workshop, through both written reports and panel debates, is contained in the last chapter of this book. In this last chapter, Carlo Carraro and Alessandra Goria outline the relevant policy questions which emerged from the scientific surveys' presentations and the debates, identify those issues which met with a general consensus and those which generated conflicting views, and finally highlight the new policy perspectives addressed and recommended.

It is worth noticing that during the workshop, despite the presence of different, sometimes conflicting stakeholders, some general ideas were supported by all parties. For example, the link between equity and efficiency of climate agreements was clearly identified and emphasized. On the one hand, an equitable agreement induces more countries to commit to emissions control, thus enhancing its effectiveness; on the other hand, a cost-effective agreement reduces the conflict on distributional issues. Another example is the importance of using the so-called Kyoto flexibility mechanisms to reduce the costs of compliance. These ideas, and the related policy perspectives, are also at the heart of most chapters of this book.

In particular, this book is aimed at providing the best available scientific and economic evidence on the links between efficiency and equity of climate policy. The book gathers 14 papers submitted for discussion at the workshop. These papers were selected because of their high quality and their relevance with respect to the workshop objectives. They certainly provide a good overview of the problems and conflicts that characterize the current debate on climate policies and on the implementation of the Kyoto Protocol in particular.

The book is divided into three parts. Part 1 is devoted to an assessment of the Kyoto Protocol. Part 2 explores possible ways of increasing the efficiency of measures designed to achieve the Kyoto targets. Part 3 analyses the equity implications of different policy options.

The three chapters of Part 1 assess the costs and benefits of the Protocol within both a static and a dynamic perspective. In 'Benefits and Costs of Kyoto', Jason Shogren provides an overview of the most contentious issues of the Kyoto Protocol from the perspective of US policy makers, constrained to act as arbiters of public opinion. He stresses that in the present state of research, weighing up the costs and benefits of Kyoto very much depends on 'what you choose to believe'. The issues represent both an intellectual and a political challenge. There are a great number of uncertainties regarding how to assess the costs of climate change, the risk of a climatic catastrophe, the structure of the market in emissions permits (which will effectively constitute a new international currency – the 'carbo'), the real effectiveness of carbon sinks such as forestry plantations (particularly important in the US case) and the potential for technological advance (which might allow countries to reduce their emissions without huge economic sacrifices). He emphasizes that the Kyoto Protocol will fail to achieve its aims unless it is successful in undergoing a shift from an initially 'deep' to a 'broad' approach, by bringing developing nations on board. Implementing the Protocol therefore requires mechanisms which involve developing countries in the process: "Think of trying to turn a battleship on a dime with a third of the crew on board".

In their chapter, Manne and Richels also recognize that approval of the Kyoto Protocol in the US will depend on its costs and that costs crucially depend on the participation of developing countries. They use the MERGE model with a general equilibrium formulation of the global economy to examine the costs of compliance in the short term, the role of 'flexibility provisions' in this, and whether the Kyoto Protocol is cost-effective in achieving its long-term goals. They show that the same benefits can be produced at lower cost when there are minimal limits on emissions trading and maximum participation by countries in emissions markets. Although the risk of 'carbon leakage' is manageable, they note the potential for 'unfair competition' from non-Annex I countries, which might profit from cheap energy to damage the industries of Annex I countries. Since Kyoto in itself will fail to stabilize global emissions, they also conclude that developing nations must be involved, with the introduction of global emissions trading schemes as soon as possible. Surprisingly, while Kyoto is more effective at reducing emissions in the short term, over the long term it is better to adopt the most cost-effective mitigation strategy from the outset. This would discourage developing countries from investing in inefficient technologies which would later have to be abandoned if any progress is to be made towards reducing global emissions.

A critical assessment of the Kyoto Protocol is also provided by Jean-Charles Hourcade and Thierry Le Pesant, even if from a quite different, I would say

'European', perspective. They trace the history of the development of the Kyoto Protocol and the political controversies over the agreement. They argue that the nature of the climate threat facing the world in fact favours the adoption of an internationally co-ordinated pricing mechanism. By contrast at Kyoto, political pressure from the US led to the adoption of a quota system where each country adopts its own strategies to reduce emissions. Each agent may end up facing very different costs – part of the reason for controversy and mutual suspicion. In fact, the difficulty of measuring compliance with the quotas has stirred up a hornets' nest of scientific/political controversies over various issues; for example, how to measure the effectiveness of carbon sinks. Hourcade and Le Pesant examine the case for caps on emissions trading, which can be seen as an attempt by Europe to prevent the US from entirely importing its emissions quotas from countries with excessive 'hot air' allocations. However, they find that this leads to excessively high costs for little extra benefit, and agree that costs must be contained as much as possible in order to encourage further adhesion to the protocol. Flexibility mechanisms may well reduce short-term abatement costs and therefore prove attractive to governments preoccupied with their political survival, but precisely because they permit continued investment in energy-inefficient capital stocks, they tend to foster solutions that prove more expensive in the long term. This may exacerbate the likelihood that such countries will find it politically impossible to comply in the future. From a dynamic inconsistency perspective, they argue for the need to strike a balance in the initial phase between excessively high carbon prices (which would discourage signatory countries from extending their commitment or other countries from joining the protocol) and excessively low carbon prices (which would fail to get countries onto the optimal long-term emissions reduction path). Like Manne and Richels, they emphasize that in the long term, the optimum reductions path requires early commitment to avoid costly investment in energy-inefficient capital stocks with low turnover. They also identify the risks of short-term thinking: ambitious reductions may be achieved rapidly in ways that may be of little value in terms of the long-term goal.

The second part of the book is devoted to a deeper exploration of the role of emissions trading. The existence of emissions quotas as agreed upon in Kyoto is taken as given. The six chapters of Part 2 try to analyse under which conditions those quotas can be achieved at the lowest cost. New proposals on the way in which emissions trading should be implemented are put forward. Applications to specific markets are explored. Indeed, an international emissions trading system is to a large extent an experimental undertaking, and a careful analysis of market design and implementation is necessary.

This analysis is provided, among others, by Peter Bohm, who focuses on the various policy options at domestic and international levels for the practical implementation of an emissions trading system. He argues that emissions trading should be extended to as many countries as possible in order to reduce costs and to limit carbon leakage, an implicit risk in any non-global approach to regulation. He also examines the case for allowing nations to borrow emissions quotas from their future allocation, arguing that this additional flexibility might make countries more willing to accept stringent controls.

One of the crucial points in domestic strategies is whether emitters, retailers or producers should have to obtain permits. Bohm argues that the impracticality of requiring individual emitters (such as motorists) to obtain permits means that permit liability is best moved 'upstream', either to the level of retailers, or better still to the level of the producers, importers and exporters of fossil fuels, which can pass the additional costs onto the consumer through pricing.

Bohm discusses whether initial permit allocations should be auctioned to firms or distributed for free on the basis of historic emissions levels ('grandfathering'). Grandfathering tends to favour already established firms, blocking the entry of new firms onto the market. Another contentious issue is that if one country gives away permits for free, its industries will have a competitive advantage compared to a country which auctions permits or which uses a carbon tax. However, Bohm argues that this effect would be compensated for by the additional efficiency benefits resulting from these alternative mechanisms, so long as the additional government revenue is used to reduce existing distortionary taxes and so provide a 'double dividend'. Bohm argues that carbon taxes and permit allocation are in fact equivalent in economic terms – for example if permits are distributed for free to certain firms via grandfathering, the equivalent carbon tax would be one which redistributed revenue to those same firms. One important difference is that carbon taxes may need to be adjusted over time in order to meet the national target, so leading to higher costs than the permit system.

Bohm argues that, even if permit systems are generally designed to integrate with the international quota system, allowing firms to trade on the world market, this does not necessarily make them the best solution: governments may in fact be as well informed as firms in taking trading decisions. Bohm defends the 'hot air' allocations of excess quotas assigned to certain countries, especially Russia and Ukraine, arguing that this was a politically necessary solution which brought these countries into the agreement, reduced the possibilities of leakage and may have encouraged other Annex B countries to accept tighter emissions quotas themselves. Bohm argues that trading must take place in an open exchange, rather than allowing bilateral trading, in order to prevent one or two large trader countries from controlling the market. Finally he criticises the 'supplementarity' case for caps on trading, arguing that this leads to higher costs for all and reduces the likelihood of obtaining the crucial political agreement of the US.

The issue of the caps on trading is explored more deeply in the chapter by Paolo Buonanno, Carlo Carraro, Efrem Castelnuovo and Marzio Galeotti. In order to give a chance for caps on trading to be welfare increasing, they adopt a model in which technical change is endogenous. In particular, both factor-augmenting technical change and emissions-reducing technical change are endogenous. In this way, it is possible to explore the validity of the argument previously raised by Hourcade and Le Pesant, namely that caps (ceilings) on emissions trading, by increasing the costs of complying with the Kyoto Protocol, foster technological innovation and the adoption of new, less emissions-intensive capital stock. This effect provides a stimulus to economic growth that may more than offset the highest abatement costs.

Unfortunately, the analysis of Buonanno *et al.* does not seem to support the conclusion of Hourcade and Le Pesant. The chapter is divided into three parts. The first identifies the new effects arising from the endogenization of technical change; the second assesses the pros and cons of the introduction of ceilings to emissions trading; the third quantifies the distributional effects of climate policy scenarios. The chapter shows that the costs of complying with the Kyoto Protocol are lower when environmental technical change is endogenous, even in the presence of emissions trading. Moreover, the introduction of endogenous technical change leads to an equilibrium in which the demand for permits is higher, but supply is also higher, because developing countries invest in R&D to increase the amount of permits they can sell in the market. As a consequence of these two effects, the price of permits is lower. As for ceilings, the analysis provides little support in favour of quantitative restrictions on emissions trading. Even if the introduction of ceilings increases the R&D efforts of buyer countries and fosters technological innovation, the overall effect on abatement costs and economic growth is negative. The reason is that the benefits from technological innovation are lower, even in the long run, than the costs of adopting a more expensive approach to climate change control. Finally, even equity is not positively affected by ceilings. Buonanno *et al.* show that flexibility mechanisms in the presence of endogenous technical change increase equity and that the highest equity levels are achieved without ceilings, at least in the long run.

The role of technical change is central also in Johan Albrecht's chapter. Albrecht explains that one way to reduce emissions is not to depend merely on carbon taxes but on the potential of new technology to reduce emissions during product *use*. He argues that existing approaches adhere too closely to existing SO_2 abatement models, whereas the range of strategies for CO_2 abatement are quite different. In the transport sector, which accounts for 25% of global CO_2 emissions, most of the emissions result from the use of vehicles rather than from their production. Empirical evidence shows that high energy taxes actually have little impact on emissions from the transport sector due to low energy price elasticities: more attention should therefore be paid to instruments which promote greater energy efficiency in the use of cars and trucks. His study models the introduction of 'tradable certificates', an allowance used to reward manufacturers for each tonne of CO_2 avoided by selling more efficient vehicles. The model shows that with this incentive, there is strong potential for significant emissions reductions over a brief time period.

Another important aspect of the implementation of the Kyoto Protocol is the possibility of early crediting of emissions reduction. This issue is analysed in the chapter by Axel Michaelowa and Marcus Stronzik. The first commitment period of the Kyoto Protocol begins in 2008, and at present there are no incentives for countries to introduce regulations in advance of this date (except in the case of the Clean Development Mechanisms, where credits accrue from 2000). In the case of individual companies for example, there is little incentive to carry out early emissions reduction since they would face the risk of being set a lower baseline when crediting begins. Michaelowa and Stronzik discuss alternative mechanisms to allow for early crediting of reductions, including early Joint Implementation and the issue of 'hot air', and early domestic emissions reduction. Proponents of such mechanisms argue that they will help to smooth out emissions reductions, but Michaelowa and Stronzik conclude that such mechanisms only make sense if accompanied by incentives for innovation (discounting low-technology projects) which take care to avoid leakage. However, they do regard such mechanisms as potentially valuable learning experiences with market-based instruments.

The theme of burden sharing is one that is essential to obtain widespread compliance with the Kyoto Protocol. In the case of the EU, Chris Böhringer, Jesper Jensen and Tom Rutherford examine the economic effects of the 'bubble' which sets an emissions target for the region as a whole, allowing distribution of the burden between member states. They use a recursively dynamic multi-sector model of six member states to examine the effects of the carbon tax. Their results show that any distributive system is likely to be politically controversial: the 'bubble' mechanism adopted at Kyoto implies significant economic transfers from Northern to Southern Europe, since the high-emitting Northern European countries face the highest carbon bills, and nations like Spain may actually benefit as their industry becomes relatively more competitive. By contrast, while uniform abatement targets would have a negligible effect on costs for the EU as a whole, they would penalize low-emitting states, such as Spain and France. Their work also illustrates the significant role that fuel shifting will have to play in reducing the costs of compliance, particularly for Northern Europe. If the changes of the optimistic 'Common Wisdom' scenario do not materialize, the costs of Kyoto could be much higher.

Nielsen and Jeppesen argue that the EU needs a co-ordinated policy to promote renewable energy use if it is to meet the Kyoto targets. Along the lines of a Dutch scheme introduced in 1998, they propose 'Green Certificates' as a cost-efficient mechanism for the electricity sector. The approach is similar to the flexibility mechanisms of Kyoto in that it combines the setting of targets by central planning with market mechanisms to ensure cost-efficiency. Another reason for the co-ordinated approach is that any mechanism must avoid distortions in the internal European electricity market (beginning in 1999): since renewable energy is at present more expensive than conventional sources, producers are unlikely to be able to make the necessary investments and remain competitive at the same time.

In this system, consumers and distributors must be able to demonstrate that they have purchased a fixed share of electricity derived from renewable sources. The certificates are awarded by state authorities to producers according to the amount of renewable energy produced. Producers are then able to recoup extra costs of generating renewable energy by selling these certificates on to consumers. The market mechanism means that renewable energy is produced in those areas which have a comparative advantage (for climatic or other reasons). Naturally, an efficiently implemented CO_2 emissions trading mechanism would obviate the need for green certificates, and also avoid interfering with national sovereignty. Nevertheless, green certificates may prove to be an effective way for a region such as the EU to regulate a specific sector. Also, the use of market mechanisms may prove to be less controversial than trying to find a political settlement to distribute abatement costs among EU nations, as in the case of the 'bubble'.

Previous chapters have mainly addressed different aspects of climate policy from an efficiency point of view (one exception is the chapter by Buonanno *et al.*). However, it is widely recognized that equity issues are also very important and, as previously said, they can hardly be separated from efficiency ones. This is why the third part of the book gathers papers which explicitly deal with both international and intergenerational equity issues.

In Adam Rose's and Brandt Stevens' chapter, 'A Dynamic Analysis of Fairness in Global Warming Policy: Kyoto, Buenos Aires and Beyond', a detailed overview of international and intergenerational equity of climate agreements is provided. Political controversy has prevented further progress from being made since the Kyoto agreement of 1997. An equitable balance will need to be struck: developed industrial countries cannot be expected to shoulder the entire burden of emissions reduction, as developing world emissions are likely to overtake them in a few decades. On the other hand, developing countries lack the surplus to divert resources from existing development and presently account for relatively little of the world's CO_2 emissions. Rose and Stevens argue that emissions trading fills the need for a system that is both cost-effective and fair. However, in the search to find a politically acceptable solution, it is important to assess the resultant costs and benefits to individual nations under different trading mechanisms. Rose and Stevens use a dynamic non-linear programming model to evaluate the impact of permit trading across time and space, examining what happens when they change the number of countries participating and the timing of their participation. They also examine what happens if they allow interperiod trading (the possibility of borrowing permits from the future or banking them in the present for later use), and if they allocate permits to developing nations according to sovereignty criteria (each nation reduces emissions by a uniform quota) or egalitarian criteria (according to population). They find that net efficiency savings are much greater when developing countries participate in trading, although for some regions, such as the former Soviet Union, the entry of developing nations onto the market prevents them from selling permits at high prices during the initial phase. Developing nations are no worse off since high initial permit allocations are made and they are able to sell their surplus: this bodes well for bringing them into the agreement. Trading across time was found to be of little extra benefit: a few countries are likely to defer emissions reductions, reducing the short-term benefits for everyone else. Early entry of developing nations provides net benefits globally because permit prices fall. The inclusion of intergenerational equity considerations failed to yield a feasible solution, and it appears that Eastern European and former Soviet Union states may require relief if this criterion is applied. They conclude that the nations to benefit most from flexible mechanisms are China and India, and, among the Annex B countries, Western Europe.

Richard Tol in his chapter, 'Equitable Cost–Benefit Analysis of Climate Change' also argues that analysis of emissions reduction strategies has tended to concentrate on efficiency at the expense of equity considerations, desirable for both moral and political motives. While equity considerations have characterized the debate over the distribution of emissions reductions targets, they have played little part in determining *how much* emissions reduction is desirable. Tol's chapter draws attention to the fact that the countries responsible for the highest CO_2 emissions are precisely those which are least vulnerable to the effects of climate change. Criticizing the 'naturalist fallacy' of commonly used models of climate change impacts, Tol uses a FUND integrated assessment model to consider three possible mechanisms that promote efficiency and equity simultaneously. While the models are not presented as practical tools for policy makers, they do suggest that equity considerations justify strong cuts in emissions. This is all the more so because the low spatial resolution of the models tends to blur out the extreme cases of vulnerability to climate change that would require even higher abatement levels. Tol also agrees that international co-operation will be essential to find an equitable solution, and that permit trading is likely to be necessary to help reduce costs.

Equity is explicitly addressed also in Bosello's and Roson's chapter. They identify two conflicting principles in the evaluation of policy: distribution of rights and distribution of costs. Although it is clear that the market principle is the most efficient solution, it is not necessarily an equitable solution. They use an integrated assessment model to examine various emissions trading schemes in terms of distributive justice. They find that trading mechanisms make implementing the protocol much less expensive, especially in the case of world trading (via the Clean Development Mechanism), where average income exceeds the business-as-usual scenario after 2050. In fact, the economic results suggest that a non-signatory nation should be allowed to trade in permits even if it refuses to adopt any emissions restraint. In some ways their analysis agrees with that of Rose and Stevens – for example, they find that widening trading to include developing nations has a detrimental effect on some regions, such as the former Soviet Union, which lose out in the initial period. While developing countries benefit from selling permits (effectively obtaining income transfers), those which benefit most from allowing non-Annex I countries to trade are the richer countries, since they are able to buy up cheap permits. Nevertheless, all scenarios show a tendency to greater income equality over time, reflecting presumptions about the higher growth rates of the developing world. However, the Pareto improvement produced by emissions trading is unequally distributed. Hence, there seem to be gains for all world regions, but some regions, in particular OECD countries, may gain more than the other ones.

As already said, developing nations must eventually be brought into the Kyoto Protocol if long-term climate targets are to be achieved. The difficulty is to reconcile this goal with their industrial development. Onigkeit and Alcamo propose that developing countries should be allowed to increase emissions until they attain a given level of income per capita (called 'graduation'). At this point, they are required to freeze emissions until per capita emissions of the industrialized nations fall to the same level. After convergence, regions continue to reduce emissions together. Under this scheme, their model suggests that global CO_2 levels can eventually be stabilized even if emissions from developing countries continue to increase up to 2030. The approach requires stringent reductions by industrialized countries in this first period, in order to offset growing population and emissions levels in the developing world. After 2030 reductions continue but at a lower rate. The mechanism is quite simple since only two main variables need be set: the level at which developing regions 'graduate' and the stringency of the reduction targets. These determine when developing regions will need to begin limiting emissions and whether this transition is smooth or abrupt. While this simplicity perhaps limits the flexibility of the approach for policy making, it should be considered as just one of a number of possible policy tools.

In climate change, one of the most important themes is intergenerational equity. How can we model the interests of future generations? In most Integrated Assessment Models, planners apply a discount rate to the welfare levels of future generations when taking decisions. This has created controversy between researchers who advocate efficient (favouring the present generation) as opposed to sustainable (favouring later generations) resource use. Can historical interest rates (which express the relationship between the value of present and future capital) indicate what extent of sustainable resource use is desirable? Gerlagh argues that part of the problem is that existing dynastic IAMs are unable to provide policy makers with reliable figures about desirable emission reduction levels. His solution is to use a complex overlapping generations model in which the discount rate is endogenous. He finds that the discount rate varies according to a number of factors, including demographic trends, social security policies and sustainability policies. Of these, the projected ageing of the population in the 21st century and the introduction of emissions permit systems have decisive impact on the discount rate. These are factors which must be taken into account by policy makers when considering the projections of economic models.

The concluding chapter of this book presents a summary of the policy conclusions derived from the debate that took place at the EFIEA workshop at which the articles published in this book were circulated and discussed. As previously said, the final chapter, written by Carlo Carraro and Alessandra Goria, also attempts to summarize a few policy perspectives that were proposed. I call them 'perspectives' because they are not normative recommendations but rather positive scenarios that suggest how climate policies could be implemented.

In particular, as far as equity is concerned, it was emphasized that practical and effective solutions in climate change policies balance equity, efficiency and political feasibility criteria. Policies focus on equitable procedures as well as on equitable outcomes. Long-term policies aim at defining a global cap, which will induce efficiency, identifying at the same time who would bear the costs and benefits of mitigation, and addressing equitable procedures.

Before any agreement on a global cap, there is, however, the need to get developing countries on board: get them involved in the whole climate change debate, and make climate change policies part of their national policies. To achieve this goal, equity and efficiency issues need to be carefully integrated. Increased equity would help increase the number of participating countries, thus enhancing the efficiency of the agreement. Increased efficiency would reduce each country's cost of controlling emissions, thus facilitating the adoption of transfer and co-operation policies that increase equity as well as the number of participating countries.

Policy integration takes into account the linkages of environmental policy with energy policies, technology development and diffusion and technological co-operation. Within the EU, policy integration occurs across sectors: transport, agriculture and energy. In particular, the integration of transport policies with other sectoral policies is promoted, for instance making land use planning an important element of transport policies. Policies also provide further incentives for cross-sectoral energy efficiency investments; in this regard, there is a strong need for a more innovative policy approach, which contemplates investment in new technologies. Integrated climate policies also evaluate and investigate the implications of liberalization in the energy market for national policies, addressing its environmental and socio-economic impact. Policy integration implies action across groups and policy targets: policies are directed to mould both producers' and consumers' behaviour, on both the supply and the demand sides.

The chapters of this book, as well as the debate at the EFIEA workshop, emphasize that an effective climate policy cannot neglect the fact that its outcome crucially depends on the actions and interactions of several players. For example, the abatement decision of a group of countries, e.g. Annex I, can be costly and ineffective if offset by emissions expansions in other countries. This is the well-known carbon leakage issue, which arises because of the free-riding strategic incentives that characterize the climate problem. Other types of interesting interactions take place between producers and consumers or between policy makers and social stakeholders. If correctly understood and managed, these interactions can help increase the dynamics of the process that may lead countries to effective control of GHGs emissions.

This opens the way to new research directions that are not covered by the chapters of this book. In particular, the focus on optimality of most analyses neglects the fact that there exist incentives that may prevent countries (players of the game) from achieving the optimal outcome. Indeed, climate change control is a public good and, as usual, the design of optimal policies cannot underscore the strong incentives of free-ride on the provision of a public good. Hence, the goal must be the joint assessment of the optimal policy and of the international regime (e.g. the number of countries committed to emissions reduction) that are likely to emerge at the equilibrium.

If the goal is to understand which international regime is likely to emerge to control GHGs emissions, the game theory is certainly the best tool to deal with it. Indeed, the game theory has extensively analysed the possibility of coalition formation in the presence of free-riding (i.e. when parties have to agree on the provision of a public good). Early contributions (Cf. Hardin and Baden, 1977) characterized the environmental game among countries as a prisoners' dilemma, inevitably leading to the so-called 'tragedy' of the common-property goods where no country reduces its emissions beyond the non-co-operative equilibrium level. But in the real world, at the same time, a large number of international environmental agreements was signed, often involving subgroups of negotiating countries and sometimes involving transfers and other links to other policies (trade, technological co-operation etc.). It was therefore necessary to develop new models which helped us understand the logic of coalition formation in the presence of spillovers, and the possibility of increasing welfare by means of appropriate mechanisms and strategies. These new models were developed in the last decade within a non-co-operative game-theoretic framework, and provide interesting indications on the likely outcomes of climate negotiations (Cf. Carraro and Siniscalco, 1993; Barrett, 1994; Barrett, 1997; Carraro, 1999; Carraro and Moriconi, 1998).

However, there is still little empirical support to these researches and results. The existing empirical literature on climate policies is large, but it assumes the exogenous formation of environmental coalitions, and assesses the effects of countries' decisions to sign the agreement on the main economic and environmental variables. However, an empirical analysis of the incentives to sign the agreement and of the negotiation process that leads to the endogenous formation of the coalition is still missing. The empirical analysis would help us to understand whether the theoretical results, usually derived in the case of symmetric countries, still hold when the negotiations take place between countries of different sizes, natural resource endowments, development stages, etc.

Empirical analyses would also help in quantifying the eventual benefits of regional agreements, the feasibility of more effective minimum participation rules, the costs of transfer mechanisms, and the trade-off between optimality and stability when issue linkage strategies are implemented (Cf. Carraro and Siniscalco, 1998). These analyses would also help in identifying which countries are most likely to join the climate coalition and through what accession rules. Finally, empirical analyses would help in figuring out which one, among the several alternatives, might be the best strategy for the EU to exert its leadership in climate negotiations.

Even if a lot of research, both theoretical and empirical, is still missing, the results outlined in this book, if coupled with a careful assessment of the political dimension of the climate problem, could provide useful indications on what can be achieved and what cannot, thus giving the EU policy makers indications of where and how to concentrate negotiating efforts and economic resources.

Note

1. The Second EFIEA Policy Workshop was held at the Fondazione ENI E. Mattei, March 4–6, 1999. The Organizing Committee was formed by Carlo Carraro (Chair), Pier Vellinga, Carlo Jaeger, Jill Jaeger, Jean Charles Hourcade, Martin O'Connor, Ferenc Toth. The Organizing Committee is very grateful to Monica Eberle and James Shaw for their very valuable help in the organization of the workshop. The Editor of this book is also grateful to Angela Liberatore and Pier Vellinga for useful comments and encouragement, to all reviewers for their contribution to the improvement of all chapters of the book, and to Monica Eberle and Roberta Ranzini for their excellent editorial assistance. Financial contribution from the EU Commission – DGXII, Environment Programme, is gratefully acknowledged.

References

Barrett, S. (1994). Self-enforcing international environmental agreements. Oxford Economic Papers. 46:878–94.

Barrett, S. (1997). Towards a theory of international cooperation. In: C. Carraro and D. Siniscalco, eds., New Directions in the Economic Theory of the Environment. Cambridge University Press, Cambridge.

Carraro, C. (1999). Environmental conflict, bargaining and cooperation, forthcoming in: J. van den Bergh, ed., Handbook of Natural Resources and the Environment. E. Elgar, Cheltenham.

Carraro, C. and F. Moriconi (1998). Endogenous formation of environmental coalitions, presented at the Coalition Theory Network Workshop on Coalition Formation: Applications to Economic Issues, Venice, 8–10 January.

Carraro, C. and D. Siniscalco (1993). Strategies for the international protection of the environment. Journal of Public Economics. 52:309–28.

Carraro, C. and D. Siniscalco (1998). International environmental agreements. Incentives and political economy. European Economic Review. 42:561–72.

Hardin, G and J. Baden (1977). Managing the Commons. Freeman & Co, New York.

PART 1

Assessing the Kyoto Protocol

CHAPTER 2

Benefits and Costs of Kyoto*

Jason Shogren

Sebastian:
What a strange drowsiness possesses them!
Antonio:
It is the quality o' the climate.
The Tempest
Act 2, Scene 1

1. Introduction

You have probably read or heard the reviews about the climate change protocol emerging from Kyoto in December 1998 – some experts in the rules of civilized engagement see the accord as a crucial first step, others see it as a serious misstep, few see it as the answer. The Protocol's short-term comeback to a long-term question has left most reviewers demanding either deeper emissions reductions or broader emissions coverage, or both. The complexities and frustrations are manifest in the comment from Bert Bolin, former chair of the Intergovernmental Panel on Climate Change: "The Kyoto conference did not achieve much with regard to limiting the buildup of greenhouse gases in the atmosphere. ... Only if the new co-operation among countries succeeds will the Kyoto conference represent a step toward the ultimate objective of the convention."[1]

But regardless of your view, the 1997 Kyoto accord did signal a new earnestness of intent toward addressing the perceived risk of climate change. Kyoto demands that developed nations turn their economies so as to hit differentiated, sub-1990-level carbon emissions targets within the next decade or so. Meanwhile, developing nations sit on the sideline uncommitted, serious in their refusal to stifle economic growth by controlling their swelling emissions. The Kyoto Protocol asks for immediate action to address an uncertain, long-term, global threat in which the nations soon to be the world's largest emitters may never participate. Think of trying to turn a battleship on a dime with a third of the crew on board. Improbable, but doable ... but for what benefit ... and at what cost?

C. Carraro (ed.), Efficiency and Equity of Climate Change Policy, 17–42.

This essay examines the benefits and costs of Kyoto. My task is to take the Kyoto Protocol at face value, consider what the accord asks for and allows, and evaluate the potential benefits and costs that might accrue if it enters into force in 2000. My attempt to remain dispassionate occasionally wavers; a remnant from lively White House discussions attended while I was serving as the senior economist for environmental policy at the Council of Economic Advisers during the run-up to Kyoto. The negotiating pressure during that time made me appreciate the force of Dales' remark that "[t]he politicians must decide what the public wants and stake their political lives on their decision; they are in a much better position to assess the benefits and costs of their action (or inaction) than any body of experts."[2]

Still, economists are not shy about saying that behaviour matters more to climate policy than many people think; and that wealth spent here is not spent somewhere else; and that with more reasonable policy, it is possible to provide more human and environmental health with less wealth. So understanding the benefits and costs can help frame the climate change debate by identifying the elements of Kyoto that inflate costs with no additional benefits.[3] We consider what the literature says about estimating the benefits and costs for the US and the world; consider the key modelling assumptions that drive these estimates – e.g. the stringency of the abatement policy, the flexibility of policy instruments such as international emissions trading systems or sinks; and the development and diffusion of technology. Not surprisingly, the benefits and costs of Kyoto depend on what one *chooses to believe* about the nature of climate protection. Most economists believe that the threat of catastrophe will have to be imminent for the Kyoto Protocol to make sense given the likely impact on the US economy.

2. The Kyoto Protocol

Achieving more financial and commercial well-being while preventing untold global catastrophe – what madman would be against that goal? Essentially this was what the representatives of some 150 countries supposedly set out to do when they met in Kyoto, Japan on December 1997 at the third Conference of the Parities (COP-3) to the United Nations Framework Convention on Climate Change (UNFCC). Their task was to create a legally binding international agreement for climate protection – the so-called Kyoto Protocol. The Protocol will enter into force 90 days after the date on which 'not less than 55 Parties to the convention, incorporating Parties included in Annex I which accounted in total for at least 55% of the total carbon dioxide emissions for 1990 of the Parties included in Annex I have deposited their instruments of ratification, acceptance, approval or accession" (Article 24). The protocol is open for signatures by Parties between 16 March 1998 and 15 March 1999. In November 1998, the US was the 60th nation to sign.

The Kyoto conference was the culmination of years of negotiations to strengthen the first international climate change treaty signed by over 160 countries at the 1992 Earth Summit in Rio de Janeiro. The original treaty, UNFCC, called on industrialized nations to voluntarily reduce their greenhouse gas emissions to 1990 levels by 2000 (see Table 1). Emissions from most nations, however, have actually risen since Rio. Since voluntary actions did not do the job, many advocates of climate protection saw the Kyoto Protocol as the way to correct this perceived misdirection.

The Kyoto Protocol takes a 'deep, then broad' stand. 'Deep' in that the emissions targets require what many observers consider a rapid reduction in carbon emissions for industrialized nations; 'then broad' in that developing nations have no obligations at this time, but are likely to join the agreement eventually once someone convinces them that it is in their best interest to join. It is this 'deep, then broad' angle that has left many experts unimpressed by the Protocol, which seems to some to be a quick political fix rather than a serious response driven by the natural sciences and economics.[4]

So what does the Kyoto Protocol say?

- *Targets and timetables* (Article 3). The Protocol set a legally binding target for 39 of the world's most developed countries to reduce greenhouse gas emissions in aggregate by 5.2% from a 1990 baseline for the period 2008–2012. The targets are differentiated by nation, ranging from an 8% reduction (the European Union) to a 10% increase (Iceland) from 1990 levels. The United States agreed to a target of 7% reduction; Japan a 6% reduction (see Table 2). Each party must show demonstrable progress towards meeting its target by 2005.
- *Nations can act jointly to hit their target* (Article 4). The Protocol allows a group of nations to form a multi-country 'bubble' in which the group has an overall target to reach. Each nation inside the bubble has its own commitment to the rest of the group. The bubble met the demand of the European Union (EU) that it should be able to comply as a group. The bubble does require the EU to adjust its commitment if its membership enlarges.
- *Greenhouse gases* (Article 3 – Annex A). The Protocol covers six greenhouse gases – carbon dioxide, methane, nitrous oxide, hydrofluorocarbons (HFCs), perfluorocarbons (PFCs), and sulphur hexafluoride (SF6) – as a 'basket.' The latter three use a 1995 baseline instead of 1990. The inclusion of the six gases allows for some flexibility in reaching the target. Reductions in one gas can be used to substitute for reductions in other gases.
- *Emissions trading* (Article 17). The Protocol allows for emissions trading among the nations to fulfil their commitments. An emissions-trading programme provides greater flexibility to a nation to achieve its target. The domestic government can issue 'greenhouse gas emissions permits' to the private sector that equal the target set by the Protocol. While the permits

Table 1. Total carbon dioxide emissions of Annex I Parties in 1990, for the purposes of Article 25 of the Kyoto Protocol

Party	Emissions (Gg)	Percentage
Australia	288,965	2.1
Austria	59,200	0.4
Belgium	113,405	0.8
Bulgaria	82,990	0.6
Canada	457,441	3.3
Czech Republic	169,514	1.2
Denmark	52,100	0.4
Estonia	37,797	0.3
Finland	53,900	0.4
France	366,536	2.7
Germany	1,012,443	7.4
Greece	82,100	0.6
Hungary	71,673	0.5
Iceland	2,172	0.0
Ireland	30,719	0.2
Italy	428,941	3.1
Japan	1,173,360	8.5
Latvia	22,976	0.2
Liechtenstein	208	0.0
Luxembourg	11,343	0.1
Monaco	71	0.0
Netherlands	167,600	1.2
New Zealand	25,530	0.2
Norway	35,533	0.3
Poland	414,930	3.0
Portugal	42,148	0.3
Romania	171,103	1.2
Russian Federation	2,388,720	17.4
Slovakia	58,278	0.4
Spain	260,654	1.9
Sweden	61,256	0.4
Switzerland	43,600	0.3
UK	584,078	4.3
US	4,957,022	36.1
Total	13,728,306	100.0

Data based on the information from the 34 Annex I Parties that submitted their first national communications on or before 11 December 1997, as compiled by the secretariat in several documents (A/AC.237/81; FCCC/CP/1996/12/Add.2 and FCCC/SB/1997/6). Some of the communications included data on CO_2 emissions by sources and removals by sinks from land-use change and forestry, but since different ways of reporting were used these data are not included.
Source: Report of the conference of the parties on its third session, held at Kyoto from 1 to 11 December 1997 (FCCC/CP/1997/7/Add.1. 18 March 1998).

Table 2. Annex B

Party	Quantified emissions limitation or reduction commitment (percentage of base year or period)
Australia	108
Austria	92
Belgium	92
Bulgaria*	92
Canada	94
Croatia*	95
Czech Republic*	92
Denmark	92
Estonia*	92
European Community	92
Finland	92
France	92
Germany	92
Greece	92
Hungary*	94
Iceland	110
Ireland	92
Italy	92
Japan	94
Latvia*	92
Liechtenstein	92
Lithuania*	92
Luxembourg	92
Monaco	92
Netherlands	92
New Zealand	100
Norway	101
Poland*	94
Portugal	92
Romania*	92
Russian Federation*	100
Slovakia*	92
Slovenia*	92
Spain	92
Sweden	92
Switzerland	92
Ukraine*	100
UK	92
US	93

* Countries that are undergoing the process of transition to a market economy.

could then be freely bought and sold domestically between firms (Article 2), it is left unclear as to whether firms can trade uninhibited across borders. The trading price forces sources to reduce its greenhouse gas emissions so long as the tax exceeded the incremental cost of emissions reduction. This would stimulate fossil-fuel users to improve energy efficiency, use less carbon-intensive fuels, and consume fewer of the goods and services produced in the carbon-intensive ways. Apparently, disagreements about international trading were almost enough to deflate the conference. In return for emissions trading, the US gave up its opposition to the EU bubble.

- *Joint Implementation/Clean Development Mechanism* (Articles 6 and 12). Joint Implementation (JI) is where one nation gets credit for implementing a project to reduce carbon emissions in another country. JI is limited to the Parties. A new device, the Clean Development Mechanism (CDM), was developed for joint projects with developing nations through the payment of a special administrative fee by developed nations. According to the Protocol, "The purpose of the clean development mechanism shall be to assist Parties not included in Annex I in achieving sustainable development and in contributing to the ultimate objective of the Convention, and to assist Parties included in Annex I in achieving compliance with their quantified emissions limitation and reduction commitments under Article 3." A small portion of proceeds from the special fee is to be used to help the poorest of the poor nations, such as the island states, adapt to climate change.
- *Carbon sinks* (Article 3). The Protocol allows for carbon sinks – land and forestry practices that remove carbon emissions from the atmosphere. Sinks could play an important role for some nations because they represent a low-cost option. Sinks are ambiguously defined in the Protocol, and will be a challenge to measure.[5] Sinks might even turn out to be an escape value for the US since little is known with certainty the net uptake of the terrestrial sinks in North America. According to the State Department, once one adjusts the US's accounting method, sinks will account for about 3% of the 7% reduction below 1990 levels as asked for by Kyoto.[6]
- *No harmonization of actions.* The Protocol allows each nation to figure out its own best strategy to meet its commitment. Not everyone sees this as a good thing – some critics have argued that the world would have been better served by a common action rather than by a common target.[7]

What did the Kyoto Protocol not achieve?

- *Developing country participation.* No agreement was reached in Kyoto on what commitment developing countries should assume to reduce their greenhouse gas emissions. Kyoto also does not include a separate article for nations to voluntarily assume binding targets. However, it is clear to everyone that climate protection requires the participation of the developing countries because by the middle of the next century they are predicted to generate the largest share of carbon emissions, but they remain unmoved by the Protocol. Developing nations have no incentive to reduce their economic growth. China, for example, is the second-largest emitter after the US, but its per capita emissions are about a seventh of those in the US. A Chinese delegate captured the sentiment underlying the opposition: "What they [developed nations] are doing is luxury emissions, what we are doing is survival emissions."[8] Substantial compensation might be required to induce their necessary participation.

- *Specifics on emissions trading and the Clean Development Mechanism.* The Protocol also left the specific rules and regulations about international emissions trading to be defined at a future date. Although both trading and the Mechanism have the potential to generate low-cost emissions reductions for developed countries and tangible benefits to the host country, two factors limit their scope – transaction costs and additionality. Transaction costs are the time, effort and other resources needed to search out, negotiate, consummate, and get governmental approvals for heterogeneous deals. How the rules are eventually defined will determine the friction in all these flexibility tools. Additionality reflects the fear that people will try to use the Clean Development Mechanism to get credit for emissions changes that would have happened anyway in spite of some reduction project. Options to address additionality range from detailed scrutiny of every project before approval to the development of simple formulae applied across all projects. An obvious trade-off exists between reliability and cost among these options. The developers of the Mechanism need to define and pre-test the institutional, administrative and financial arrangements, the guidelines on the criteria for eligibility and certification, and verification and monitoring of emissions reductions.
- *Specifics on compliance and enforcement.* Both emissions trading and the Clean Development Mechanism require measures to assess compliance and to hold participants responsible for non-compliance. Under the Framework Convention and the Kyoto Protocol, the Annex I capped countries are ultimately responsible for achieving their emissions targets, whether they are net buyers or sellers of permits or credits. Presumably the Protocol Parties will then reassign this responsibility to the domestic private sector through strategies to monitor emissions and spot-check specific investment projects. The Protocol says enforcement procedures to deal with non-compliance will be established at the first meeting of the Parties to the Protocol. But until the details are fleshed out and this uncertainty is resolved, those firms falling under a trading system will remain sceptical about the workability of the scheme. For instance, whether the buyer or seller should be liable for the penalty of trading a permit that should not have been traded remains a major question mark.

3. Benefits from Kyoto

The potential benefits from Kyoto are captured by the avoided damages from climate change.[9] Potential climate risks avoided include more severe weather patterns, hobbled ecosystems, less biodiversity, less potable water, loss of coastal areas from rising sea levels, rises in mean temperature, and more infectious diseases such as malaria, yellow fever and cholera. On the plus side, climate change might benefit agriculture and forestry by increasing productivity with longer growing seasons and more fertilization. These gains (or losses) can be categorized into four broad sets, increasingly difficult to quantify – the avoided losses to market goods and services, non-market goods, secondary impacts, and catastrophes.

Traditionally, people have judged the benefits of climate protection as the reduction in human and environmental risks from the *business-as-usual* (BAU) baseline. Under BAU, modellers have estimated that carbon concentrations might be expected to double pre-industrial levels within the next half century, with mean temperatures predicted to rise by about 1° Celsius by 2050, and 2.5° by 2100. With Kyoto, the expected rate of slowing temperature rise is minor – between 4% and 14% of BAU baseline, with global-mean warming reductions between 0.04° and 0.10° Celsius by 2050, and 0.08° and 0.28° Celsius by 2100.[10]

Researchers have estimated the total impact on Gross World Product from climate change around 1–2%. The impact on Gross Domestic Product (GDP) in the US has been estimated to be around $\pm 1\%$. Most industries in the developed nations are separate from climate – less than 3% of US livelihoods, for instance, are earned in agriculture and other climate-sensitive outdoor activities.[11] And even if we include the potential non-market damages, Nordhaus has argued that the market and non-market benefits to the US are probably at most about 2% of GDP.[12] At the margin, the damages avoided have been estimated to be about US$25 tC, the extra benefit obtained from reducing carbon emissions by one tonne.[13] These impacts are not trivial, but the impact on economic output is not likely to cause the next global depression either.

Thomas Schelling views climate protection as a political problem, whose costs will be relatively low: "a few trillion dollars over the next 30 or 40 years, out of an OECD GDP rising from US$15 trillion to US$30–40 trillion annually."[14] Although this may be achieved, he wonders whether this is really the right question. He points out that climate policy really amounts to a wealth transfer from today's industrialized nations to the future generations in the developing nations. The benefits from Kyoto are most likely to accrue to the future generations in developing nations because their economies depend more on favourable climate for agriculture, forestry and fishing. He wonders whether it would be better to invest in development today than pay for climate relief tomorrow.

Two topics in non-market valuation are likely to trigger major debates about the likely magnitude of potential benefits – human health and ecosystem endangered species services. First, consider health. There are numerous lists of the potential threats to human health, including old scourges like cholera, plague, yellow and dengue fever, tuberculosis, and malaria, and 30 diseases new to medicine, like E. coli, hatavirus and HIV.[15] But there is less talk about the odds that the events will come to pass. What are the odds, what is the variability around these odds, how credible are these odds?[16]

Despite the warnings, the cause and effects are still uncertain. It is not clear whether malaria rates will increase if mosquitos should fail to adapt to changes in temperature, humidity or precipitation.[17] And attached to these threats must be advances in technology and nutrition and medical care. Adaptation may be the key to prevention. Once we acknowledge that adaptation plays a key role, we must account for the fact that risk depends on private and collective choices. The economic variables that drive adaptation, such as relative price and wealth, must be considered in the estimation of health risk. Given the relative marginal effectiveness of different self-protection actions, how people confront risk differs across individuals and situations, even though the natural phenomena that trigger these actions apply equally to everyone.

More information is needed to determine the bias in assessing health risks solely in terms of natural science information, given that the sources of systematic variation are relative prices, incomes, and other economic and social parameters. If economists are going to be effective in the debate over setting health-based standards, they must insert themselves into the econometrics of epidemiology.[18] Biases associated with measurement and specification errors are prevalent. When account is taken of self-protection behaviours, ambient concentrations are not synonymous with exposure. Economic variables affect behaviour which affects the risks faced by people, and exclusion of these variables from risk assessment biases predictions. Studies have shown that people persistently below the poverty line are far more likely to become sick than are wealthy people for a variety of reasons including habits, lifestyle, less medical screening, and the ability to self-protect. The evidence suggests that behavioural choices frequently associated with poverty (e.g. high discount rates) are the most significant threat to health. Wealth equals health, in warmer or colder climates alike.

Estimating the social value of endangered species and ecosystem services is also a challenge. When considering endangered species values, people disagree about the usefulness of the primary tool to reveal the monetary value of these preferences – contingent evaluation surveys. These public opinion surveys use a sequence of questions to put a monetary value on personal preferences. But since people are responding to a survey rather than facing their own budget constraint and actually spending their own money, no market discipline exists to challenge their statements. For instance, if one summed the stated preferences from various endangered species surveys as a crude measure of benefits, the average person was willing to pay about US$1000 to protect 18 different species. Multiplying US$1000 by the number of US households suggests that we would be willing to pay over 1% of GDP to preserve less than 2% of the endangered species.[19] Many will find these values to be suspiciously high. Despite the challenge in measuring the value of preservation, determining a plausible range for these values is needed for helpful judgments about the potential for climate change benefits.

One might claim that climate protection might avoid damage to the global ecosystem services costing in the range of US$33 trillion.[20] This estimate is meaningless, however, since willingness to pay is constrained by the world's ability to pay – a world GDP of about US$18 trillion.[21] And if one accounts for the fact that people make some contribution to total world income, the potential maximum ecosystem benefits are easily cut in half.[22] But it is questionable whether these numbers really mean anything at all given the false baseline of an all-or-nothing outcome with climate change. The most likely changes will not be a binomial Armageddon vs. Eden revelation.

Another way to amplify the benefits of the Kyoto Protocol is to consider the potential secondary impacts that come from discouraging coal and other fossil fuel consumption. The Kyoto Protocol would reduce emissions of such air pollutants as carbon monoxide, sulphur and nitrogen oxides, and toxic trace pollutants in exhaust gases. By reducing BTUs generated by fossil fuel consumption, emissions of these pollutants would inevitably fall, reducing damage to health, visibility, materials and crops. Studies in Europe and the US have estimated that the non-climate benefits might be as large as or larger than the benefits from avoiding climate change. The estimated secondary benefits range from US$3 to US$78 per tonne carbon reduction (US$1992) based on a pre-1997 ozone/particulate matter National Ambient Air Quality Standards (NAAQS) baseline.[23] The benefits from reducing air pollution damages could offset 30–100% of carbon abatement costs given the pre-NAAQS baseline.

But should these secondary impacts count when considering the level of climate protection? One argument against their inclusion is the fear of double counting – the already existing NAAQS rules that were tightened in 1997 might produce most of the potential health benefits, not climate change policy. The Administration has estimated that extra benefits from climate change given a post-NAAQS baseline are about a billion dollars in 2010.[24] If our current air quality policies are effective, they should be capturing all the positive net social benefits. Claiming these net benefits for climate change just says that our other environmental policies are flawed, and that we should be focusing on attention on improving them. Climate change policy should not serve as a catch-all bail-out for all our perceived social ills.

Finally, researchers, policy makers and politicians have raised the spectre of catastrophe and surprise. Modellers often presume climate change will be gradual – a slow and steady rise in temperature or precipitation. But many people warn that this steady flow ignores the real risk of a sudden rupture or a straw-camel-back scenario, e.g. a catastrophe such as a structural change in ocean currents like the Gulf Stream, the melting of the Western Antarctic ice sheet, waves of environmental refugees. These threats are enough to scare most people into action.

While careful not to make any causal link, some point to El Niño as an illustrative example of the damage one might expect with climate change. Stuart Eizenstat, Undersecretary of State for Economic, Business, and Agricultural Affairs, said, "For a preview of the type of severe weather ... look at the devastation wrought by this winter's El Niño."[25] But according to FEMA, the El Niño winter was no costlier to US taxpayers than the previous two winters in the US – US$289 million in 1997–98 compared to US$294 million in 1996–97 and US$280 million in 1995–96.[26]

The problem is that researchers do not have any reasonable estimates of the odds that these events will come to pass. The best these researchers are willing to say is that these severe events are 'uncertain'.[27] But all 'all severity – no probability' scenarios can lead people to knee-jerk reactions. The challenge is to get researchers to quantity these odds for more informed policy judgments. It matters whether the odds are 10% or 0.01%.

In addition, the BAU baseline is not the only benchmark in town for judging the benefits of Kyoto. In fact, the BAU path is less credible after Kyoto since the developed world has already agreed to do something to address climate change.[28] According to Article 2 of the Framework Convention on Climate Change, the objective is to stabilize "greenhouse gas concentrations in the atmosphere at a level that would prevent dangerous anthropogenic interference with the climate system." Numerous different emissions reduction scenarios can work to stabilize concentrations at some level, including the 'broad, then deep' pathway recommended by many researchers and policy makers: broad participation by both developed and developing countries, and a gradual emissions reduction path to achieve a desired long-term concentration target.[29]

Delay is not denial, some say.[30] The 'broad, then deep' represents the view that we can allow emissions to grow at least for the next few decades before serious reductions are necessary.[31] And while advocates of this policy run the risk of being dubbed irrelevant to the process, the moderate reduction baseline calls for an emissions path that peaks around 2020 and would result in essentially the same concentration level in 2100 as Kyoto, at a fraction of the cost. Implementation of this benchmark would require an initial, modest increase in the price of carbon emissions, and a credible commitment to increase the price over time. Shifting emissions reductions into the future allows time for a gradual adaptation of the energy capital stock, developing low-cost and low-carbon technology substitutes, removing carbon from the atmosphere via the carbon cycle, and, since the economy yields a positive return on capital, future reductions can be made with a smaller commitment of today's resources.

The benefits generated by this moderate benchmark are nearly identical to those produced by the Kyoto Protocol, even accounting for secondary and catastrophic impacts. The temperature difference between the Kyoto Protocol and the moderate baseline is less than 0.1° Celsius at any time over the next century. This small difference reflects the long lags in emission flows and concentration stocks over the century, and that the most serious emissions will come later in the century from the expanding economies of the developing nations. The net result is that Kyoto did not gain any benefits over the next best alternative.

4. Costs of Kyoto

As expected, the costs of Kyoto are all over the board. Some studies suggest the US could meet its target at negligible or modest cost; others call Kyoto an 'economic disarmament' driven by rank political opportunism. And to assume the costs are somewhere in between puts the theorist at risk of being compared to a cigarette manufacturer stealing the life out of children or to a fool who has bestowed long-lasting economic advantages to his international competitors. It's wild out there.

On the modest side sits the Clinton Administration's report on the Kyoto Protocol that emerged from the Council of Economic Advisers (CEA). Again relative to the BAU baseline, the Administration report states the costs to the US to meet its Kyoto target are "likely to be modest if those reductions are undertaken in an efficient manner employing the flexibility measures of emissions trading (both domestic and international), joint implementation, and the Clean Development Mechanism." By modest, the Administration means an annual GDP drop of less than 0.5% – roughly some US$10 billion dollars; no expected negative effect on the trade deficit; increase gasoline prices by about 5 cents a gallon; lower electricity rates; and no "significant aggregate employment effect."[32] The marginal costs in this case fall around US$10–20 tC. Essentially, a 'broad and deep' baseline is built right into the Administration's estimates of costs, and these estimates might be plausible if all goes exactly right with the world; a big 'if'.

In contrast, forecasts from the economic consulting firms, WEFA Inc. and DRI/McGraw Hill Inc., estimate the US GDP could take an annual hit of nearly 3% of GDP, or about US$250 billion a year – with intra-nation emissions trading. Plus the trade deficit would increase by tens of billions of dollars; gasoline prices would increase by nearly 50 cents a gallon; electricity prices would nearly double; and two million US jobs would disappear.[33] The impact on energy use is analogous to that felt during the decade of OPEC price shocks. The marginal costs here are upwards of around US$200–300 tC. These estimates obviously do not see all going right with Kyoto. Rather they see trouble in premature assumptions of rapid technological improvement, and in the likely friction imposed on any emissions trading or technology transfer system by the institutions that eventually might see the light of day.

Other observers are more in the middle of the pain distribution.[34] Kyoto will not destroy national economies, but it will not be painless either, they say.

Researchers at Resources for the Future, for instance, estimate that Kyoto could cost about 1% of GDP annually, at a worst 2%.[35] Plus energy costs would increase for an average US household (about US$2,500 per year) by about 25%, this includes a gasoline price rise of 30 cents a gallon. These estimates fall within the range of those cost estimates produced by Charles Rivers Associates – about 1.4–2.0% losses in 2020.[36] The Australian government estimated global losses at about 1% annually globally at 2020.[37] Prior to Kyoto, the Energy Modeling Forum (EMF) estimated the short-term costs of a 1990 by 2010 strategy at 0.5–1.5% annual drop in US GDP; and the medium-term costs between 0.6–2.6% drop in GDP by the year 2020.[38] The post-Kyoto EMF estimates have yet to be released.

Another interesting story emerges when we compare the costs of Kyoto relative to the more moderate 'broad, then deep' emissions path baseline. Nordhaus and Boyer have used the RICE-98 model to address the relative benefits and costs of the Kyoto Protocol to a moderate or 'optimal' baseline.[39] While preliminary, their results suggest that while both paths yield nearly identical emissions reductions, the Kyoto accord without global trading could be 8–14 times more expensive than the moderate path.

In general and for all models, cost estimates are likely to be on the low side for several reasons – models presume the most efficient possible climate control programme, even though today only one such programme is ongoing; models assume the control programme is announced early and maintained indefinitely, even though a government will be hard pressed to maintain consistent control over the decades; many models focus on long-term equilibrium and ignore the short-run adjustments such as the oil shocks of the 1970s. There are good reasons to believe that any international or domestic emissions trading programme or Clean Development Mechanism will have significant transaction costs due to market friction. Some economists think these factors would raise cost estimates by a factor of one to four. This factor would cause Kyoto to reduce GDP by 1–10% from baseline. For comparison, the US now spends about 2% of GDP on all environmental programmes combined.[40]

Finally, what about the common charge that modellers habitually overestimate the costs of environmental regulations, and thus it is likely that the costs of Kyoto are also too high by definition. This blanket assertion is not supported by the facts. Granted economists did predict the control costs for sulphur dioxide could be US$1500 per tonne, when today a tonne actually costs US$100. And good reasons exist for this gap – unanticipated technology breakthroughs, railroad deregulation, many permits given out for free. Ray Squitieri at the Department of Treasury has compiled the actual evidence. He finds that costs are just as often on the low side. For instance, predictions straddle actual costs for asbestos, coke ovens, vinyl chloride regulation; numerous unpredicted changes in technology and the economy occurred to lower the cost of CFCs, cotton dust, SO_2 control; recall that the National Ambient Air

Quality Standards were estimated to be achieved by 1977, but these standards are still not attained in 75 areas with 75 million people.

5. What you choose to believe

How one sees the benefits and costs of Kyoto depends on what one chooses to believe about the nature of three elements that underlie climate protection – the cusp of a catastrophe, the degree of flexibility, and the origins of technological advance.[41]

The Cusp of Catastrophe

What you choose to believe about the benefits of Kyoto depends on how you perceive the risk of catastrophe. If you believe disaster is imminent, emissions reductions cannot come soon or fast enough. If you don't, it is hard to justify the likely costs of the Kyoto Protocol without global trading. Reliable information is needed to guide people from their diffuse prior beliefs about the odds of a catastrophe. Whether the information will be forthcoming between now and potential ratification fights is unclear.

Numerous unanswered questions persist about the structure of atmospheric systems and their potential thresholds. As Schelling and many others note, uncertainties abound. We do not know which regions will get warmer or cooler; which will get wetter or drier; which will get stormier or calmer. Climate policy debates eventually reach the point in which the modeller is asked whether he or she has accounted for the likelihood that a change in the ecosystem will be discontinuous – a catastrophe. Most modellers acknowledge their models do not address the potential for discontinuous shocks, like a sudden shift in the Gulf Stream, or an unravelling of the web of life due to the loss of some keystone species.

Some observers view the increased temperature over the past century as well within the bounds of natural variability. Others such as Vice President Gore have been known to point out that the evidence of global warming keeps "piling up, month after month, week after week. How long is it going to take before these people in Congress get the message?"[42] John Holdren, a member of the President's scientific advisory committee, agrees: "Every day the evidence becomes more persuasive that global warming is underway." But daily weather and seasonal means are highly variable making it a real challenge to discern trends; that is, separating signals from natural noise.

But does this mean that society is on the cusp of catastrophe? Not necessarily – the doomsayers have a terrible track record. In addition, numerous risk perception studies have revealed that people commonly overestimate the chance that they will suffer from a low probability/high severity event, e.g. a nuclear power accident.[43] When the outcome is potentially very bad, people inflate the chance that the outcome will be realized. Policy makers are not immune to this human fallibility either. Overestimation of risk can multiply throughout the general public. For example, people pondering the storage of nuclear waste can transform images of a fortified storage facility containing sanitized, air-tight receptacles into an abandoned dump site teeming with rusty, leaking vats of toxic material. The images induce vivid perceptions, both of which can persist in a community, causing considerable disagreement about how to regulate the risk.

Experience tells people little about low-probability risks like climate change. They must rely on outside sources of information to help them make judgments about the likelihood that a bad event will actually come to pass. And if that outside information stresses severity without giving some notion of the odds, people systematically bias their risk perceptions upward. Other evidence suggests that when people are given good news with the bad, the bad news often dominates.[44]

Of course, sticking one's head in the sand is not particularly useful either. Climate protection is viewed as hedging against uncertainty – i.e. planet insurance.[45] As such, it is important that we understand the range of potential impacts, not just the expected value.[46] Peck and Teisberg explore how risks of large losses affect the estimated costs of given climate protection strategies.[47] Working from Nordhaus's survey of expert opinions about the odds of losses,[48] they create eight states of nature by assigning either a high or a low value to three key parameters that reflect uncertainty – the probability of a loss, the climate sensitivity, and the utility discount rate. The results suggest that the optimal policy under uncertainty is about the same as the policy for the lowest-loss state of nature. This occurs in part because the lowest-loss state receives the greatest likelihood of coming to pass. Also a probabilistic damage function based on the mean relative pessimism from the same expert opinion survey suggests that under the extreme damage scenario opined by natural scientists, the climate protection is six times more stringent than that of Nordhaus's original prediction.[49]

Again, using the same opinion survey as the starting point, a third integrated model constructs an empirical relationship between carbon emissions accumulation and the odds of a catastrophe.[50] More carbon emissions without a disaster in one year lead to a greater chance of a catastrophe the next year. The results suggest that the odds of a catastrophe occurring by about 2100 range from 4% to 63% depending on how emissions affect the odds of a loss. While this range does not reduce our uncertainty much about the odds of disaster, the framework is a worthy one deserving more attention in future modelling efforts. Moreover, commissioning a newer, updated expert opinion survey seems most worthwhile given the lessons learned over the last decade about the workings of the climate system.

Finally, uncertainty about the underlying physical processes and random variability requires policy makers to decide whether to control for potential catastrophe now or wait for more information about the climate. Kyoto reduces emissions, and consequently causes learning to slow down because less information now exists about the cause and effect. An integrated assessment model with endogenous learning shows that uncertainties about the climate system are resolved in 90–160 years, far longer than most people expect.[51] A trade-off exists between emissions policy and learning – the more emissions reductions, the less learning about the system under study. Policies of climate protection based on a presumption of complete information can be out by as much as 25% in either direction once learning is accounted for. Elements of truth exist in both sides of the debate. Policy makers would serve a more useful role as arbiters that help reconcile risk perceptions, rather than inflame the differences. People with a sane expectation of rationality should demand no less.

The Degree of Flexibility

What you choose to believe about the costs of Kyoto can be framed by focusing on if and how you think the flexibility provisions will come to pass, what we assume about future policy regimes, and how many nations will actually be included in any exchange system. The costs to meet a policy depend on a firm's legal ability to use low-cost carbon reductions and how quickly society wants to change the energy systems and capital structure of the US economy. A stringent, inflexible carbon policy will induce greater economic burden than a loose, flexible policy. Obviously more flexibility and more trading partners can reduce costs, as a firm can search out the lowest-cost alternative. One can expect the opposite with inflexible rules and few trading partners. It is estimated that any agreement without the cost flexibility provided by trading will at least double the US costs.

Flexibility can be measured as the ability to reduce carbon at the lowest cost, either domestically and internationally, including so-called 'when and where' flexibility – which assumes a world emissions budget could be spent optimally over space and time to capture all potential intra- and inter-temporal efficiencies. This would allow the banking and borrowing of allowable carbon emissions. Providing a firm or nation more flexibility to reach a given target and timetable also will reduce costs. Almost all studies of the costs of Kyoto already presume that carbon reductions are implemented domestically with well-designed, cost-effective policy tools like a carbon tax or tradable permits system. International flexibility is also key for US firms to meet domestic targets at lower cost by financing the upgrading of inefficient energy facilities in developing countries. Limiting such opportunities for flexibility – due to poor policy design, inherent problems in administering international policies, or lack of interest on the part of other countries – will mean potentially much higher costs. The key is to distribute emissions internationally so as to minimize the costs of climate policy.

The Kyoto Protocol established a narrow coalition of developed nations which now must reach out to the developing countries to join later. This approach begins with a narrow participation by a limited set of nations in a relatively ambitious agreement that involves considerable costs and hence requires fairly sophisticated policy instruments. People have to judge the odds that developing nations will eventually join the Protocol. The most contentious issue in Kyoto was how to encourage or pressure developing nations to commit to emissions reductions. The US pushed for a voluntary system, but the EU maintained that discussions on developing countries should begin only after the developed nations took the lead. China, Brazil and other nations stopped several efforts in Kyoto to create a voluntary opt-in process for the developing nations to adopt binding commitments. China also successfully opposed putting the issue on the agenda for COP-4 in Buenos Aires in November. The difference in outlook is obvious – the developing nations are looking into the future and seeing the other side of the Environmental Kuznets Curve; the developed nations are looking at today and seeing the steep climb up the same Curve. Rich nations got rich through carbon, poor nations want the same opportunity. Who can blame them.

Kyoto runs the risk that the developing countries will never join later because the costs of doing so will be too high. By increasing the relative costs of carbon in the narrow coalition, carbon-intensive industries will be tempted to move to developing countries, thereby making these economies even more carbon dependent as they try to grow their way past the real health problems they face now.[52] Their addiction to carbon-based growth increases the costs of joining the treaty. The suppliers of carbon-intensive energy will look for existing markets and will create new markets. If developing countries do not alter their emissions path, global emissions levels will continue to increase even if all the developed nations completely eliminate all their emissions. Without China or India and the rest of the developing world, the Kyoto Protocol will not work. The Administration has indicated that they would not submit the Protocol to the Senate until 'meaningful participation' from developing nations was obtained. Unskilled as I am in affairs of state, my guess is that the debate over whether climate change policy is really back-door foreign aid will stall serious proposals to pay for meaningful participation over the next decade.

We also need to work through the details of how the system would be designed before alternative policies can be usefully evaluated.[53] The Kyoto Protocol does not have a set position as to what either a domestic or an international trading system would look like. This is serious since we would essentially be creating a new global currency – the 'carbo'. The global economy would be under the thumb of the institution that controlled the supply of the carbo. Would we give control over to a quasi-independent institution like the Federal Reserve? Would we turn over the carbo supply to Alan Greenspan, Chairman of the US Federal Reserve Board? Society would want to turn over the supply to someone who understands how to keep the economy from stagnating, which is not likely to be the administrator of a federal environmental ministry.

What is curious about carbon emissions trading is that its biggest supporters have often been the environmental activists rather than economists. The environmental community prefers emissions trading over carbon taxes because the quantity of carbon flowing into the atmosphere is fixed, thereby shifting risk from the environment to the economy in the form of price uncertainty. Some advocates have pushed for relatively high transaction costs that would limit the cost-savings of a trading system; others have argued for a flexible system that allows for banking and borrowing of permits. But many economists have questioned the feasibility of carbon trading because the international market is likely to be thin as most nations have indicated inaudible interest in the system, and the costs of monitoring and enforcing the system are likely to be high.

Defining the rules for flexibility incentive systems is wide open. One way to explore the nature of trading is to testbed alternative systems in laboratory markets prior to actual field implementation.[54] Experimental economists could play an important role in reducing the associated uncertainty. This holds for joint implementation and the Clean Development Mechanism as well. A testbed carbon emissions trading system designed in laboratory markets can evaluate the institutional factors that will influence the effectiveness of carbon trading. Experimentalists can consider how flexibility in trading, imperfect information, multi-gas trading, links between domestic and international trading, and other factors affect the potential efficiency of trading.

A serious effort is vital to understand what aspects of emissions trading can reduce the costs of climate change policy. The effort should first design and parameterize a market that reflects the costs and productivity of the countries or regions expected to participate in an emissions trading initiative as suggested in the Kyoto Protocol. Researchers can then use the parameters of the nations to design World Wide Web market experiments; to testbed various trading proposals that have been described to meet the Kyoto Protocol objectives using the market parameterization; to evaluate the robustness of emissions trading market institutions to market 'frictions' – impediments to efficient, cost-minimizing market outcomes; and to evaluate how the scope of the market affects market performance.

A further consideration is that carbon sinks are a wild card in the search for flexible, low-cost solutions. Recall that a carbon sink is a process that destroys or absorbs greenhouse gases, such as the absorption of atmospheric carbon dioxide by terrestrial (e.g. trees) and oceanic biota. The main anthropogenic sink is tree planting and other forest management actions. Soils and other types of vegetation also provide a potential sink. It is estimated that forests around the world contain about 830 Pg of carbon in their vegetation and soil, with about 1.5 times as much in soil as in vegetation. For the US, forests are an important terrestrial sink given that they cover about 750 million acres. Land use changes in the US have increased the uptake of carbon to an estimated 200 MMTCE (million metric tonnes of carbon equivalent).

A few studies have found that carbon sequestration through sinks could cost as little as US$25/tonne in the US for 150 MMTCE.[55] But serious uncertainties remain about how to measure and account for estimates of net carbon. For example, how forest management activities affect soil carbon is unknown, and since forest soils contain over 50% of the total stored forest carbon in the US, this difference can have a significant impact on estimates. Also, some researchers have shown that sinks are not as effective as predicted when one accounts for the interaction of forest reserves and the timber market. The more land that is set aside for carbon sinks, the quicker the cycle of harvesting on other forestland, and the less total net carbon sequestration.[56] Some fear that these ambiguities about sinks could divert attention from first-order priorities to second-order technicalities.[57]

The costs of climate protection are also amplified by pre-existing distortions created by the existing tax system for labour and capital income. Labour and capital taxes distort behaviour because they reduce employment and investment levels below what they would have been otherwise. Now add on a carbon tax (or permit) that discourages consumption and production, and you further reduce employment and investment, which then exacerbates the labour and capital tax distortions. One estimate is that this amplified distortion will inflate the control costs by some 400%.[58] One could reduce these extra costs by channelling the revenue from the carbon tax, if any existed, to reduce the labour and capital taxes, and thereby reduce the pre-existing distortions. Revenue recycling could shave control costs to 75%. But the political reality is that the odds of a tax or permit system that raises revenues to be recycled are unlikely. Carbon permits most likely would be given away for free to producers because, as it has been put by those wary of trying to sell a tax hike to Congress given the disastrous BTU tax experience: "If it don't bring no revenue to the government, it ain't no tax."

Furthermore, costs will increase if proposals are acted on to restrict flexibility further by requiring that nations do some fixed percentage of emissions cuts at home. The EU environmental ministers met after Kyoto to define a strategy to restrict US efforts to use emissions trading to reduce the costs of hitting its target. Seeing trading as flight from responsibility, the EU drafted text for the next negotiations to prevent 'loopholes' in the Protocol. Although no consensus was reached, some ministers argued that 50% of emissions reductions must come from domestic cuts. Peter Jorgensen, spokesman for the EU commission, stated: "As it is the leading emitter of greenhouse gases, the United States needs to take tough domestic measures ... [the US] is going to try and buy its way out of its Kyoto commitments, and we are determined to prevent that from happening."[59] This quantity constraint of 50% do-it-domestic will inflate the costs of hitting the target. "The burdens of global leadership are sometimes heavy, indeed."[60]

The Origins of Technological Advance

What you choose to believe about the origins of technological advance will colour the costs of Kyoto. For any given target and set of policy provisions, costs decline when consumers and firms have more plentiful low-cost substitutes for high-carbon technologies. Engineering studies suggest 20–25% of existing carbon emissions could be eliminated at low cost if people switched to new technologies like compact fluorescent light bulbs, improved thermal insulation, heating and cooling systems, and energy-efficient appliances. Engineers argue that the origins of technological advance are firmly rooted in non-price responses – people do the right thing for the right reason, and once they understand the potential benefits of low-carbon technologies, they will just switch.

Economists disagree. They see the origins of advance as driven by changes in relative price. Even if new technologies are available, people do not switch unless prices induce them to switch. They are unwilling to experiment with new devices at current prices. People behave as if their time horizons are short, perhaps reflecting their uncertainty about future energy prices and the reliability of the technology. Plus factors other than energy efficiency also matter to consumers, i.e. quality and features, and the time and effort required to learn about a new technology and how it works.

The difference in views on the origin is revealed in the debate about the autonomous rate of improvement in the energy-to-GDP ratio underlying all models of climate economics. Modellers debate the appropriate rate of 'autonomous energy efficiency improvement' (AEEI), the approximation of the rate of change in the energy/GDP ratio when energy prices remain relatively flat. Although historical evidence suggests that the autonomous change is about 0.5% per year, some people argue the announcement of the Kyoto Protocol will prompt businesses to accelerate the implementation of energy-efficient methods of production. This 'announcement effect' has been argued to increase the autonomous rate to 2% per year or more. These significant leaps are hard to justify, however, after examining the evidence. The Climate Change Action Plan (CCAP) thus far has spent a half a billion dollars thus far to reduce carbon emissions in the US by about 14 MMTCE. Given we have to reduce emissions by over 500 MMTCE, one would have to believe in the statement that we are on the cusp of a very steep S-shape diffusion curve.[61]

Economists remain wary of claims that alternative energy-saving technologies are readily available at no extra cost to consumers. People do not always take advantage of cost-effective, energy-efficient technologies that, in the long run, are good for both the pocket and the environment. At current prices, many consumers may not be willing to experiment with compact fluorescent light-bulbs, improved thermal insulation, better heating and cooling systems, and energy-efficient appliances. Several studies have estimated that when consumers buy air conditioners, space-heaters, water-heaters, and refrigerators the pay-off predicted by the engineers does not come through.[62] Also, their implicit time horizon can be much shorter than the time horizon reflected in market interest rates, and even when they are presented with estimates of the likely future cost savings they pay more attention to immediate outlays. As our experience with the oil shocks of the 1970s shows, choices do change when prices rise. Economists see the most effective way to curb excessive energy consumption to be to raise its price to reflect the harmful effects on the environment of burning fossil fuels.

However, at the White House conference on climate change last October, President Clinton made it clear that he did not think Americans would not tolerate higher energy taxes to combat climate change.[63] His answer was to promote new energy-efficient technologies rather than impose steep energy price increases. The President's plan to devote US$6 billion for research subsidies and tax credits over the next five years. The Congressional Budget Office estimated that the US currently spends nearly US$5 billion annually on programmes directly or indirectly related to climate protection. The CBO also concluded that the effect of current programmes and tax policies "on total emissions was unclear ... [but] it would probably be small. ... Since most of the funds are spent to learn more about [climate change] and to improve energy efficiency in the future, the short-term effect is minimal."[64]

The theory of non-price-policy response to technology adoption must recognize the fact that preferences can prevent adoption to the levels predicted by engineering studies. High adoption rates will be realized once prices go high enough to eliminate the barrier, e.g. high carbon prices can provide the stimulus for some technologies to clear the barrier. Policies could be designed to try and eliminate barriers, but firms have found that these technologies are not cost-effective for them. Subsidies that increase their adoption will generate benefits which are outweighed by the policies' costs.

Although economists do not see the average person switching energy sources just for the sake of switching, they do accept that the search for profits can create R&D breakthroughs that reduce the costs of backstop technologies.[65] These breakthroughs change the relative prices that can induce energy users to switch from fossil fuels to solar. Based on a Hotelling model of scarcity rents, for example, Chakravorty *et al.* show that if historical rates of cost reduction in the production of solar energy are maintained – 30–50% per decade, more than 90% of the world's coal will never be used.[66] The world will make the transition to the backstop technology: to solar from coal and oil even without a carbon tax. Global temperatures will increase by 1.5–2° Celsius by around 2050, and will then decline steady back to pre-industrial levels.

Finally, what you choose to believe about technology depends on how you think about the interaction of the technologies of risk reduction. Climate change discussions usually separate human responses to potential threats into two broad categories – mitigation and adaptation. By mitigating, humans reduce the *odds* of a bad event happening; by adapting, they reduce the *consequences* when a bad event actually does occur. But for the most part, the climate change literature has modelled mitigation and adaptation separately. This is unfortunate since significant interactions are likely to exist between how people choose to mitigate and adapt. These risk reduction strategies probably complement or negate each other. Understanding the interaction between the two can help formulate what you think of Kyoto. The benefits of mitigation will be lower the more people can adapt to the climate.

People can privately and collectively affect the threats they confront. This realization may have profound impact on the formal evaluation of climate change policy. First, researchers need to rethink the traditional risk assessment–risk management bifurcation currently applied to the research and management of climate change risk. Second, we should also acknowledge that both private and collective risk-reduction actions be considered in benefit–cost analysis and programme evaluations to avoid the under-valuation of risk reductions. Third, researchers should also consider the implications of when a person passes a risk along to someone else across time and space.

6. A choice of vision

A moderate 'broad, then deep' approach with gradual emissions reductions made sense before Kyoto and it still makes sense today.[67] Kyoto provides no additional benefits for the extra cost it will impose relative to this alternative benchmark. One might claim that this additional cost is the price of building international trust and securing a politically credible 'first step'. This might well be the case. But it is not obvious to everyone that this step would not have occurred anyway, or that the billions of redirected dollars could not be spent more usefully on a dozen other environmental, health or safety issues.

But whether or not one thinks this matters depends on what Thomas Sowell calls your choice of vision, people with an unconstrained vision believe we all have a vast untapped morality buried within, waiting to emerge with the right direction. Thus solutions like the Kyoto Protocol are primary; the trade-offs involved are secondary. As Sowell puts it, "Every closer approximation to the ideal should be preferred. Costs are regrettable, but by no means decisive."[68] People with a constrained vision, however, weigh ideals against the costs of achieving them. Real incentives will be needed to get people to take on the goals of Kyoto intentionally. The concrete messages from economics are that the catastrophes have to be exceedingly likely for Kyoto to make sense. But for now and the foreseeable future, the uncertainties of climate change leave enough latitude so that whether you choose to believe the benefits justify the costs of Kyoto rests on your choice of vision.

Where to now? If climate change is really just about the developed world picking up the tab for benefits accruing to future generations in the developing world, there must be a better way – a more direct way – to do this than Kyoto. But if climate change is about more than this, we need to consider questions to forge a larger middle ground on benefits and costs. Find the price to induce the developing countries to come on board. Define the odds for catastrophe and surprise concepts. Quickly examine the nature of sinks to sort out whether the costs of measurement, verification and enforcement problems exceed the benefits, thereby reducing the odds of this discussion degenerating into a sideshow distraction. Commission testbed studies on emissions markets because effective institutional design requires no less; it worked for the spectrum markets, it can work for emissions markets as well. Construct real case studies to understand what institution-building exercises work across developed and developing nations. Carefully evaluate the trade-off frontier of flexibility and stringency within political constraints. Give more consideration to the incentives for technological progress created by different climate policies over the long term, including the opportunity cost of inducing innovation in climate protection versus other deserving goals. Finally, be vigilant that we are not only talking to ourselves as we sharpen the benefit and cost estimates as events unfold and new research results come forth.

Notes

* This paper draws on work with M. Toman forthcoming in: P. Portney and R. Stavins, eds, Climate Change Policy, Public Policies for Environmental Protection 2nd edition, Washington, D.C., Resources for the Future. All views herein are mine. Thanks to Joe Aldy, Bob Hahn, Sally Kane, Randy Lutter, David Montgomery, Al McGartland, Ray Squitieri, Rob Stavins, Bob Tuccillo, Jonathan Weiner, and participants at the AEI conference on Climate Change Policy for their helpful comments.

1. Bolin, B. The Kyoto negotiations on climate change: a science perspective. Science. 16 January 1998;279:330–31.
2. Dales, J. Pollution, Property, and Prices. University of Toronto Press. 1968.
3. Several technical reviews of the costs and benefits of climate change policy exist. See the Intergovernment Panel on Climate Change (IPCC) Working Group III. J. Bruce, H. Lee and E. Haites, eds, Climate Change 1995: Economics and Social Dimensions of Climate Change. The Contribution of Working Group III to the Second Assessment Report of the Intergovernment Panel on Climate Change. Cambridge, Cambridge University Press, 1996; and its critique, W. Nordhaus, ed., Economics and Policy Issues in Climate Change. Washington, D.C., Resources for the Future, 1998. Others question whether benefit-cost analysis is useful at all considering the uncertainties involved – see for example the papers by Bolin and Tietenberg in the climate change policy forum in Environmental and Development Economics. 1998;3:347–409.
4. See, for example, Jacoby, H., R. Prinn and R. Schmalensee, Kyoto's unfinished business. Foreign Affairs. July/August 1998:54–66.
5. See Sedjo, R., B. Sohngen and P. Jagger, Carbon sinks in a post Kyoto world, Climate Issues Brief No. 12, Washington, D.C., Resources for the Future. October 1998.
6. US Department of State, The Kyoto Protocol on climate change: State Department fact sheet. 15 January 1998.
7. Cooper, R. Toward a real global warming treaty. Foreign Affairs. March/April 1998;66–79.
8. Quoted in Huber, S. and C. Douglass, Two Perspectives on Global Climate Change. A Briefing Book. Center for the Study of American Business, Washington University of St. Louis, July 1998.
9. Benefits can be more than just avoided damage if one includes indirect or secondary benefits such as reduced congestion or air pollution. We address the validity of this point shortly.
10. Wigley, T. The Kyoto Protocol: CO_2, CH_4 and climate implications. Geophysical Research Letters. 1998;25:2285–88.

11. These aggregate estimates also mask significant variability in impacts across regions, economic sectors, and industries.
12. Nordhaus, R. To slow or not to slow: the economics of the greenhouse effects. Economic Journal. 1991;101:920–37.
13. Pearce, D. Economic development and climate change. Environment and Development Economics. 1998;3:389–92.
14. Schelling, T. The costs of combating global warming. Foreign Affairs. November/December 1997;8–14; Schelling, T. Some economics of global warming. American Economic Review. 1992;82:1–14.
15. See for instance Epstein, P. Climate, ecology, and human health. Consequences. 1997;3:3–19.
16. Moore, T. (1998) makes the case that moderately warmer weather is more conducive to human health, such that climate change might reduce mortality rates in the US by about 40,000 per year. See Moore, T. Health and amenity effects of global warming. Economic Inquiry. 1998;36:471–88.
17. Krupnick, A. Climate change, health risks, and economics. Weathervane, www.weathervane.rff.org May 1998.
18. See Crocker, T. and J. Shogren, Endogenous risk and environmental program evaluation. In: Knaap, G. and T.J. Kim, eds, Environmental Program Evaluation. A Primer. Urbana, IL: University of Illinois Press. 1997;255–69; Duncan, G. Does poverty affect the life chances of children? American Sociological Review. 1998;63:406–23; Kremer, M. Integrating behavioral choice into epidemiological models of AIDS. Quarterly Journal of Economics. 1996;111:549–74; Korenman, S. and J. Miller, Effects of long-term poverty on physical health of children in the national longitudinal survey of youth. Photocopy, 1997.
19. Brown, G. and J. Shogren, Economics of the endangered species act. Journal of Economic Perspectives. 1998;3–20.
20. Costanza, R., R. D'Arge, R. De Groot, The value of the world's ecosystem services and natural capital. Nature. 1997;387:253–60.
21. See for example Smith, V.K., Mispriced planet. Regulation. Summer 1997;16–17. On reflection, most economists would agree with M. Toman's point that the US$33 trillion figure is 'a serious underestimate of infinite', see Nature. 1998;395:430.
22. See Alexander, A., J.A. List, M. Margolis and D'Arge, R.C., A method for valuing global ecosystem services. Ecological Economics. 1998;27:161–70.
23. Burtraw, D. and M. Toman, The benefits of reduced air pollutants in the US from greenhouse gas mitigation policies. Climate Issues Brief No. 7, Resources for the Future, October 1997.
24. See the PM NAAQS and Ozone NAAQS Regulatory Impact Analyses, Office of Air Quality Protection and Standards, US Environmental Protection Agency. http://www.epa.gov/ttncaaa1/t1ria.html
25. Eizenstat, S., Stick with Kyoto. Foreign Affairs. 1998;77:119–21.
26. Allen, J., El Niño's price tag sets no disaster record. Washington Post, 5 April 1998.
27. Barron, E., Written testimony, committee on environment and public works, US Senate, 11 July 1997; Barron, E., Climate models: how reliable are their predictions? Consequences. 1995;1:16–27.
28. In fact, some companies are being quite aggressive. British Petroleum, the world's third-largest petroleum company, has taken on emissions trading itself, and has opened a new solar manufacturing facility. See Percy, S., Making progress beyond Kyoto, how workable is the Kyoto Protocol? Weathervane. www.weathervane.rff.org March 1998.
29. The 'broad, then deep' path seems to be consistent with one side of the domestic US political strategy reflected in the non-binding Byrd–Hagel resolution, passed in the Senate 95-0, that stated that the US should accept no climate agreement that did not demand comparable sacrifices of all participants.
30. Schmalensee, R., Greenhouse policy architectures and institutions. MIT Joint Program on the Science and Policy of Global Climate Change. Report No. 13, November 1996; Schneider, S.

and L. Goulder, Achieving carbon dioxide targets cost-effectively: what needs to be done now? Nature. 1997;389:13–14.

31. Wigley, T., R. Richels and J. Edmonds, Economic and environmental choices in the stabilization of atmospheric CO_2 concentrations. Nature. 1996;379:240–43; Manne, A. and R. Richels, On stabilizing CO_2 concentrations: cost-effective emission reduction strategies. EMF 14, April 1997.
32. The pre-Kyoto results from the President's Interagency Analysis Team (IAT) are within this range as well. An exception is that the IAT estimates that reducing emissions to 1990 levels by 2010 would cost 900,000 jobs in 2005, and 400,000 jobs by 2010. Also see Yellen, J., The economics of the Kyoto Protocol, statement before the Committee on Agriculture, Nutrition, and Forestry. US Senate. 5 March 1998.
33. Also see the report by the US Energy Information Administration, *Impacts of the Kyoto Protocol on US energy markets and economic activity*. October 1998. The report finds that full US compliance with the Kyoto Protocol could raise gasoline prices 53% and electric rates 86% by the year 2010. The report predicts a drop in US GNP of about 4% by 2010 under a worst-case scenario.
34. See for example Grubb, M. et al., The costs of limiting fossil-fuel CO_2 emissions: a survey and analysis. Annual Review of Energy and the Environment. 1993;18:397–478; Edmonds, J. et al., Return to 1990: the costs of mitigating United States carbon emissions in the post-2000 period. Battelle, October 1997; McKibbon, W. and P. Wilcoxen, A better way to slow global climate change. Brookings Institute, March 1997.
35. Hamilton, M. and C. Chandler, Cures that involve a world of pain. Washington Post. 13 November 1997.
36. Montgomery, D., Global impacts of a climate change treaty. In: Adlen, J. ed., The costs of Kyoto. Washington, D.C., Competitive Enterprise Institute, 1997, 57–72.
37. Brown, S. et al., The economic impact of international climate change policy. ABARE 1997.
38. These estimates are robust across different modelling runs. The EMF compares a diverse group of economic models employing different methodologies. Standardizing these models by assuming common exogenous parameters yielded similar results. This suggests that the choice of method is secondary to the choice of values for population growth, per capita income, energy intensity, and technical progress.
39. Nordhaus, W. and J. Boyer, Requiem for Kyoto: an economic analysis of the Kyoto Protocol. Yale University Photocopy, 29 June 1998.
40. Discussions with Ray Squitieri, US Department of Treasury.
41. Also see Repetto, R. and D. Austin, The costs of climate protection. WRI 1997.
42. Warrick, J., "People are sweltering", says Gore, tying record heat to global warming. Washington Post, 15 July 1998.
43. Lichtenstein, S. et al., The judged frequency of lethal events. Journal of Experimental Psychology. 1978;4:551–78; Viscusi, W.K., Fatal tradeoffs. Oxford, Oxford University Press, 1992.
44. For example, Fox et al., Consumer preferences for irradiation given alternative information. Working paper, Kansas State University. July 1998.
45. See for example, Blinder, A., Needed: planet insurance. New York Times. 22 October 1997.
46. The economics literature on catastrophes include Cropper, M., Regulating activities with catastrophic environmental consequences. Journal of Environmental Economics and Management. 1976;3:1–15; Clarke, H. and W. Reed, Consumption/pollution trade-offs in an environment vulnerable to pollution-related catastrophic collapse. Journal of Economic Dynamics and Control. 1994;18:991–1010; Eismont, O. and H. Welsch, Optimal greenhouse gas emissions under various assessments of climate change ambiguity. Environmental and Resource Economics. 1996;8:129–40.
47. Peck, S. and T. Teisberg, Uncertainty and the value of information with stochastic losses from global warming. Risk Analysis. 1996;16:227–35.
48. Nordhaus, W., Expert opinion on climatic change. American Scientist. 1994;82:45–51.

49. Roughgarden, T., Quantifying the damage of climate change: implications for the DICE model. Stanford University. 14 March 1997.
50. Gjerde, J., S. Grepperud and S. Kverndokk, Optimal climate policy under the possibility of a catastrophe. Statistics Norway. February 1998.
51. Kelly, D. and C. Kolstad, Bayesian learning and accumulation of stock externalities. UC-Santa Barbara. 1996.
52. One guess is that carbon leakage would probably be between 10% and 50%. See Rutherford, T., International competitiveness and national plans. University of Colorado, 1995.
53. See the recent ideas by Hahn, R. and R. Stavins, Thoughts on designing an international greenhouse gas trading system. JFK School of Government, Harvard University. September 1998.
54. See Shogren, J., Lessons from the lab. Regulation. January 1998, 6.
55. See for example Stavins, R., The costs of carbon sequestration: a revealed-preference approach. American Economic Review. 1999;89:994–1009.
56. Discussion with Brent Sohngen, Ohio State University, August 1998.
57. Jacoby, H., R. Prinn and R. Schmalensee, Kyoto's unfinished business. Foreign Affairs. July/August 1998;54–66.
58. Parry, I.W.H., Reducing carbon emissions: interactions with the tax system raise costs. Resources. Summer 1997;128:9–12.
59. Europe Union Starts planning strategy to stress domestic action under Kyoto Pact. Daily Environment Report. 31 March 1998;1060–2976.
60 Firor, J., US needs to lead by example in reducing emissions; how workable is the Kyoto Protocol? Weathervane, www.weathervane.rff.org March 1998.
61. United States Climate Action Report – 1997, Department of State, July 1997.
62. For example, see Metcalf, G. and K. Hassett. Measuring the energy savings from home improvements investments: evidence from monthly billing data. NBER working paper 6074. June 1997.
63. Anonymous, Clinton opposes higher energy taxes to curb global warming. New York Times. 7 October 1997.
64. Congressional Budget Office. Climate change: the policy challenge and current programs. CBO. August 1998.
65. Goulder, L. and S. Schneider, Induced technological change and the attractiveness of CO_2 abatement policies. Resource and Energy Economics. 1999;21:211–53.
66. Chakravorty, U., J. Roumasset and K. Tse, Endogenous substitution among energy resources and global warming. Journal of Political Economy. 105:1201–34.
67. Hahn suggests a better first step would be a series of institution-building exercises aimed at engaging developing nations in climate change policy. (Hahn, R., ed. The economics and politics of climate change. Washington, D.C., American Enterprise Institute, 1998). This approach could make sense, especially if one addresses the historical fact that once a problem is transformed into institutional rules, interest groups that invested in these rules will fight to keep them, regardless of the inefficiency. See Yandle, B., Bootleggers, baptists, and global warming. PERC Policy Series PS-14. November 1998.
68. Sowell. T., A conflict of visions. Ideological origins of political struggles. New York, William Morrow and Company, 1987, 34.

CHAPTER 3

The Kyoto Protocol: A Cost-Effective Strategy for Meeting Environmental Objectives?*

Alan S. Manne and Richard G. Richels

1. Introduction

The Kyoto Protocol represents a milestone in climate policy.[1] For the first time, negotiators have attempted to lay out emissions reduction targets for the early part of the 21st century. The goal is for Annex 1 (developed countries plus economies in transition) to reduce their aggregate anthropogenic carbon dioxide equivalent emissions by at least 5% below 1990 levels in the commitment period 2008–2012. The Protocol, however, has yet to enter into force. To do so will require ratification by 55 countries representing 55% of total Annex 1 CO_2 emissions in 1990.

As each country considers ratification, important questions will arise. High up on the US list is the issue of economic costs. The Senate, for example, has stated that "any Protocol should be accompanied by a detailed financial analysis of impacts on the economy."[2] Not surprisingly, US negotiators had hardly returned from Kyoto before the first hearings were scheduled on Capitol Hill. Although the issue of costs is but one of many important considerations, policy makers are keenly interested in the economic implications of ratification.

This paper is intended to help clarify our understanding of compliance costs. The focus is on three questions, which we believe to be of particular relevance: What are the near-term costs of implementation? How significant are the so-called 'flexibility provisions'? And, perhaps most importantly, is the Protocol cost-effective in the context of the long-term goals of the Framework Convention?[3]

Unfortunately, the answers to these questions will not come easily. It has always been difficult to calculate the economic costs of implementing climate policy. Kyoto has done little to simplify matters. Indeed, it raises at least as many questions as it resolves. These questions fall into two categories: those related to the near-term implementation of the Protocol and those related to the evolution of climate policy over the longer term.

C. Carraro (ed.), Efficiency and Equity of Climate Change Policy, 43–61.

The Protocol is unclear on a number of topics. These include the rules governing emissions trading, joint implementation (JI), the Clean Development Mechanism (CDM), and the treatment of carbon sinks. In addition, there is a weak knowledge base regarding the costs of sink enhancement and of controlling several of the relevant trace gases. Until these issues are clarified, analyses will be highly speculative.

Calculating the costs of Kyoto is also complicated by the issue of 'what happens next?' Energy sector investments are typically long-lived. Today's investment decisions are not only influenced by what happens during the next decade, but also by what happens thereafter. In order to estimate the costs of implementing emissions cuts in the first commitment period, assumptions are required concerning the longer-term requirements. Unfortunately, the international negotiation process offers little guidance on this issue. This further complicates the process of analysis.

We do not wish to suggest that economic analysis is premature at the present time. Uncertainty is rarely an excuse for paralysis. It does mean, however, that we must be careful to highlight the tentative nature of the projections and focus, to the greatest extent possible, on the insights for decision making. Here, sensitivity analysis can be particularly useful. For example, in the case of several of the flexibility provisions (emissions trading, joint implementation and the Clean Development Mechanism), we explore a variety of scenarios regarding constraints on the purchase of carbon emissions rights. While the exact magnitude of the benefits will continue to be debated, the insights nevertheless appear to be quite robust.

We also examine the Protocol in the context of the longer-term goal of the Framework Convention, i.e. the stabilization of greenhouse gas concentrations in the earth's atmosphere. A particular concentration goal can be reached through a variety of emissions pathways. Considerable effort has been devoted to trying to understand the characteristics of cost-effective pathways.[4] It is interesting to examine Kyoto in the context of this work. The price tag for moving forward may be formidable. Consistent with the Framework Convention, it is essential that "policies and measures to deal with climate change should be cost-effective so as to ensure global benefits at the lowest possible costs."[5]

2. The model

This analysis is based on MERGE (a model for evaluating the regional and global effects of greenhouse gas reduction policies).[6,7] MERGE is an intertemporal market equilibrium model. It combines a bottom-up representation of the energy supply sector together with a top-down perspective on the remainder of the economy. Savings and investment decisions are modelled as though each of the regions maximized the discounted utility of its consumption subject to an intertemporal wealth constraint. Each region's wealth includes not only capital, labour and exhaustible resources, but also its negotiated international share in carbon emissions rights.

For the present version of the model, known as MERGE 3.0, we have adopted 10-year time intervals through 2050 and 25-year intervals through 2100. Geographically, the world is divided into nine geopolitical regions: (1) the US, (2) OECDE (Western Europe), (3) Japan, (4) CANZ (Canada, Australia and New Zealand), (5) EEFSU (Eastern Europe and the Former Soviet Union), (6) China, (7) India, (8) MOPEC (Mexico and OPEC) and, (9) ROW (the rest of world). Note that the OECD (regions 1 through 4) together with EEFSU constitute Annex 1 of the Framework Convention.

Particularly relevant for the present analyses, MERGE provides a general equilibrium formulation of the global economy. We model the possibility of international trade in carbon emissions rights. This is sometimes known as 'where' flexibility. It would allow regions with high marginal abatement costs to purchase emissions rights from regions with low marginal abatement costs. In addition, MERGE can be used to examine the related issue of 'when' flexibility – intertemporal transfers of carbon emissions rights.

We also model international trade in oil, natural gas, and energy-intensive basic materials. We are therefore able to examine issues related to 'carbon leakage'. Such leakage can occur through a variety of pathways. For example, Annex 1 emissions reductions will result in lower oil demand, which in turn will lead to a decline in the international price of oil. As a result, non-Annex 1 countries may increase their oil imports and emit more than they would otherwise.

The present version of the model includes the notion of endogenous technical diffusion. Specifically, in the electric sector, the near-term adoption of high-cost carbon-free technologies leads to accelerated future introduction of lower-cost versions. The model also includes both price-induced and non-price conservation. For most regions and time periods, the AEEI (autonomous energy efficiency improvement) rate is taken to be 40% of the rate of GDP growth. By 2100, this leads to regional energy–GDP ratios that are much closer to each other than they were in 1990.

In calibrating MERGE for the present analysis, several supply- and demand-side parameters were adjusted so that the global emissions baseline would approximate the Intergovernmental Panel on Climate Change (IPCC) central case 'no policy' scenario (IS92a).[8] Figure 1 shows carbon emissions for each region in the reference case scenario. For more on the model and its key assumptions, see our website:

http://www-leland.stanford.edu/group/MERGE/

3. Treatment of sinks and non-CO_2 greenhouse gases

Few issues have engendered as much confusion as that of carbon sinks. Key questions include their definition, the extent to which they are included in the

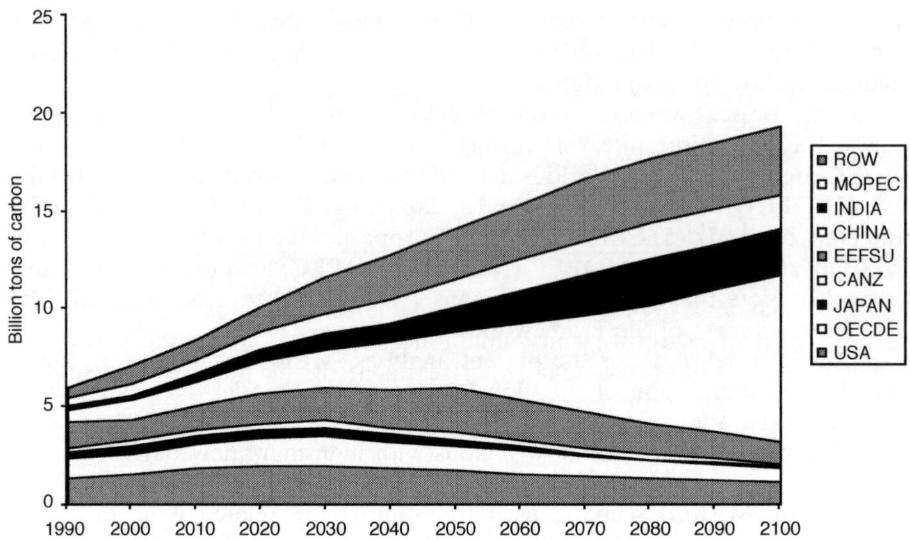

Figure 1. Regional carbon emissions – reference case.

Protocol, the amount currently being sequestered, their time profile, and the costs of sink enhancement.

The Protocol states that Annex 1 commitments can be met by "the net changes in greenhouse gas emissions from sources and removal by sinks resulting from direct human-induced land use change and forestry activities limited to afforestation, reforestation, and deforestation since 1990, measured as verifiable changes in stocks in each commitment period."[9] The confusion results from alternative interpretations regarding the treatment of soil carbon, an issue flagged for further study in the Protocol. Their inclusion may result in large increases in the international legal definition of sink potential.

The quality of the data is uneven. The supply curves for sink enhancement are particularly questionable. The degree of confidence concerning current and predicted future levels of carbon sequestration varies enormously across regions of the globe. Not surprisingly, information is most reliable (albeit still poor) for Annex 1 countries. Comparatively little effort has been made to collect such data elsewhere.

We have adopted the values shown in Table 1 as placeholders. To provide some perspective, in order for the US to reduce industrial carbon emissions by 7% below 1990 levels in 2010, it would have to reduce emissions by approximately 550 million tonnes below its reference trajectory. Sink enhancement would satisfy 9% of this obligation. For the purposes of the present analysis, we assume that this sink enhancement is costless.

CO_2 is by far the most important of the greenhouse gases. In addition, the

Table 1. Sink enhancement

	Million metric tonnes of carbon annually
USA	50
OECDE	17
Japan	0
CANZ	50
EEFSU	34

Protocol includes five other trace gases (methane, nitrous oxide, hydrofluorocarbons, perfluorocarbons, and sulphur hexafluoride). Given the scarcity of reliable emissions and cost data, the treatment of the non-CO_2 greenhouse gases is also problematic. For the purposes of the present analysis, we assume that each gas is reduced proportionately. With this proportionality assumption, the inclusion of the non-CO_2 greenhouse gases does not affect the requirements for CO_2 reductions.

As with our treatment of sinks, we do not include the costs of abating the non-CO_2 greenhouse gases in our estimates of the costs of complying with the Protocol. Clearly, an important next step would be to develop supply curves for the cost of abating non-CO_2 greenhouse gases and for sink enhancement. Neither of these costs is included in the present version of MERGE.

4. 'Kyoto forever'

We begin with an examination of a 'Kyoto Forever' scenario. This is a case in which the Kyoto constraints on Annex 1 countries are maintained throughout the 21st century. With regard to non-Annex 1 emissions, we assume they will continue to be bounded by their business-as-usual baseline (Figure 1). The latter constraint is imposed in order to prevent carbon leakage. Later on, we will explore the impact of relaxing this constraint.

Numerous studies have shown that global mitigation costs can be reduced substantially by allowing emissions reductions to take place wherever it is cheapest to do so – regardless of geographical location.[10] The Kyoto Protocol includes several provisions allowing for a limited amount of 'where' flexibility. These include emissions trading and joint implementation among Annex 1 countries. They also include provisions for a Clean Development Mechanism (CDM) that is intended to facilitate joint implementation between Annex 1 and non-Annex 1 countries.

As with the definition of sinks, the Protocol leaves many critical details unresolved. For example, it remains unclear whether there will be limits on the extent to which a country can rely upon the purchase of emissions rights to satisfy its obligations. The Protocol states that "the Conference of the Parties shall define the relevant principles, modalities, rules and guidelines ... ."[11] Similar

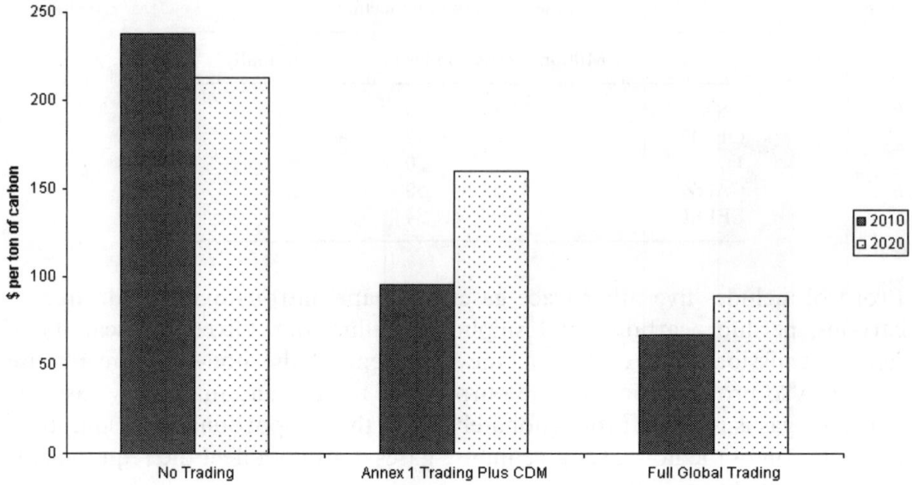

Figure 2. Incremental value of carbon emissions rights in US under Kyoto Forever.

ambiguity surrounds the Clean Development Mechanism. Again, the elaboration of 'modalities and procedures' is left to a future meeting of the Conference of the Parties.[12]

In this section, we explore three scenarios: (1) no trading, (2) Annex 1 trading plus CDM, and (3) full global trading. These three options are representative of alternative implementations of the Kyoto Protocol. Each has its own advocates and opponents, but we do not consider them equally likely. In our opinion, there is little likelihood of enticing all major countries to participate in a global market in emissions rights during the initial commitment period (2008–2012).

The full global trading scenario places an upper bound on the CDM's potential to reduce GDP losses. In calculating the potential size of the contribution from a CDM, we therefore calculate this upper bound on the export of emissions rights from non-Annex 1 regions. Because of the difficulties in implementation of the CDM, however, we assume that only 15% of the potential would be available for purchase through this mechanism. This is a highly subjective estimate. Given the complexities of the CDM, however, we are not inclined to assign a higher value.

Figure 2 reports the incremental value of carbon emissions rights to the US in 2010 and 2020. We focus first on 2010. In the most constrained scenario, the US must satisfy its emissions reduction requirements within its own geographical boundaries. In this case, the value of emissions rights approaches US$240 per tonne. With Annex 1 trading plus CDM, the value drops to slightly less than US$100 per tonne. As might be expected, the value of emissions rights is lowest with full global trading. Here, it falls below US$70 per tonne.

For the two scenarios in which trading is permitted, the value of emissions

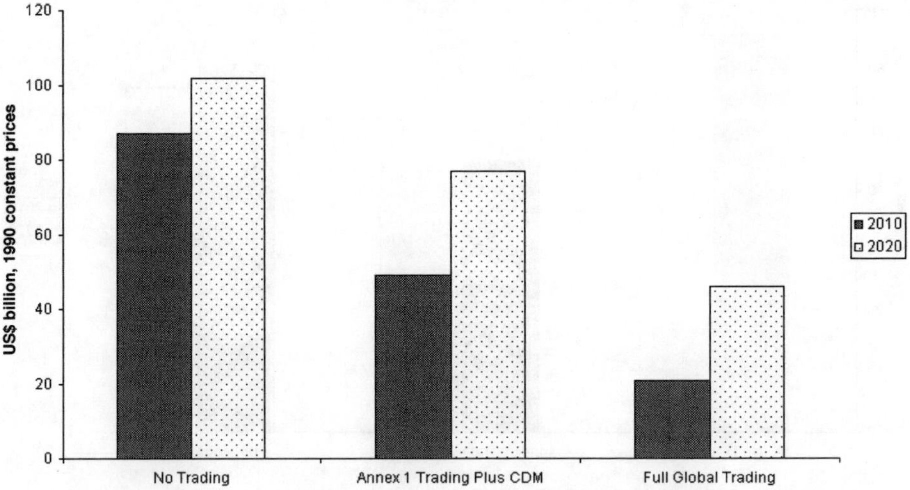

Figure 3. Annual US GDP losses under Kyoto Forever (US$ billion).

rights increases in 2020. This is because EEFSU's projected emissions lie below its negotiated constraint for 2010. It has been allocated more emissions rights than it needs to satisfy its internal obligations. By 2020, however, EEFSU's economic growth is expected to be such that it no longer enjoys an excess of emissions rights. As a result, there is more competition for emissions rights in the international marketplace, and there is an increase in their price.

Another way to view the costs of abatement is to show the GDP losses. Figure 3 contains those for the US. Losses are highest in the absence of trade. Here, they exceed US$80 billion in 2010. This is approximately 1% of US GDP. To the extent that trade is introduced, losses decline. In the most optimistic scenario (full global trade), losses are approximately US$20 billion or one-quarter of 1% of GDP in 2010.

Of the three scenarios, 'Annex 1 trading plus the CDM' is most consistent with the Protocol as it currently stands. However, the US Senate has stated that the US should not be a signatory to the Protocol if it does not mandate specific commitments for developing countries.[13] If this were to result in full global emissions trading, we move in the direction of the right-most bar of Figure 3.

There is, however, strong sentiment among many parties to the Framework Convention to limit substantially the extent to which Annex 1 countries can meet their obligation through the purchase of emissions rights. Several influential developing countries have expressed strong opposition to the concept altogether. Figure 3 shows the costs of the no trading scenario. We now turn to the case where trading is permitted, but with limitations on the purchase of emissions rights.

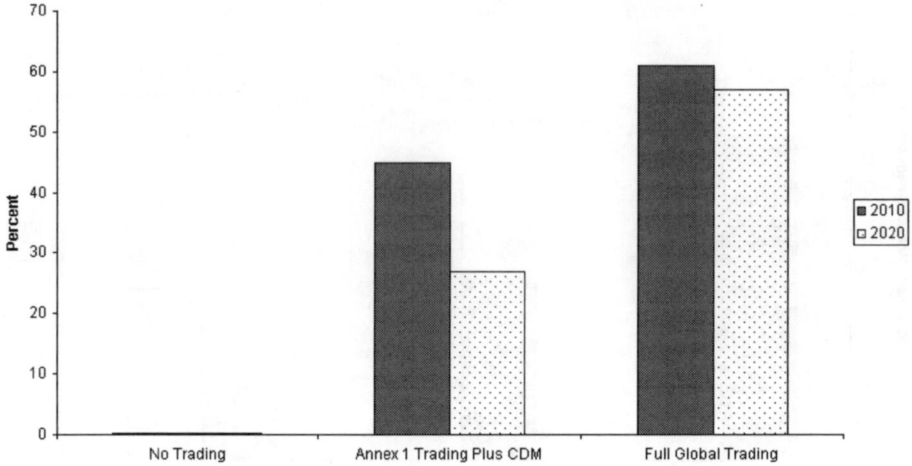

Figure 4. Percentage of US obligation satisfied through the purchase of emissions rights.

5. Limits on the purchase of carbon emissions rights

Figure 4 shows our estimates of the percentage of the US emissions reduction obligation that would be satisfied through the purchase of emissions rights under base case assumptions. With full global trading (the least-cost of our three scenarios), trading is used to satisfy more than 50% of the US obligation. But suppose that limits are placed on the purchase of emissions rights? For example, suppose that international negotiators agree that Annex 1 buyers can satisfy only one-third of their obligation through this means. What would be the impact on GDP losses?

Figure 5 compares three cases. All assume full global participation in an international market for carbon emissions rights, but only the first assumes no limits on the amount a country can buy. The second and third cases are based upon the one-third limitation. We further make the distinction between a buyers' market and a sellers' market. With the former, sellers of emissions rights are price takers. Buyers exert sufficient market power to hold the international price to the marginal cost of abatement in the selling countries. However, since a country is only able to satisfy one-third of its obligation through the purchase of emissions rights, it must eventually rely on its own domestic marginal abatement capabilities to meet its obligations. Hence, there is an important distinction between the international price and the domestic price. Conversely, with a sellers' market, buyers face but one price. Here, the rents accrue to the sellers.

Figure 6 shows the GDP losses associated with the three scenarios. Note that losses in 2010 are two and one-half to three times higher with the constraint on the purchase of carbon emissions rights. That is, the benefits from 'where'

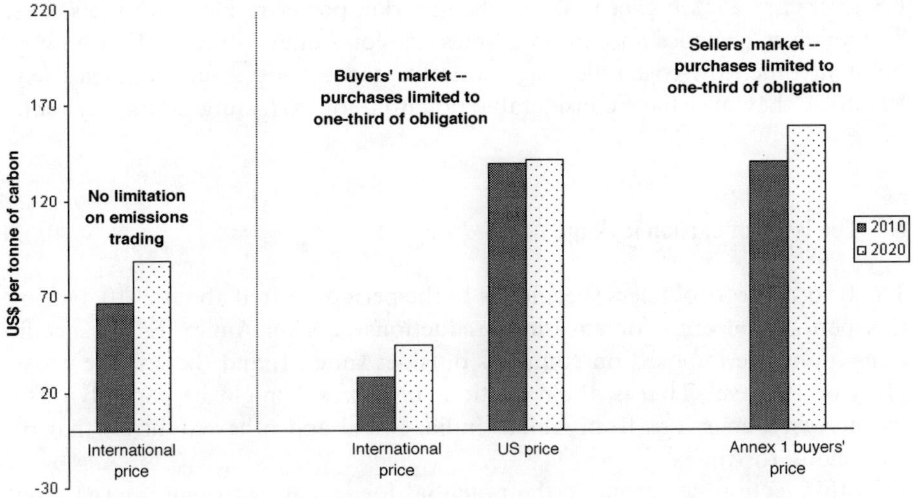

Figure 5. Incremental value of carbon emissions rights with and without limits on the purchase of emissions rights – Kyoto Forever.

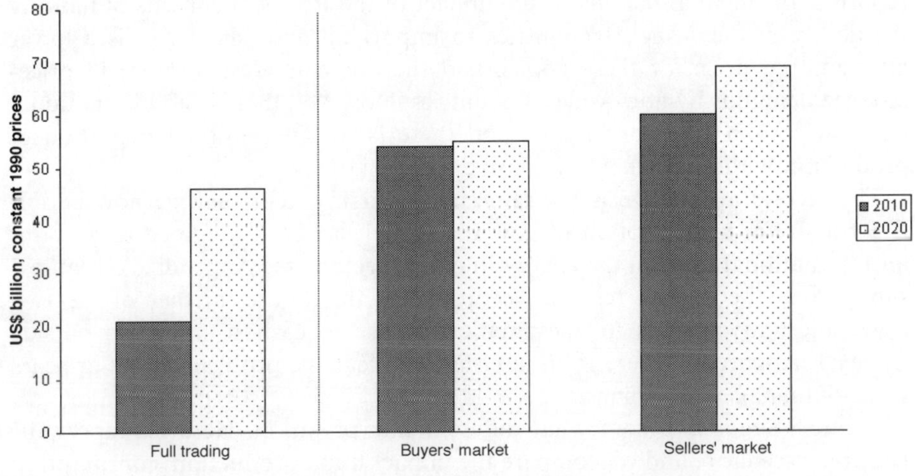

Figure 6. Annual US GDP losses with full trading – Annex 1 may satisfy only one-third of obligation through the purchase of emissions rights – Kyoto Forever.

flexibility are greatly diminished. The message is clear. Developing country participation in the market for carbon emissions rights is a necessary, but by no means a sufficient condition for reaping the full benefits of 'where' flexibility. To achieve a cost-effective solution, buyers must also be unconstrained in the manner in which they fulfil their obligation.

Also note that the distribution of the rents makes a difference to GDP losses.

US losses are 25% higher in 2020 when market power resides with the sellers. The analysis provides an additional message for Annex 1 buyers. If at a given point in time, low-cost sellers are concentrated among a few countries (e.g. EEFSU), they may have considerable potential for extracting monopoly rents.

6. The issue of carbon leakage

The Kyoto Protocol refers specifically to the period centred about 2010. During this period, the onus for emissions reductions falls on Annex 1. No specific obligations are imposed on countries outside Annex 1, and there is the possibility of 'leakage'. That is, the reductions in Annex 1 might be partially offset by increased emissions from China, India, Brazil and other countries that do *not* belong to Annex 1.

In this section, we examine the potential for leakage through international fuel markets and through the migration of energy-intensive industries. We therefore drop the assumption that non-Annex 1 countries are constrained to their reference-case emissions. Two variants on the reference scenario are reported. In the first, the only trade impact of the Protocol consists of limiting the ability of the Annex 1 countries to import oil and gas. There is a lower international price for these goods, and there is a modest increase in price-induced demands by non-Annex 1 countries. However, there is no international trade in carbon emissions rights, and there is no international migration of production within the energy-intensive sectors (EIS).

The second alternative is the same as the first, except that we now permit EIS trade. For a description of how the model has been modified to account for international trade in the energy-intensive sectors, see Appendix A. Figure 7 summarizes the overall results. According to this figure, neither of the two trade alternatives leads to a dramatic increase in carbon emissions outside Annex 1. Apparently there is an international leakage problem, but it appears to be of manageable dimensions.

Figure 8 suggests a somewhat different interpretation. Here we report the EIS trade scenario, and we compare the impact upon production–consumption ratios in each region. Under the reference case, these ratios are close to unity (the horizontal line) in most regions. The bars in Figure 8 show that the Protocol could lead to serious competitive problems for EIS producers in the US, Japan and OECD Europe. The Protocol would lead to significant reductions in their output and employment, and there would be offsetting increases in regions with low energy costs. One can easily anticipate calls for protection against 'unfair competition'. In its present form, the Protocol could lead to acrimonious conflicts between those who advocate free international trade and those who advocate a low-carbon global environment.

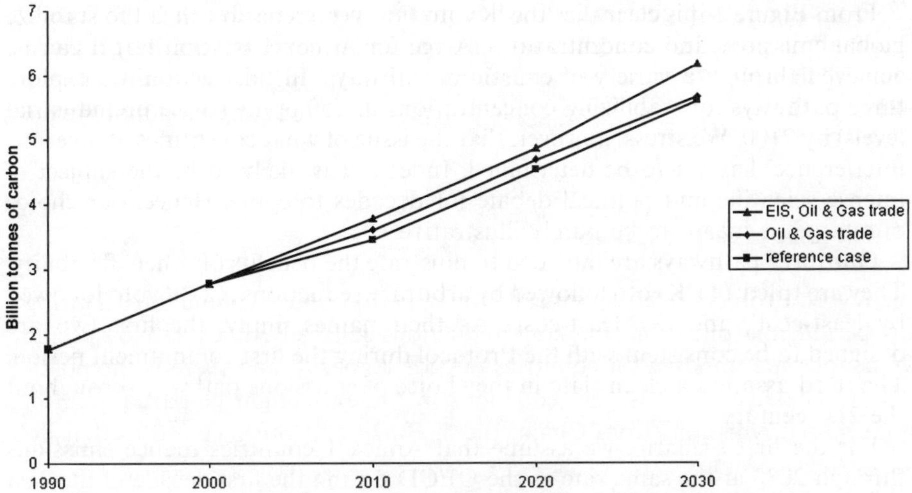

Figure 7. Carbon emissions outside Annex 1 – alternative leakage scenarios.

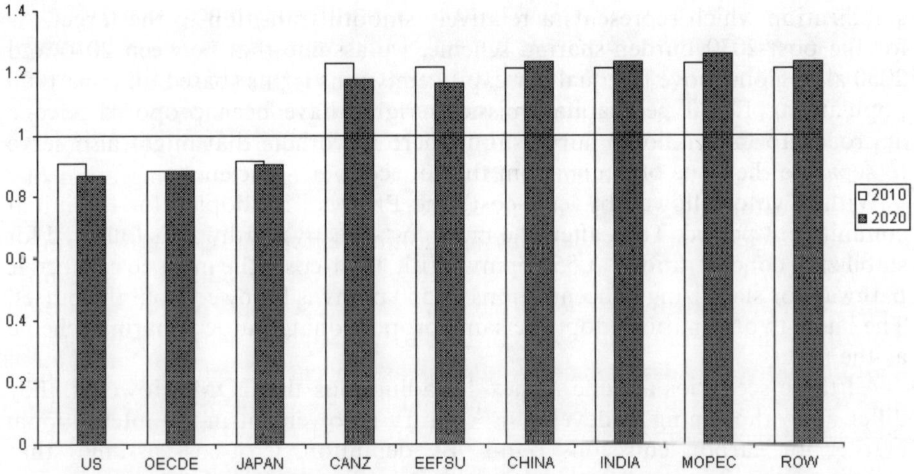

Figure 8. Ratios of domestic EIS supplies to demands – Kyoto with leakage.

7. Evaluating Kyoto in the context of the longer-term goal

The objective of the Framework Convention is "the stabilization of greenhouse gas concentrations at a level that would prevent dangerous anthropogenic interference with the climate system."[14] The drafters of the Protocol focused exclusively on the initial steps to be taken by Annex 1 countries. Little attention was paid to the ultimate goal. We now examine the Protocol in the context of a long-term stabilization objective.

From Figure 1 it is clear that the 'Kyoto Forever' scenario will fail to stabilize global emissions and concentrations. A particular concentration target can be achieved through a variety of emissions pathways. In this section, we explore three pathways for stabilizing concentrations at 550 ppmv (twice preindustrial levels) by 2100. We stress, however, that the issue of what constitutes 'dangerous interference' has yet to be determined. Indeed, it is likely to be the subject of intense scientific and political debate for decades to come. Hence, our choice of a target is meant to be purely illustrative.

Our three pathways are intended to illustrate the benefits of 'when' flexibility. They are titled: (1) 'Kyoto followed by arbitrary reductions', (2) 'Kyoto followed by least-cost', and (3) 'least-cost'. As their names imply, the first two are designed to be consistent with the Protocol during the first commitment period. The third assumes a clean slate in the choice of emissions pathway throughout the 21st century.

For the first scenario, we assume that Annex 1 countries reduce emissions through 2030 at the same rate as the OECD during the first decade of the 21st century (2% per year). During this period, non-Annex 1 countries are permitted to emit up to their reference case levels. By 2020, emissions in the developing nations are larger than those in Annex 1. We then choose a pathway to stabilization which represents a relatively smooth transition to the target. As for the post-2030 burden-sharing scheme, we assume that between 2030 and 2050 all regions move to equal per capita emissions rights (based on their 1990 population). Equal per capita emissions rights have been proposed as one approach to international fairness, but there are others that might also serve to separate the issue of equity from that of economic efficiency.

With 'Kyoto followed by least-cost', the Protocol is adopted for the initial commitment period. Thereafter, the most cost-effective pathway is followed for stabilizing concentrations at 550 ppmv. With 'least-cost', the most cost-effective pathway for stabilizing concentrations at 550 ppmv is followed from the outset. The latter two scenarios adopt the same proportionate burden-sharing scheme as the first.

All three scenarios assume Annex 1 trading plus the CDM. However, they differ as to the timing of developing country involvement in the international market for carbon emissions rights. By definition, least-cost assumes that emissions reductions will be made where it is cheapest to do so, regardless of the geographical location. Hence, in the least-cost scenario, we assume global emissions trading from the outset. In the case of 'Kyoto followed by least-cost', we assume that global emissions trading is delayed until after the first commitment period. With 'Kyoto followed by arbitrary reductions', global emissions trading does not begin until 2030, the year that developing countries agree to lower their emissions below business-as-usual.

Global carbon emissions

Figure 9 shows global carbon emissions for the reference case and the three stabilization scenarios. Following a least-cost strategy from the outset results

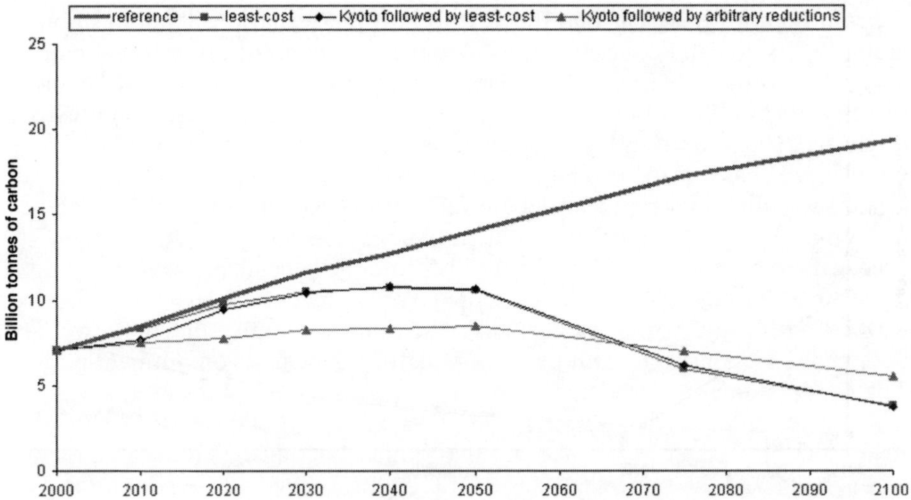

Figure 9. Global carbon emissions – reference case and three alternative emissions pathways for stabilizing concentrations at 550 ppmv.

in an emissions pathway that tracks the reference path through 2010 and then departs at an increasing rate thereafter. There are several reasons why a gradual transition to a less carbon-intensive economy is preferable to one involving sharper near-term reductions.

Concentrations at a given point in time are determined more by cumulative, rather than year-by-year, emissions. Indeed, a concentration target defines an approximate carbon budget, i.e. an amount of carbon that can be emitted between now and the date at which the target is to be reached. At issue is the optimal allocation of the budget. Reasons for relying more heavily on the budget in the early years include: (1) providing more time for the economic turnover of existing plant and equipment, (2) providing more time to develop low-cost substitutes to carbon-intensive technologies, (3) providing more time to remove carbon from the atmosphere via the carbon cycle, and (4) the effect of time on discounting on mitigation costs.[15]

We next turn to the two scenarios where we adopt the Protocol for the first commitment period. Notice that the two emissions pathways behave quite differently post-2010. 'Kyoto followed by least-cost' follows the least-cost pathway once the Protocol's constraints are relaxed. 'Kyoto followed by arbitrary reductions', on the other hand, bears no resemblance to the least-cost pathway. What is striking about Figure 9 is that with a 550 ppmv target, the Protocol is inconsistent with the most cost-effective mitigation pathway, i.e. 'least-cost'. Indeed, it appears that the ultimate target would have to be considerably lower than 550 ppmv for the Protocol to be justified in terms of cost-effectiveness.

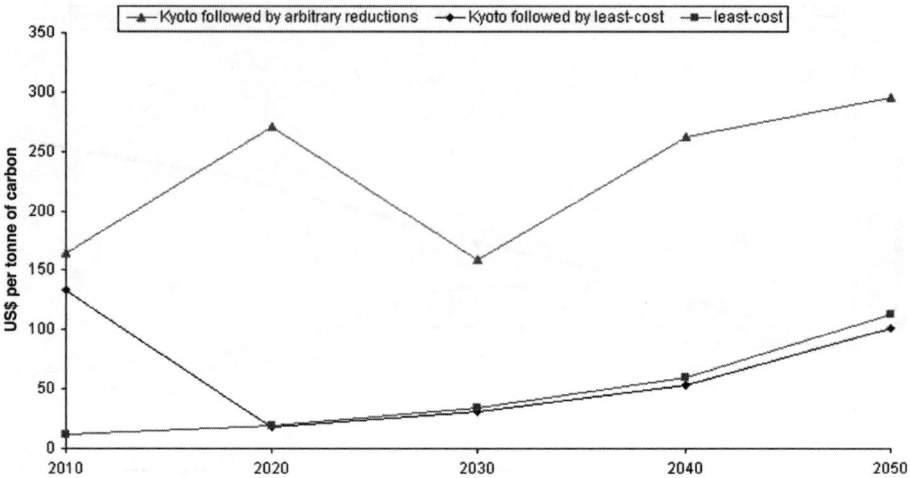

Figure 10. Incremental value of carbon emissions rights under three alternative emissions pathways for stabilizing concentrations at 550 ppmv – global trading.

Near-term losses

It is instructive to look at the incremental value of emissions rights for the three stabilization scenarios (Figure 10). With the least-cost path, the value is relatively low in the early years (US$11 per tonne of carbon in 2010), and it rises gradually over time. With 'Kyoto followed by least-cost', the value is US$130 per tonne in 2010 and then tracks the least-cost path thereafter. In the case labelled 'Kyoto followed by arbitrary reductions', the incremental value of emissions rights starts at about US$160 per tonne and it remains high.

Figure 11 shows US GDP losses in 2010 and 2020 under the three stabilization scenarios. Notice that GDP losses in 2010 differ for the two scenarios involving the initial adoption of the Protocol. Because of the long-lived nature of energy investments, investors are concerned both with what happens in the initial commitment period and with what happens thereafter. In the case of the more rapid transition away from the baseline ('Kyoto followed by arbitrary reductions'), investors will be forced to invest more heavily in high-cost substitutes in the early years. With 'Kyoto followed by least-cost', they will have more flexibility.

It is striking by how much GDP losses can be lowered under the 'least-cost' scenario. This strategy involves a more gradual transition away from the baseline in the early years. It relieves much of the pressure for premature retirement of existing plant and equipment and for dependence on high-cost substitutes (on both the supply side and the demand side of the energy sector). Relative to the reference case, the US also receives some benefits as an oil importer. Recall that a carbon constraint decreases the overall demand for oil and lowers its price on the international market.

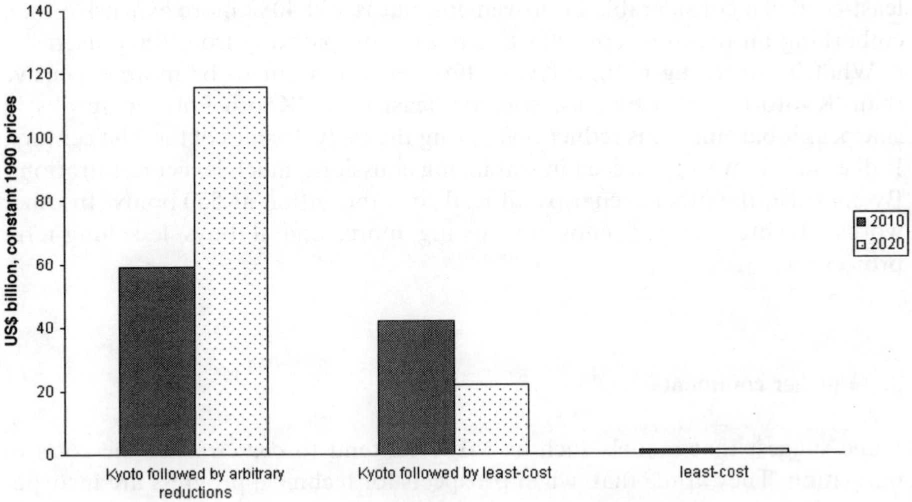

Figure 11. **US GDP losses under alternative 550 ppmv stabilization scenarios.**

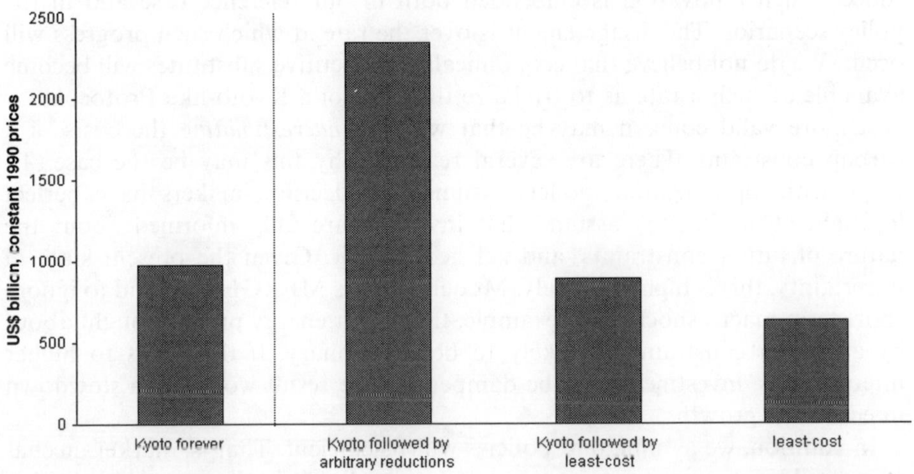

Figure 12. **Global consumption losses through 2100 discounted to 1990 at 5% – Kyoto Forever vs. three scenarios for stabilizing concentrations at 550 ppmv.**

Global losses

Finally, it is instructive to examine losses from a global perspective (Figure 12). For the purposes of the present comparison, we focus on the present value of consumption losses over the 21st century discounted to 1990 at 5%. The relative magnitude of the cumulative losses for the three stabilization scenarios comes as no surprise given the previous discussion. 'Kyoto followed by arbitrary reductions' is by far the most expensive of the three paths. 'Kyoto followed by least-cost' is a considerable improvement, but is still 40% more expensive than embarking on the most cost-effective mitigation pathway from the outset.

What is surprising is that 'Kyoto Forever' turns out to be more expensive than 'Kyoto followed by least cost' or 'least cost'. 'Kyoto Forever' results in sharper global emissions reductions during the early decades of the 21st century. It does not, however, succeed in stabilizing emissions, much less concentrations. By contrast, the other scenarios all lead to stabilization at 550 ppmv. In other words, 'Kyoto Forever' ends up costing more, and it buys less long-term protection.

8. Further comments

Some suggest that models such as MERGE tend to *overestimate* the costs of mitigation. They argue that, when prospects for technical progress are incorporated, the costs of a carbon constraint, even a sharp near-term constraint, will be minimal. We, too, are optimistic about the outlook for technical innovation. Indeed, such innovation is embedded both in our reference case and in the policy scenarios. The disagreement is over the rate at which such progress will occur. We do not believe that economically competitive substitutes will become available at such a rate as to trivialize the costs of a Kyoto-like Protocol.

A more valid concern may be that we are *underestimating* the costs of a carbon constraint. There are several reasons why this may be the case. To begin with, optimization models assume that decision makers have perfect foresight. That is, they assume that investors are fully informed about the nature of future constraints, and act accordingly. Given the present state of uncertainty, this is highly unlikely. Models such as MERGE also tend to ignore short-term macro shocks. For example, the higher energy prices brought about by a carbon constraint are likely to be inflationary. If this leads to higher interest rates, investment may be dampened. The result would be a slowdown in economic growth.

In addition, we assume that policies will be efficient. That is, market mechanisms will be chosen over 'command and control' approaches to accomplishing environmental objectives. Whereas, in recent years, there has been an increasing trend toward market mechanisms, the approach to be taken with climate policy is by no means assured. Moreover, even if such a commitment were made, we have no assurances that the requisite international institutions will be available when needed.

Although it is easy to quibble over the numbers, the real value of analyses lies more in insights than in numerical values. Indeed, we believe that the current exercise has yielded several insights that may be of value to those charged with interpreting the current proposal. Here, we summarize what we have learned:

- First, it is extremely unlikely that a 'Kyoto Forever' scenario will stabilize emissions – much less concentrations. Non-Annex 1 emissions are quickly overtaking those of the OECD and the economies in transition. Hence, meeting the stabilization goal of the Framework Convention will eventually require the participation of developing countries.
- International co-operation through trade in emissions rights is essential if we are to reduce mitigation costs. The magnitude of the savings will depend on several factors. These include the number of countries participating in the trading market, the shape of each country's marginal abatement cost curve, and the extent to which buyers can satisfy their obligation through the purchase of emissions rights.
- With regard to the latter, limitations on the purchase of emissions rights may be especially costly. In the example explored here, limiting purchases to one-third of a country's obligation increased GDP losses by a factor of at least two and one-half in the year 2010. If proponents of such limitations are successful, they may seriously reduce the benefits from 'where' flexibility.
- The issue of monopoly power in markets for emissions rights may turn out to be important. This is most likely to occur if trading is limited to Annex 1 and the majority of inexpensive emissions rights are concentrated in a small number of countries. If these countries were successful in organizing a sellers' cartel, they might be able to extract sizable rents.
- The near-term costs of the Protocol will depend on expectations regarding the future. Energy investments are typically long-lived. Today's investment decisions are influenced not only by what happens during the next decade, but also by what happens thereafter. Hence, analyses which focus solely on 2010 may be underestimating the costs of Kyoto.
- Finally, and perhaps most importantly, unless the concentration target for CO_2 is well below 550 ppmv, the Protocol appears to be inconsistent with a cost-effective long-term strategy for stabilizing CO_2 concentrations. Rather than requiring sharp near-term reductions, it appears that a more sensible strategy would be to make the transition at the point of capital stock turnover. This would eliminate the need for premature retirement of existing plant and equipment, and would provide the time that is needed to develop low-cost, low-carbon substitutes.

Note

* This research results from our involvement in Stanford University's Energy Modeling Forum 16 Study. The authors are indebted to James Deaker and Robert Parkin for their research assistance. We have benefited from discussions with Jae Edmonds, Henry Jacoby, David Montgomery, William Nordhaus, Stephen Peck, Thomas Rutherford and John Weyant. Funding was provided by EPRI. The results presented here are solely those of the individual authors.

References

1. Conference of the Parties (1997). Kyoto Protocol to the United Nations Framework Convention on Climate Change. Third Session Kyoto. 1–10 December.
2. 105th Congress, 1st Session (1997). S. Res. 98.
3. Intergovernmental Negotiating Committee for A Framework Convention on Climate Change (1992). Fifth Session, Second Part, New York. 30 April–9 May.
4. IPCC (1997). Stabilization of atmospheric greenhouse gases: physical, biological and socio-economic implications. Technical Paper III.
5. Intergovernmental Negotiating Committee for A Framework Convention on Climate Change (1992). op. cit.
6. Manne, A., Mendelsohn, R. and Richels, R. (1995). MERGE: a model for evaluating regional and global effects of GHG reduction policies. *Energy Policy*. 3:1.
7. Manne, A. and Richels, R. (1995). The greenhouse debate: economic efficiency, burden sharing and hedging strategies. *The Energy Journal*. 16:4.
8. IPCC (1994). Climate Change 1994. Cambridge University Press.
9. Conference of the Parties (1997). Kyoto Protocol to the United Nations Framework Convention on Climate Change. *op. cit.*
10, Richels, R., Edmonds, J., Gruenspecht, H. and Wigley, T. (1996). The Berlin mandate: the design of cost effective mitigation strategies. EMF-14, Working Paper 3. Stanford, CA, Stanford University.
11. Conference of the Parties (1997). Kyoto Protocol to the United Nations Framework Convention on Climate Change. *op. cit.*
12. Conference of the Parties (1997). Kyoto Protocol to the United Nations Framework Convention on Climate Change. *op. cit.*
13. 105th Congress, 1st Session (1997). S. Res. 98.
14. Intergovernmental Negotiating Committee for A Framework Convention on Climate Change (1992). *op. cit.*
15. Wigley, T., Richels, R. and Edmonds, J. (1996). Economic and environmental choices in the stabilization of atmospheric CO_2 concentrations. *Nature*. 379:18 January.

Appendix 1. Modelling international trade in the energy-intensive sectors

MERGE 3.0 has recently been modified to include the possibility of trade in EIS (energy-intensive sectors). EIS is an aggregate including ferrous and non-ferrous metals, chemicals, non-metallic minerals, paper, pulp and print. This aggregate does not include the energy-intensive industry of petroleum refining. The model may be run either with or without EIS trade.

The new feature is introduced in a way that preserves the basic simplifying characteristics of the ETA-MACRO submodel. That is, energy, capital and labor are substitutes that enter into an aggregate production function. They produce an index of the value of all non-energy goods which may be used for consumption, investment and interindustry payments for energy costs.

It is assumed that trade will continue to represent a relatively small amount of each region's total internal demand for EIS. The GTAP (General Trade Analysis Program, 1992) database is employed to estimate each region's EIS demands. In all other respects, the model is the same as MERGE 3.0.

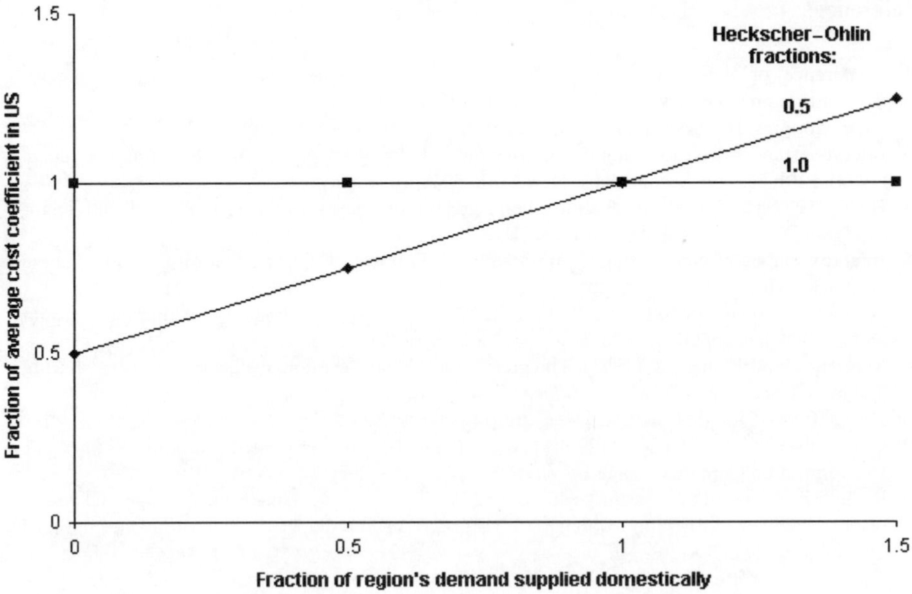

Figure A.1. EIS supply curves – marginal cost of non-energy inputs to EIS.

For projecting the impact of the Kyoto Protocol, each region is taken to be self-sufficient at base year energy prices. Changes in the location of production are attributed primarily to changes in the cost of energy. At base year prices in the US, 85% of the cost of EIS consisted of non-energy inputs (labour, shipping, capital, iron ore, etc.), and 15% of the cost consisted of energy inputs (half electric and half non electric). Under these conditions, a doubling of energy prices would imply only a 15% increase in the cost of EIS. This is why it is assumed that the demand for EIS is inelastic with respect to the price of energy. For projecting future demands, the income elasticity is taken to be 0.5.

For modelling purposes, we have supposed that the marginal supplies of EIS in all regions are determined by the same international technology that prevails in the US. Each region has the same energy–EIS production ratio. For non-energy inputs, each supply curve is linear. Its positive slope serves the same purpose as an Armington elasticity describing substitution between foreign and domestic goods. This is the way in which we avoid penny-switching as a characteristic solution mode.

The slope of the non-energy supply curve is described as a Heckscher–Ohlin fraction. If this fraction is unity, EIS is viewed as a perfectly homogeneous commodity. Small changes in energy costs will then lead to large changes in the international location of production. If this fraction is less than unity, the supply function is less elastic, and the changes in location will be less dramatic (see Figure A.1.)

CHAPTER 4

Negotiating Targets, Negotiating Flex-Mex: the Economic Background of a US–EU Controversy

Jean-Charles Hourcade and Thierry Le Pesant[1]

Introduction

Some observers qualified Kyoto's Conference as a 'flexibility feast', but this feast almost ended in a failure when the EU on one side and US and the rest of the JUSCANZ[2] on the other demonstrated rather different views about the enforcement of the flexibility mechanisms; and when the G77 rejected the concept of joint implementation (JI) and resisted to see the word 'emissions trading' written in the Protocol.

Hence, the last-minute invention of the Clean Development Mechanism (CDM) saved the Kyoto Protocol, but this could not but conceal major points of contention concerning the principles, modalities, rules and guidelines governing the enforcement of flexibility mechanisms. More specifically two sorts of preconditions to the use of these mechanisms were set:

- The first is encapsulated in a G77 'non-paper' at the SBSTA meeting at Bonn in June 1998: "*Until the question of emissions rights and entitlements has not been addressed equitably, it will not be possible to have emissions trading*".[3] The question of entitlements to trade remained a major point of discussion at COP4 in Buenos Aires (November 1998) and was not really addressed at COP 5 in Bonn (October 1999).
- The second is the European interpretation of Article 17 of the Kyoto Protocol which states that "*the Parties included in Annex B may participate in emissions trading for the purposes of fulfilling their commitments under Article 3. Any such trading shall be supplemental to domestic actions for the purpose of meeting emissions limitation reduction commitments under that Article*". Whereas the US and many JUSCANZ countries interpret the last sentence as trivial, the EU, following the statement of its Council of Environment Ministers of October 6th 1998,[4] recalls that "*a concrete ceiling on the use of flexible mechanisms has to be defined ... in quantitative and qualitative terms based on equitable criteria.*"

C. Carraro (ed.), Efficiency and Equity of Climate Change Policy, 63–90.

The first precondition raises the most intellectually and politically difficult problems involving equity issues. But as will be demonstrated, it may become one of the drivers of the supplementarity quarrel beyond 2012.

The rationale for trading systems is very clear from an economic perspective; they aim at minimizing the overall economic and social costs of meeting commitments, and to hedge against uncertainties about the domestic cost of commitments. The economic rationality for preconditions other than a fair organization of the market is less obvious. In a first best world indeed, any Party, observing the international price of the carbon, will set up any domestic action which marginal cost will be lower than this price and the supplementarity criteria will be automatically fulfilled.

In fact the supplementarity preconditions convey a set of very different concerns:

- distrust vis-à-vis market mechanisms and the possibility of fair and transparent GHGs trading,
- political suspicion that flexibility mechanisms will be used by the richest countries with the highest levels of emissions to avoid the political and economic cost of domestic measures,
- concerns about the 'hot air' during the first budget period and about possible 'loopholes' enlarged by the CDM,
- concerns about the compatibility between GHGs trading and existing energy taxation future carbon taxes,
- concerns about simple 'cap and trade' mechanisms not sending signals likely to put countries on an emissions path compatible with minimizing the cost of meeting the long-run of the Framework convention on Climate Change (FCCC).

In this paper, we will first come back to the choices for co-ordination of climate policies from both an analytical and a historical point of view. We will then demonstrate the importance of the dynamic inconsistency problem: on the one hand, too-low prices in the first budget period may not give the appropriate incentive; on the other hand, too-high carbon prices may lower the capacity of countries to accept ambitious commitments for the second budget period (Bohm, 1999). Finally, we will examine the question of the involvement of developing countries in climate policies in the light of this dynamic inconsistency.

1. From Rio to Kyoto: pervading economic ambiguities

Co-ordinating climate policies requires three questions to be answered simultaneously: (a) how stringent should the abatement policy be? (b) what types of economic instruments are likely to maximize the cost-efficiency of action? (c) will the burden of action be shared in an equitable and fair manner?

The negotiation process on any global environmental issue can be carried out in order to reach one of the three alternative agreements: quantitative emissions reduction targets; a common set of policy signals; or simple reporting procedures. The latter hypothesis having been ruled out very quickly in the case of climate change, the remaining alternatives offer a classic 'price versus quantity' dilemma. Despite its theoretical character, this discussion helps in understanding neglected fundamental structures of any attempt at co-ordination under uncertainty and provides a useful insight into the EU/US controversy about the 'supplementarity condition'.

Indeed, before Rio (1992), the European Commission and the majority of the EU member states historically favoured an approach by co-ordinated policies and measures including a carbon-energy tax; it is only with the Berlin Mandate that the approach by means of quantitative targets was adopted. However, it would be misleading to make a retrospective analysis of such a move; it was hardly the result of a conscious strategy by the key actors. More specifically the linkage between quantitative limits to GHGs emissions and the use of trading systems was not clearly understood and assumed by all stakeholders, which explains part of the stumbling-blocks on the road to COP 6.

1.1. Co-ordination under uncertainty: prices versus quantities

Current policy controversies demonstrate that the 'price versus quantity' dilemma well known by professional economists remains largely ignored by policy makers. This dilemma stems from the fact that, because of the uncertainty about the response of the economic system to public incentives, it is impossible to ascertain what level of carbon tax[5] is required in a given country to meet a given abatement target. This is simply a case of the impossibility of planning both in quantity and prices which is one of the area of consensus amongst economists, after the historical failure of translating in the real world the theoretical possibility to decentralise central quantitative objectives through prices demonstrated by some Eastern European economists (Kantorovitch 1976; Liptak, 1965).

In the real world, a trade-off has to be made between a price co-ordination and a quantity co-ordination. In a price co-ordination approach, the costs of climate policy are controlled (the tax represents the marginal willingness to pay for precaution), but the outcome is not fully predictable; in a quantity co-ordination approach, the outcome is certain, but not the costs incurred by each participant. This explains why the use of a tradable permit system is a necessary complement to any binding commitment on quantitative emissions limits. Indeed, since abatement costs are unpredictable *ex ante* it would be economically and politically risky for a country to commit itself seriously (i.e. without envisaging deliberate non-compliance) without any hedging against adverse consequences.

In purely economic terms, Weitzman (1974) demonstrated that the trade-off

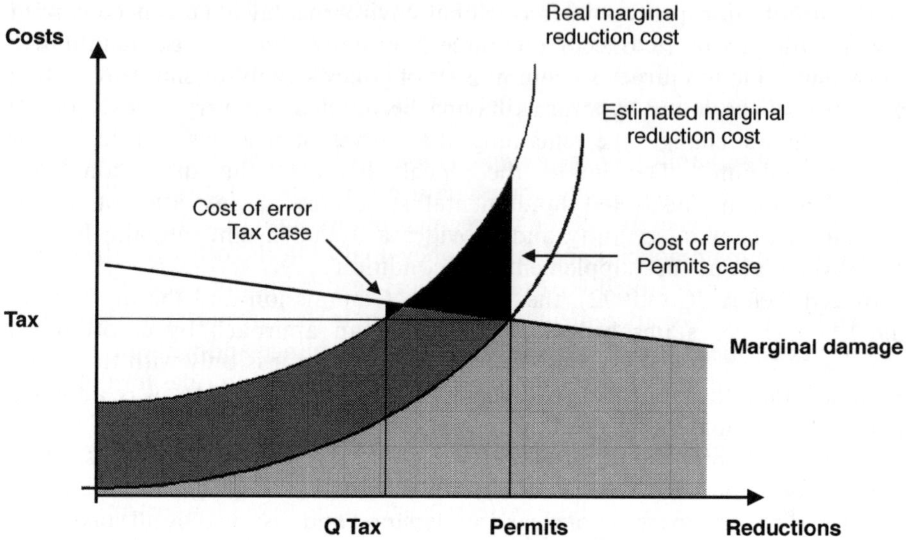

Figure 1. Deadweight loss of tax versus permits in case of flat marginal damages.

between price and quantity co-ordination depends upon the shape and the degree of uncertainty of the marginal abatement cost curve and of the marginal damage curve. When the marginal abatement cost curve is very steep and the marginal damage curve rather flat, price co-ordination is preferable, unless the degree of uncertainty about the marginal damage curve is dramatically higher; the conclusion is the reverse if the marginal damage curve is higher. Indeed, as abatement costs are highly controversial, this means that in the case of wrong cost forecasts, the cost of an error leads to higher taxes in case of a steep marginal damage curve, and higher quotas in the case of a flat marginal damage curve (Pizer, 1997), as shown in Figure 1. Building on Weitzman's result, Newell and Pizer have demonstrated that global warming being a stock pollution, the benefit curve of marginal avoided emissions is necessarily flat (Newell and Pizer, 1998).

The assumption of a very high probability level of a climate catastrophe such as the slowing down of the thermoalyne circulation in the North Atlantic does not suffice to reverse this result unless there is a significant probability that this catastrophe may occur. Because existing scientific information provides no evidence about such, co-ordination through prices (i.e. taxes) should be the first choice.

1.2. Increasing ambiguities about co-ordination patterns

Beyond this purely analytical result, arguments in favour of a price-based co-ordination are based on the intrinsic deadlocks of a quota approach in terms of procedural efficiency (political acceptability, difficulty in revealing the information about preferences and margins for action, including no-regret potentials). These difficulties can be grouped in three broad categories which have played a role in the process between Berlin and Kyoto:

- the non-observability of the *ex-ante* emissions baselines and of the domestic abatement costs may incite some countries to demonstrate high baselines, to minimize the existence of no-regret options and to deny the linkages between climate policies and other policy goals. On the other hand, governments may be tempted to exert political leadership by announcing very ambitious GHGs abatement targets which in fact result from structural policies decided for reasons other than climate change and without adopting real additional measures. Given the difficulty in justifying rules for differentiating targets there is a tendency to converge on quasi flat-rate abatements which are consistent with neither efficiency nor equity principles;
- the necessity of setting up large-scale GHGs trading systems at the world level: past experience of these systems has been analysed for cost-minimising the reduction of a single pollutant (unleaded petrol, SO_2) in homogeneous sectors and in a regional or a national context, mainly in the US with pre-existing strict command and control policies. There are many reasons why it is hazardous to apply the outcome of such systems (Hourçade and Baron 1993, Godard and Henry 1998) to the case of GHGs: first, several gases are involved and many uncertainties remain about how to measure them in a reliable way and about what common metrics should be used; second, these gases are emitted by a various range of activities characterized by very different economic logic and management routines; third, the trading must take place between countries having very different institutional and legal traditions; fourth, the general equilibrium effects depend on the way in which the quotas are reallocated to the industry at the domestic level (auction versus gratis allocation). Overall it was predictable from the outset that the setting up of such systems would require the resolving of some harmonization problems in the absence of a pre-existing legal framework;
- the equity issue: meeting the long-term targets of the Climate Convention requires the involvement of developing countries in mitigation policies; but this involvement should be guided by the principle of common but differentiated responsibilities; to translate this principle in terms of primary emissions rights may trigger a *regressum ad infinitum* of controversies due to the combination of scientific uncertainties and debates about legitimate ethical criteria.[6] As demonstrated by the fierce debate between the World Resources Institute (1990) and A. Agarwal and S. Narain (1991), any attempt to measure the total emissions of a country is confronted with the wide range of uncertainty concerning emissions from sources other than CO_2 from the formal energy sector and the fragility of the conversions factor (Global Warming Potential). This is all the more likely to stem the burning

debates that emissions of CH_4 and CO_2 from fuel-wood and deforestation are linked to the satisfaction of basic needs.

This set of arguments explains why the European Commission and some key European countries attempted to impose co-ordination through a price signal in the form of carbon tax or carbon energy tax in 1992 and, to a lesser extent, between 1992 and 1995. The other approach, which was for a long time officially supported by no major country but only by environmental NGOs, was systematized by M. Grubb (1989). The basic argument was that trading systems were the only way to organize financial and technological transfers to the developing countries in order to help them cut down their emissions. It was not based on a response to the 'pure efficiency' counter argument but on a set of political economy considerations providing a different view on the procedural and political constraints affecting the negotiations. At this stage, we will not elaborate on the fact that some of the political economy arguments against a quota approach are weaker in the case of repeated games with a learning institutional process. We will rather present four main arguments which have played a real role against a co-ordination through carbon taxes:[7]

- the unrealistic political aim of reaching a consensus on a co-ordinated carbon tax given the cultural and institutional context in the US;
- the sovereignty principle which would lead many countries (the US but also European countries such as the UK) to reject any restriction on the liberty of their parliaments to decide for their fiscal regime;
- the opposition of energy-intensive industry and of the energy sector, including OPEC countries;
- the fact that a quantitative target approach linked with trading systems will provide a means for the developing world embarking on mitigation efforts, while the alternative approach of raising domestic energy prices in these countries, however consistent it may be for the sustainability of their development, would create acceptability problems over the short term because of negative distributional impacts.

The pros and cons of each alternative solution are well established and have played a role in the process at different points in time; but a lot of the implementation problems of the Kyoto Protocol come from the fact that the very logic of each approach has not been perceived equally in various regions of the world, the Kyoto Protocol resulting from an ambiguous compromise between these two approaches.

The Berlin Mandate contained indeed a lot of ambiguities. In the US and in countries such as Canada, Australia and New Zealand the interpretation was univocal; the adoption of the quota approach was linked to the perspective of international trading systems, and a country which was committed to reducing its GHGs emissions would logically try and achieve this objective where it is cheaper to do so; moreover the idea of trading was politically acceptable and was perceived as providing the basis for a compromise with the industry. The dominant interpretation in Europe was far from being along these lines. On this side of the Atlantic, some countries and ideological forces which supported the quota approach understood the Berlin Mandate as a way to force the US to implement domestic policies and measures to reduce their emissions. They did not perceive the adoption of the quota approach meaning, at least for their partners, the abandoning of any attempt to co-ordinate GHGs abatement measures internationally.[8]

It is remarkable that very little work on emissions trading systems was done in the European countries. During the preparation of Kyoto, the European Commission worked principally on a set of P&M to be either adopted at the EU level or imposed internationally. But there was never a strong political push in favour of some items of this list and no explicit ranking of these items with regard to the required level of co-ordination. In other words the intellectual preparation for the Kyoto Conference by the European Union was focused on how to distribute targets within the EU bubble, but did not succeed in proposing P&M adapted to the new negotiation contexts. In the meantime, the US and other OECD countries prepared for the Kyoto Conference as if there would be no objection to the immediate use of trading systems.

1.3. Kyoto's unfinished business[9]

Kyoto targets are interpreted in very different ways by the various stakeholders: in the US, the opponents to any mitigation policy argue that they result from a pure political process and will entail dramatic costs for society; on the other hand, many environmental non-governmental organizations express the concern that the adopted targets may not be ambitious enough to enforce a real precautionary principle vis-à-vis climate risks.

In fact, 2012 is too short a time horizon for climate policies, and no judgment about Kyoto targets can be made without considering them as a simple passing point. The question is whether they are compatible with a long-term emissions pathway suitable for stabilizing GHGs concentration below a tight ceiling. The lesson from the 'time flexibility' debate initiated by Wigley, Edmonds and Richels (WRE) in *Nature* (1996) is indeed that drastic short-term abatements are not the least-cost pathway to meet a given concentration target. But since the ultimate target compatible with the long-term objective of the FCCC is currently unknown (the 550 ppmv target proposed by the EU is scientifically arbitrary) the danger in optimizing the response throughout the next century is to make too costly the shift towards more ambitious targets (say 450 ppmv) if this is proved to be necessary in the light of new information. Because of the inertia of economic systems, any acceleration in abatements inflates the costs drastically, and after one or two decades of 'no action', it may then be socially preferable to renounce climate control.

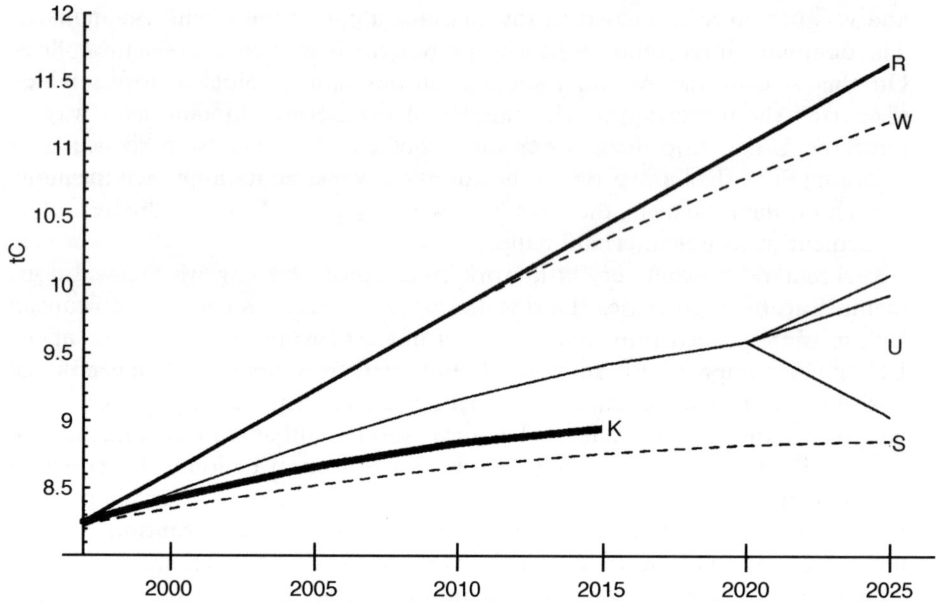

Figure 2. Different future CO_2 emissions strategies. R: reference emission profile following IS92a; W: WRE; U: HGH; K curve: Kyoto targets; S: 550 ppmv stabilization.

Figure 2 compares the aggregate Kyoto target (K) to two emissions trajectories published in *Nature* during the pre-Kyoto timing debate. Rather unsurprisingly, this target represents a far more ambitious departure from current trends than that suggested by WRE for a 550 ppmv concentration ceiling; but it is also lower than the trajectory proposed by Ha-Duong, Grubb and Hourcade (HGH) who calculated the expected value of 450 ppmv, 550 ppmv and 650 ppmv and projected the costs of switching from a 550 ppmv to a 450 ppmv objective. Too much importance should not be placed on this result; it can reflect a high risk-aversion to climate change, but it can also come from the fact that key players have incorporated an allowance for many 'no-regret' cases in their evaluation of mitigation. However, it demonstrates that Kyoto targets safeguard enough flexibility for switching towards a target as ambitious as 450 ppmv.

Far more controversial is the distribution of emissions targets among countries. In principle this distribution should indeed be set so as to equate to marginal welfare losses in each country, but, given the non-observability of emissions baselines, of abatement costs and of country-specific 'no-regret' cases, it is impossible to follow such a rule. For example, the US can argue that their emissions baseline will in 2012 be 30% over the 1990 level while the figures presented in the preparatory work of the European Commission suggested only a 12% increase. But the EU scenarios incorporated in their baseline (i.e.

at zero cost) many measures already adopted in the EU (or on the road towards adoption) which were not explicitly proposed for reasons of climate policies but which would result in lower GHGs emissions trends. Conversely the required abatement in the US can be suspected not to represent a good indicator of the effort since the likelihood of no-regret cases should be higher in this country that in the EU, given current discrepancies in per capita energy consumption.

Another very controversial aspect of Kyoto targets, which is one critical trigger of the 'supplementarity quarrel' is obviously the risks of excess assignments of emissions rights in Russia and Ukraine if the current post-communism transition crisis does not comes to an end in the next few years. There is indeed a widely held view that if an international GHGs trading system operates, the inflow of the hot air would result in the trading of fictitious reductions.[10] What has to be emphasized at this stage is the similarity between the perception of this risk and the question of the developing countries in a climate policy, a similarity based on the difficulty of setting emissions baselines from which emissions abatements can be calculated. Kyoto experts from most of the OECD countries have been inclined to think that, since no new commitments would be required from developing countries (i.e. no binding commitments in addition to the reporting obligations contained in the FCCC), the G77 and China would not pose major problems and that the proposal of 'activities implemented jointly' would be a modest but practical way to interest them in climate policies. The importance of AIJ was underlined when the US Senate declared that the Kyoto Protocol would not be ratified if developing countries (at least the majority of them) did not demonstrate a meaningful participation in climate mitigation.

However, the last-minute refusal of AIJ by the G77 and China marked the end of the 'flexibility feast', the neologism Clean Development Mechanism, which attempted to reconcile the need for flexibility for Annex B countries and the demands of new means for enhancing sustainable development by developing countries. But the ambiguities around the CDM are still pending, as emphasized in this blunt statement by the 'father' of the concept: "Though I facilitated approval of this proposal, I did not like it. My reservation was that the CDM is considered as a form of joint implementation but I do not understand how commitments can be implemented jointly if only one Party involved is committed to limit or reduce emissions and the other Party is free from the quantitative point of view .... This problem can become serious if baselines from individual CDM projects are set without regard to the total emissions of greenhouse gases in the country where the project is located" (Estrada–Oyuela, 1998).

This explains why it is suspected that the CDM will generate inflows of fictitious abatements in the system, the so-called tropical air, except if much progress is made on measuring project baselines, low-cost monitoring systems or any form of incentive framework designed to reduce the uncertainty about the reality of abatements. However, in terms of failure to secure real abatements, the problem posed by the CDM is quantitatively unimportant compared with the question of the rules that should govern the quota allocation beyond 2012 and that have to be clarified in 2005 for including developing countries in climate policies. The link between the trading systems, the CDM and the target setting over the long run was made explicit by the G77 and China at Buenos Aires. This request expresses both the concern of developing countries to be excluded from the new technological markets covered in the Kyoto Protocol and the impossibility for them to commit to binding emissions limits without prior clarification of the rules of the game.

The controversial judgment on the Kyoto framework reveals the reality of the neglected challenges behind the 'price versus quantity' dilemma and resulted in the supplementarity quarrel.

2. The economic rationale behind the supplementarity quarrel

The idea of international emissions trading is that a Party that faces higher costs or political difficulty in achieving its initial 'assigned amount' can exchange with a Party that is willing and able to do more. As explained before, this 'where flexibility' is intrinsically linked to a co-ordination through emissions quotas. It is necessary to face uncertainties about commitment costs, it would minimize the burden of emissions targets and it might be beneficial to the environment since each Party would lower the obstacle to adopting more ambitious targets for the second abatement period (Bohm, 1999).

The insistence of the EU on limiting the use of the flexibility mechanisms is then often viewed as a purely political move to force the US to take domestic measures instead of meet their targets only through the import of low-cost emissions reduction units or as the result of the ideological reluctance of a significant number of the public to accept the very idea of trading systems for managing a global environmental good. This feeling was exacerbated by the compliance and the 'hot air' issues. The fact that large excess assignments from Russia an Ukraine may be transferred at very low cost to the wealthiest countries was seen as violating the spirit of the Kyoto agreement.

However, the supplementarity condition is somewhat paradoxical from a purely static economic viewpoint. Indeed, the supplementarity criteria will be automatically satisfied, as any Party could set up a domestic plan of action whose marginal costs would be lower than this price. The higher the price of the permit, the more domestic actions undertaken by Parties as shown in Figure 3 for the US.

The economic rationale behind the supplementarity relates to two problems which have in common the fact that the short-term minimization of abatement costs may not be an appropriate indicator of a cost-effective climate strategy over the long term:

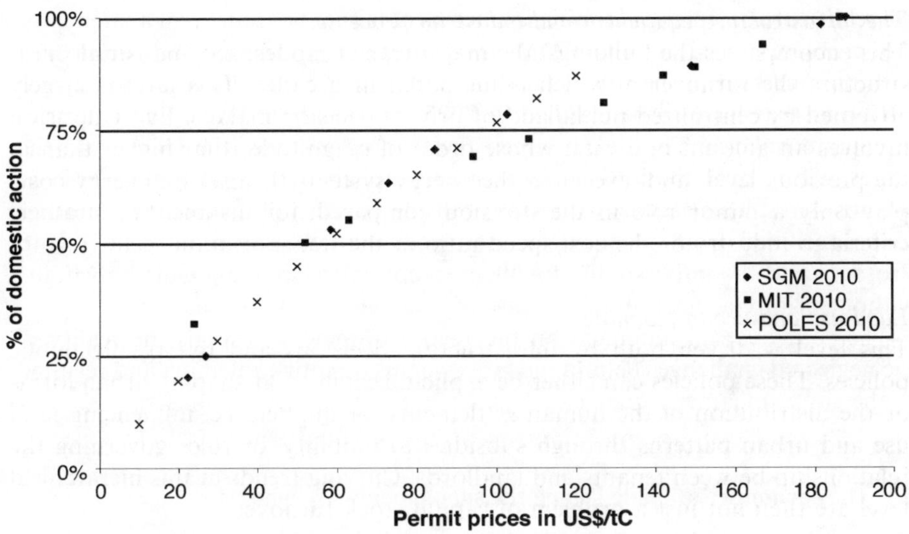

Figure 3. The influence of permit price on domestic efforts, an example with the US.

- the dynamic inconsistency problem: the flexibility mechanisms may discourage the adoption of measures designed to curb down the emissions in the very long term, some governments preferring to escape the resulting political cost in the short term;
- the risk that the trading system will dismantle the policies and measures currently implemented in Europe (mainly the energy taxation) and that have proved to be efficient over the long term to manage a rather low energy intensive development style.

2.1. *Inertia, price signals and dynamic inconsistency*

The analysis of the dynamic effects of trading systems must take account of the controversies about the role of price signals as a driver of technical change and the importance of inertia of technical systems and consumption patterns pointed out in the debate about timing of action.

Mark Jaccard (1997) portrays a three-level hierarchy of the decisions governing the dynamics of capital stock:

The end-use equipment
The decision is made by private decision makers (households, a division in a company). The turnover of capital stock ranges from a few years to two decades. At this level, the cost of delivering a given energy service is the key selection criterion.

The infrastructure equipment and industrial processes
This encompasses the buildings, the major transit modes, and industrial infrastructure the turnover of which is measured in decades. This level is largely governed by centralized public and/or private decision makers. Every decision involves an amount of capital whose order of magnitude is far higher than in the previous level, and, except in the energy systems themselves, energy costs play only a minor role in the decision compared, for instance, to strategic criteria in industry or the cost/speed ratio in the transportation sector.

Land use and urban planning
This level is driven both by infrastructure decisions and by specific public policies. These policies can either be explicit, i.e. aimed at shaping urban forms or the distribution of the human settlements, or implicit, i.e. influencing land use and urban patterns through subsidies to mobility, or rules governing the relationship between tenants and landlords. Curving trends at this hierarchical level are then not just a problem of capital stock turnover.

Inertia in the economic system results mainly from the interactions between these three levels. For example, urban forms determine not only the transportation needs but also the relative share of journeys made on foot, on bicycles, by public transport or by private car. The centring of activities around the proximity of infrastructures, the required investment, the nature of skills and the amount of vested interests generates dynamics which are hard to curve overnight. We can easily imagine, in the transportation sector for example, two very different equilibriums with relatively similar total costs but very different carbon contents and very high costs for shifting from the adopted one to the other. Lecocq *et al.* (1998) demonstrated that the cost of a delay in mitigation actions in these sectors (which might result from too-low permit prices) would lead to a great increase in the costs supported by the flexible sector as shown in Figure 4.

Such a context of deflated prices is all the more likely in that excess assignments in Russia and Ukraine, the possibility of negative costs in many Annex B countries (IPCC 1996), and possible 'tropical air' in CDM projects, may keep the price of carbon rather low in the first budget period and mask the real long-term cost of the carbon constraint. Simulations by SGM (Batelle) and EPPA (MIT) give a price of carbon around US$25/tC[11] for a permit market extended to developing countries (Ellerman D., Decaux A., 1998; Edmonds J. A. *et al.*, 1998).

One can then imagine a situation in which, up to 2012, OECD countries have no incentive to adopt appropriate measures to control emissions in the transportation sector and will be trapped in carbon demanding trends in lifestyles and technology which will make it more difficult to accept more ambitious targets beyond 2012. The risk of dynamic inconsistency of choices if one resorts only to 'cap and trade' policies is then all the more serious in that while the

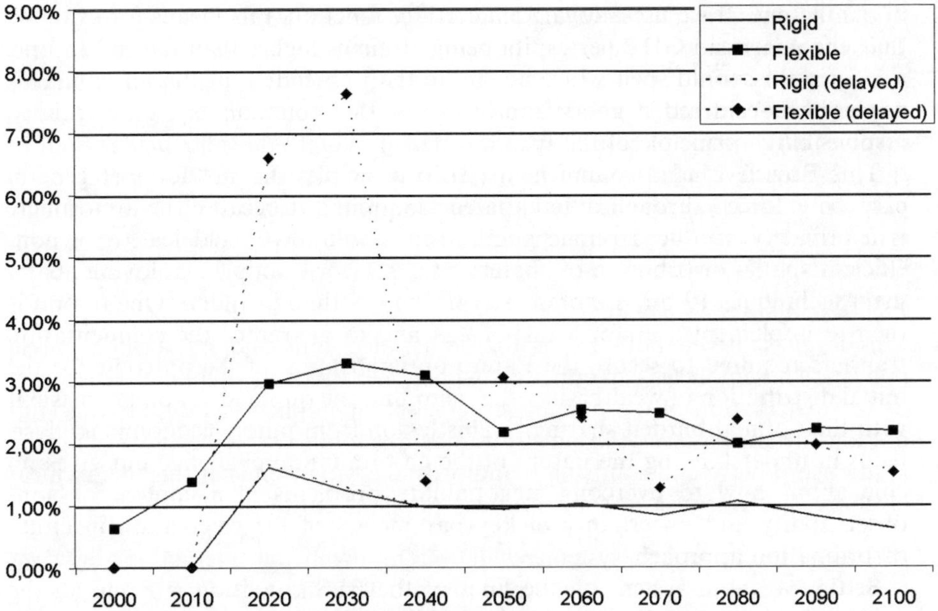

Figure 4. The influence of a delayed action on overall abatement costs, an exemple using STARTS model (From Lecocq *et al.*, 1998.)

question of the Russian hot air attracts attention today, over the long term, it might prove less important than the question of the developing countries' involvement.

2.2. *Target-setting rules beyond 2012: the joining up by developing countries may aggravate the risk of dynamic inconsistencies*

To clarify the rules governing the quota allocation it is necessary for developing countries making a start in climate policies not only from the viewpoint of the 'South' but also from the viewpoint of the 'North'. But since any solution has to be accepted by countries far more heterogeneous than those in the Annex B, this raises a sensitive equity–efficiency dilemma which may turn into dangerous stumbling-blocks on the way to the enforcement of a quota co-ordination approach beyond the first commitment period.

As long as the initial wealth distribution is not questioned (this premise is consistent with political realism since it is unlikely that a large reallocation of wealth would be operated only to mitigate global warming) the answer of public economics is that equity and efficiency can be separated: the amount of public good to be provided should be such that its marginal cost equates to the sum of marginal willingness to pay for it, and each agent should be charged as a function of its marginal willingness to pay, which in turns depends on its income. If we use a logarithmic utility function, this means for example that, the American GDP per capita being 19 times higher than the Indian one, a US citizen should spend 19 times more than an Indian at the margin. This might be considered a good translation of the 'common but differentiated responsibility' principle of the FCCC.

This Bowen–Lindahl–Samuelson (BLS) principle can, in the most general case, be enforced through differentiated taxation; but because the atmosphere is a privately produced public good, such a solution would lead to a non-efficient spatial distribution of abatement costs (for example employing abatement techniques 19 times more costly in the US than in India). One option is then to implement a uniform carbon tax and to guarantee the compensation transfers required to secure the Pareto optimalization of the outcome for the initial distribution of wealth; the other is to find the quota allocation consistent with the optimal burden sharing.[12] This lesson from purely economic analysis helps in understanding the nature of the co-ordination challenge but gives no clue about how to overcome negotiability problems in a context of non-observability and uncertainty of key parameters of the calculation in either co-ordination approach.

Setting a quota system on the lines of the BLS condition is made all the more difficult in that climate integrity is a 'future' public good whose valuation depends on expectations about the 'states of the world' in the long term and involves both inter- and intra-generational equity considerations.[13] In a cost–benefit framework where climate damage is to be considered, a 'dynastic solidarity' view where each country considers solely the damage which would occur to its descendants requires an agreement about the geographical distribution of climate damage. The last parameter introduces such a level of uncertainty that it will result in endless controversies and strategic manipulation of information. This parameter is neutral only in a 'universal solidarity' perspective where the citizens of each country are concerned about the impact of climate change whatever the location of the impact; in this case the burden sharing is identical to the burden sharing in a cost–efficiency framework. But this is not to say that an agreement on the remaining parameters (the emissions baseline, abatement cost curves, country-specific economic and environmental side-effects) can be easily found.

To prevent equity from becoming such an endless matter of controversy, it is then necessary to find robust rules that can be viewed by each actor as leading to an acceptable outcome given its own vision of equity and its own vision of abatement costs. These rules should not leave the way open for strategic manipulations at each point of time and should be computed on the basis of observable parameters that are accepted as a proxy of real costs in both absolute and relative terms. Many different rules of this sort have been proposed so far for resolving the entitlement issue (Berk *et al.*, 1999; Rose *et al.*, 1998); per capita allocation (Agarwal and Narain, 1998); convergence towards the same per capita emissions level or towards the same per GDP emissions level (Colombier, 1998); cumulated carbon emissions per capita or cumulated impact on radiative forcing (the Brazilian proposal); or declining carbon/income elasticity as a function of the income level (Jacoby, 1998). In fact none of these rules will be accepted and apt to support a general agreement unless they lead to burden sharing acceptable by all parties and are robust against the various competing views about the future.[14]

In a preliminary exercise (Lecocq *et al.*, 1999) we tried to test two of the most simple of these rules for reaching a 550 ppmv concentration target (a convergence towards a per capita equality and Jacoby's entitlement rule), under four scenarios combining high growth (H) and low growth (L) assumptions (3% and 2.2% annual world growth rate up to 2100, with optimist (a) and pessimist (b) income elasticity of carbon demand (in an endogenous technical change specification).[15] The main lesson of the exercise is the heterogeneity of the results for regions: (i) FSU (Former Soviet Union) has dramatically different results depending upon the rule and should univocally support the Jacoby rule; (ii) the adoption of this rule would create difficulties with China which appears to be a big loser in some scenarios; (iii) the net impact on ROW (Rest of the World) and IND (India) are more sensitive to the baseline in the convergence rule.

None of these rules seems to be robust enough for assumptions about growth and that there will be a tendency of each negotiating Party to advocate in favour of the rule which maximizes its own interest, and to use to this respect the growth and development patterns assumptions the most efficient to support its argument. Hence, because of this context of pervading uncertainty, a simple rule of thumb may not apt to reduce the tensions due to the contradictions of interest about burden sharing.

This is all the more important in that entitlement rules will have wide implications on the dynamics of carbon prices. Permit prices are given in Figures 5 and 6 for the convergence and Jacoby entitlement rules respectively. Prices in 2010 are comparable (between US$50 and US$70 per tonne of carbon), but medium-term permit dynamics differ quite significantly. In the convergence case, carbon prices rise significantly in 2020 (from US$75 to US$130 per tC) and fall in 2030, while under the Jacoby rule they rise slowly but continuously (US$55–105 in 2020 and US$75–180 per tC in 2030). This is due to the different step-in procedures. In the convergence case, developing countries enter jointly in 2030, bringing a massive amount of cheap abatement opportunities into the market.[16] In the Jacoby case on the other hand, they join the market more progressively: China enters in 2010 in all cases, ROW in 2020 (H) or in 2030 (L) and India in 2020 (H cases) or in 2030 (L cases). The progressive step-in induced by the Jacoby rule generates a smoother price signal. Nevertheless, this 'progressivity' itself has to be controlled: if for example the step-in occurs only beyond a given per capita GDP (in the case of Portugal for example), low growth baselines in the South will lead to delayed entries, and thus to a very heavy load on Annex B countries in the interim.[17]

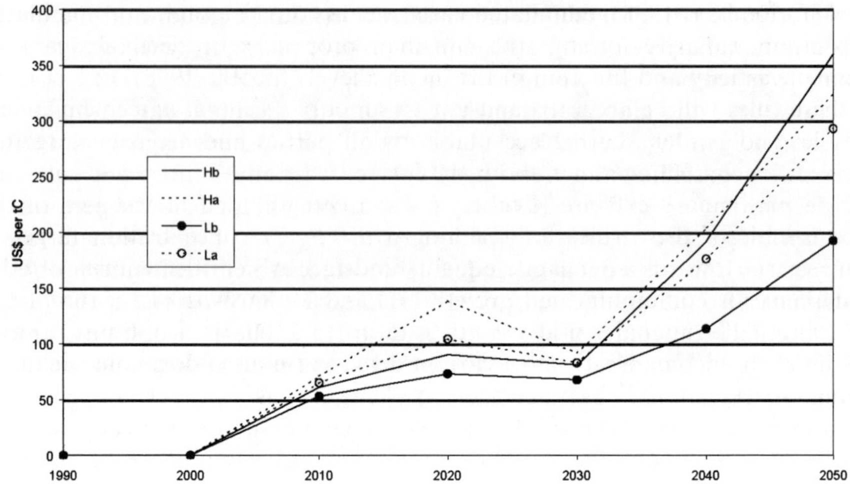

Figure 5. Permit prices – per capita convergence rule – Lecocq *et al.*, 1999.

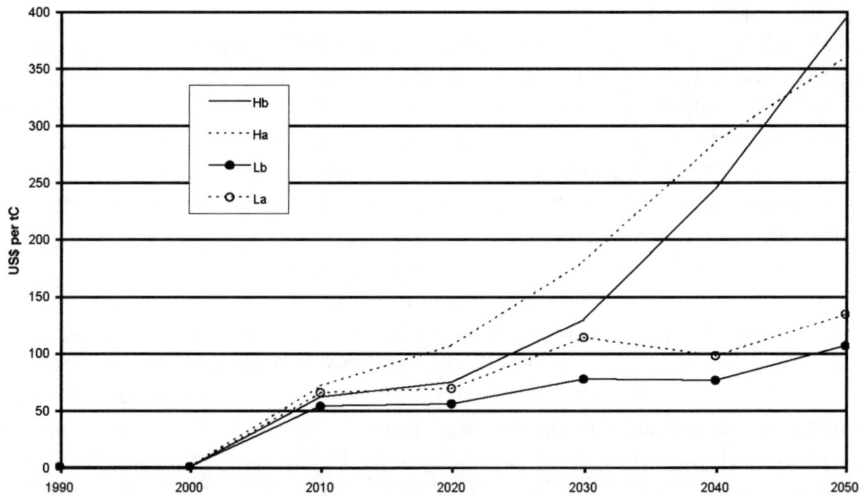

Figure 6. Permit prices – Jacoby rule – Lecocq *et al.*, 1999.

These results tend to suggest that because all these entitlement rules are likely to prove inequitable *ex post*, a quantity co-ordination approach is confronted to a real squeeze:

- on the one hand, faced with the challenge of encouraging developing countries to participate, there will be a tendency to accept a rather lax interpretation of the criteria and the risk of introducing waves of 'tropical air' in the system. The resulting deflation and increased volatility of carbon prices will reinforce the risk of dynamic inconsistency in climate policy and the doubt about the capacity of trading systems to provide the right signals in due time;

- on the other hand, by comparison with the no-trade case, carbon trading divides by 3 the global abatement costs and is the only way to make the system politically acceptable.[18] Since it is difficult to hedge against surprises regarding baselines and abatement cost curves, any entitlement rule can indeed be proved to be inequitable *ex post* and might incite lower-performing countries to stop their participation to the FCCC or to negotiate lower targets for the subsequent period. Carbon trading helps mitigate problems stemming from 'misallocated quotas', because the implications of the wedge between the outcome of the rules at date t and the first-best entitlement consistent with the real state of the world at the same date are all the more reduced that global abatement costs are low.

3. Translating the supplementarity condition

3.1. A quantitative concrete ceiling, a self-defeating approach?

One approach to make the concept of 'supplementarity' operational is to set quantitative limits on the assigned amounts (ERUs) obtained through trade. The formulae proposed by the EU Council of Ministers of 13 May 1999 to set the 'concrete ceiling' on all three Kyoto mechanisms are:[19]

(a) *Net acquisitions by an Annex B Party for all three Kyoto mechanisms together must not exceed the higher of the two following alternatives:*

- 5% of its base year emissions multiplied by 5 plus its assigned amount divided by 2, where the 'assigned amount' is the total allowed in each of the five years 2008–2012 inclusive.

Or

- 50% of the difference between its annual actual emissions in any year of the period 1994–2002, multiplied by 5, and its assigned amount.

(b) *Net transfers by an Annex B Party for all three Kyoto mechanisms together must not exceed:*

- 5% of its base year emissions multiplied by 5 plus its assigned amount divided by 2.

However, both ceilings may be increased to the extent that an Annex B Party achieves emissions reductions higher than this ceiling through domestic action undertaken after 1993.

If we let aside this last sentence, because it can be viewed as preparing a diplomatic compromise, it remains that setting quantitative limits on the use of IET presents the following weaknesses from a purely economic perspective:

- this formula results in lowly differentiated quantitative limit on transaction which would not be consistent with the principle of 'common but differentiated responsibilities';
- any such lowly differentiated concrete ceiling places a higher burden on the countries whose marginal abatement cost curve is higher (as shown in Figure 7). It will then result in a far higher burden on Japan than on the EU and a higher burden on the EU than on the US. Most models indeed confirm the intuition that there is a higher amount of low cost abatement potential in the US than in the EU.[20]
- a concrete ceiling on trading provides high trading incomes to the sellers of 'hot air'; the entire tradable share of hot air would be sold anyway, and this at a higher price than in an unregulated market;[21] the non-traded hot air would be banked and sold at higher prices during the following budget periods.

To restrict the use of ITEPS is therefore a self-defeating response to a legitimate concern. Figure 7 represents the increased cost due to a ceiling on trading (based on the Austrian proposal made in 1998, extrapolated from MIT abatement curves (Ellerman *et al.*, 1998)[22] and the various cases of market power (either to the buyer or to the seller). It shows that marginal abatement costs among countries are very different in case of ceiling leading to important increase in overall costs (more in Japan and the EU than in the US).

Moreover, a concrete ceiling will increase the volatility of carbon prices, because of the huge gap between the upper and lower bounds of such prices; the rise and fall of carbon prices would in fact be dictated by the change in market power, with the dramatic implications already experienced in the oil market.

3.2. Achieving supplementarity without quantitative limits to trading

There are other ways to achieve the supplementarity condition, but attention must be paid to the counterproductive effects. For example, restricting the international trading to high energy intensive industry exposed to international competition would force each country to undertake domestic action on transportation. But given the uncertainty on the transportation sector, this would place a high risk on countries where an increase in transportation would represent a higher share of the increase in total emissions. Another example is a minimum price that could be imposed for any retrocession of permits from governments to industry or any transfer to other governments; but this would open the way to the creation of black markets.

In fact, the only way to respond to the supplementarity condition is to come back to its origin, the weaknesses of any co-ordination approach, and to consider some form of hybrid 'quotas plus prices' system. Interestingly, such a combination of price signals and quotas is worked currently in the US (Kopp

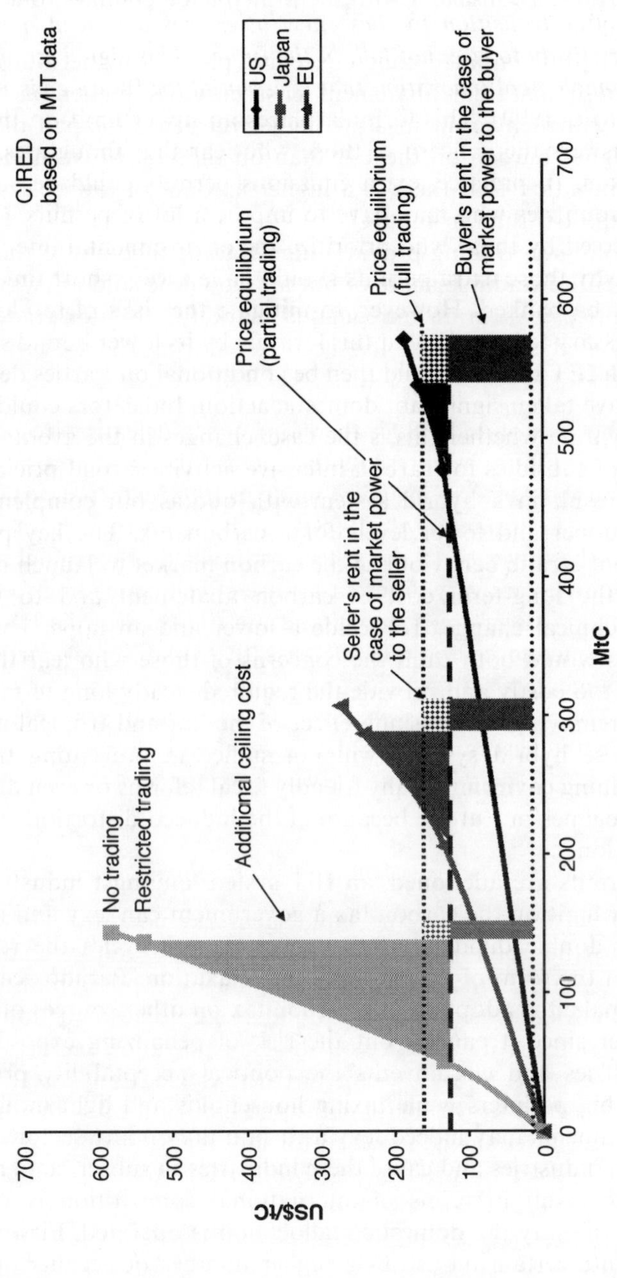

Figure 7. An example of the influence of a concrete ceiling.[23] Hourcade and Le Pesant 1998.

et al., 1999) for a domestic early action programme which places a 'safety valve' on the carbon-price: "*By beginning with a low ceiling price that rises over time, we provide a softer transition to these new incentives and send a signal that energy prices are likely to rise, not fall, in the future. This signal strengthens the incentive for technological innovation that is essential for future emissions reductions*" (Kopp *et al.*, 1999). This formula places an upper limit on the price of permits and answers the concern of those who fear that thoughtless commitments were taken. In practice, extra emissions permits could be sold at this price to those countries who may have to import a lot of permits. Obviously, this will be rejected by those who prioritize the environmental integrity of the system; this is why those extra permits should have a very short time duration and should not be banked. However, to mitigate the risks of too-low carbon prices, it is necessary to complement this formula by its lower bound symmetric. The access to an IET system should then be conditional on Parties demonstrating that they have taken significant domestic action. Indicators could be developed to demonstrate whether this is the case: changes in the average 'price of carbon'; levels of subsidies for carbon intensive activities; road pricing.

This would result in a hybrid system with quotas but complemented by some form of upper and lower levels for a carbon tax. The key problem is indeed to prevent erratic behaviour of the carbon market to launch misleading signals about the long-term cost of carbon abatement and to freeze the dynamics of technical change. To decide a lower and an upper level for the price of permits would both calm the concerns of those who fear that Kyoto targets may be too costly and provide the required steady long-term signal on innovation by removing the masking effect of the hot and tropical air.

However, these hybrid systems will not suffice in preventing the risk of GHGs undermining environmentally friendly fiscal reforms or even dismantling existing fiscal regimes in Europe because of the induced distortions in international competition.

In fact, if permits are auctioned, an IET system amongst industries simply places an upper limit on the carbon tax a government can levy and retains the potential for a double dividend, if the government recycles the revenues of these permits in the form of decreased general taxation. Paradoxically such a system would make the adoption of a carbon tax on other sources of emissions politically easier since it cancels out the risk of penalizing exposed energy-intensive industries and circumvents the political acceptability problems of exempting the big polluters while taxing households and light industries and services. Governments may indeed levy a carbon tax on any sector other than heavily exposed industries and cause these industries to subscribe to an ITEPS.

However, the result in terms of international competition is very much conditional on the way the domestic reallocation is operated. First, a government can allocate certain industries a higher number of permits than others for strategic reasons or because of domestic pressures. Second, if some countries put permits up for auction while the others give them away gratis, the heavy industries of the former will be penalized because of the 'wealth effect': they will be obliged to pay for all their emissions and not solely for the emissions exceeding their quotas. Obviously, in a general equilibrium framework, it can be demonstrated that gratis allocation creates both barriers to entry for newcomers and a rent which is not socialized; whereas the auctioned permit can lead to a double dividend which could favour low-emitting labour-intensive firms (Cramton and Kerr, 1998). The problem is then restricted to a minor part of the industry. The existence of emissions trading systems, far from weakening the domestic P&M, can facilitate their adoption if the reallocation rules of these permits are harmonized. One way to deal with this issue would be to ensure that the rules, modalities and guidelines for IET ensure that domestic reallocation rules are compatible with basic World Trade Organization principles and do not result in implicit subsidies and/or barriers to new entrants. These conditions would not violate national sovereignty.

3.3. CDM: the knife-edge path of the double additionality

The hybrid system suggested above allows for a corridor for constraints on carbon prices, and limits the risks of dynamic inconsistencies in climate policies stemming from the embarkation of developing countries on a quota co-ordination approach. But it does not suffice to address all the concerns about CDM which will be the first of the 'flexibility mechanism' to enter into force and will probably remain the principal way of abating GHGs emissions in those of the developing countries which are not in position to accept binding targets for their emissions. The very semantic innovation of CDM (where the D matters) as a substitution for the concept of AIJ, indicates that the challenge in designing the CDM is to secure two forms of additionality:

- *an environmental additionality*: it is important to maximize carbon abatements that would not have occurred otherwise;
- *a development additionality*: the CDM should provide the means of enhancing sustainable development in poor countries, and this independently from the impacts of climate change.

While most points of the Kyoto Protocol emphasizes this second dimension, its flexibility dimension is the only part that is very clear so far, based on emissions reductions from specific projects. However, such a project-based approach is opened to criticisms:

- related to the environmental additionality. One of the most obvious problems stems from the difficulty in observing the emissions baseline and the level of reduction for each certified project; this is all the more dangerous in that there is a common interest of the host country and of the investor to agree on overestimated amounts of GHGs reductions;
- related to the long term control of emissions trends. A purely project-based

approach might not be suitable to provide a support to structural measures (energy pricing, removal of subsidies for fossil fuel consumption, transportation and infrastructure policies) necessary to help developing countries to shift towards less carbon-intensive development patterns over the long term;
- related to the development additionality. There will be no surplus for the host country if the funds transferred correspond strictly to the coverage by companies of Annex B of the expenditure relating to a specific abatement project. The CDM may then generate still fewer positive externalities on the development in that the resulting activities may not correspond to the development priorities of the country and, in the worst case, may contradict such priorities and capture a part of overseas aid.

To put a concrete ceiling on the use of such a mechanism meets the 'environmental additionality' but only over the short term and, by not answering the lasting questions of the 'South' about how to reconcile environment and development, risks endangering the climate control over the long term. Here again, because the problems come from the deadlocks of a quota approach due to the difficulty of measuring baselines and costs (at a micro level instead of at a macro level such as in the case of trading systems) the response should be found in the design of the system and its articulation with domestic P&M in these countries.

Some countries might be willing to commit themselves to domestic policies and measures which would both contribute to enhancing their development and result into a control of their GHGs emissions (energy efficiency keeping the energy demand within the financing capacity without lowering the level of delivered energy services, or mitigation of local environmental problems). But these 'no regrets' options are inhibited by internal and external constraints (capital shortages, lack of technology, market failures) and no explicit support other than the financial mechanism of the Convention is provided in the Protocol. It is necessary to design the CDM in such a way that it operates a leverage effect to enhance the policies and measures adopted by non-Annex B countries as a function of their development objectives and would result in a lower level of GHGs emissions. The issue is how to couple two sets of items of the Kyoto Protocol: those related to the CDM and those related to the P&M.

To achieve such a coupling between CDM and development policies, it is first necessary to recognize that there is ultimately 'something' to pay for including developing countries in sound climate policies; this 'something' may be either a financial transfer from the North, or the risk of some 'tropical air' in the system. While the first 'cost' may yield a positive outcome for both environment and development, it is desirable to minimize the latter.

This debate should be articulated *de facto* along two lines:

- *the correspondence between CDM and development policies*: technically, it is easy to demonstrate that the CDM could operate a leverage effect on 'no-regrets' P&M, infrastructure programmes or renewable energy dissemination with a long-term structural effect on the development and innovation; indeed the incremental revenue from carbon credits would make the projects made profitable by domestic policies more attractive to industry (Mathy and de Gouvello, 1999);

- *the potential for the industry of developing countries* to be credited with a part of the ERU. Both a fairness and an environmental efficiency issue are embedded in this debate. Indeed, the CDM will generate rents (the difference between the marginal abatement cost of a project and the international price of carbon modulo a risk premium or a transaction cost); these rents will be collected by the investors of Annex B. This is why the option of allowing the domestic industry of non-Annex B countries to receive carbon credits from investments in CDM operations, and to bank or sell them on the international market, has to be seriously envisaged. This is a way of mobilizing initiatives by private and public economic actors of developing countries. The counter-argument of Annex B countries may be that the rent is due to their voluntary commitment to GHGs emissions limits and that it is economically and environmentally unacceptable to allow a country to receive ERUs from projects while no effort is made elsewhere in the economy. A cynical reaction would be to allow emissions to increase and maximize the opportunities for CDM projects.

One response to controversies at both these levels would be to allow domestic economic agents to receive carbon credits from CDM projects in which their domestic industries invest as soon as, on the basis of the country reports provisioned in the FCCC under Article 4, the CDM executive board, by delegation of the COP, will have certified that the domestic P&M (general or in specific sectors) would result into lower GHGs emissions. This, *per se*, does not reintroduce an implicit conditionality; this is not indeed a prerequisite to access to CDM since it does not rule out the possibility of attracting CDM in sectors not concerned by P&M or in countries not having adopted policy interventions in line with the objectives of the FCCC.

4. Conclusion

We have tried to demonstrate in this paper that the supplementarity dispute between the EU and the US is logically tied to the specific difficulties raised by any quota co-ordination approach as soon as one considers the uncertainty regarding emissions baselines and abatement costs. It is this very uncertainty which makes flexibility mechanisms necessary to hedge against *ex post* inequity in quota allocation, but it is also this very uncertainty which explains the dynamic inconsistency problems raised by the enforcement of such systems in the real world.

Because any *ex ante* quantitative limit to trade tends to aggravate the problem (by putting more burden on countries with high marginal abatement costs and by increasing the volatility of carbon prices), we suggest that an appropriate response to the supplementarity condition could be found in hybrid co-ordination approaches where the 'cap and trade' mechanism is bounded by a lower and an upper limit to carbon prices and where the CDM is used as a leverage to domestic policies and measures in developing countries. We are conscious that this requires far more precise investigation to find out the appropriate institutional design, but our object in this text is to go over some basic economic elements of the discussion which have been too much overlooked in the past decade.

Notes

1. Centre International de Recherche sur l'Environnement et le Développement, Unité Mixte – Centre National de la Recherche Scientifique – Ecole des Hautes Etudes en Sciences Sociales, Corresponding author: lepesant@centre-cired.fr. Research support from the EC DGXII under contract ENV4-CT98-0810 is gratefully acknowledged.
2. Specific terms are listed in the glossary at the end of the chapter.
3. FCCC/SB/1998/MISC.1/Add.3/Rev.1 Preparatory work for the fourth session of the Conference of the Parties on the items listed in decision 1/CP.3, paragraph 5, Indonesia (on behalf of the Group of 77 and China).
4. General Secretariat of the Council, Community strategy on Climate Change. Meeting document CONS/ENV/98/7 SN/109/98.
5. We will not elaborate here on the relative role of price and non-price signals; in the real world, both are required and their combination depends upon the range of activity and the technology. However, the non-price signals are so country specific (trade regulation for example), or might encompass such a wide range of techniques (emissions standards) that they cannot be harmonized internationally. The only category of P&M that can be easily harmonized at the world level are price signals (carbon tax, carbon-energy tax, subsidies on clean technologies).
6. In other words, the uncertainties surrounding the equity discussion in quantitative terms appear to be so important that they may generate a self-defeating negotiation process and inhibit any stable agreement. As we put it six years ago, "*This overview shows the* regressum ad infinitum *of scientific controversies induced by a focus of the negotiation on quota allocation. This does not mean that equity must not be considered in the negotiation process; it means that it will be impossible to agree on a numerical criterion of equity. The conclusion is that negotiations aiming at the allocation of initial quota would place each country in a position where it would necessarily adopt non-cooperative argumentation. Because of the amount of controversies and value judgements involved, it would be difficult to put some rational in the equity debate; this would exacerbate the division lines instead of easing a progressive convergence. An international tradable permit system does not overcome this preliminary difficulty, and negotiation would prove self-defeating*" (Hourcade, 1994).
7. It should be clear that it was not the perspective of a carbon tax at the national level that was rejected but that of an international carbon tax.
8. It behoves each government to select the most appropriate policies to meet its target and it is difficult to justify why a government should accept constraints on this selection; either the co-ordinated policies lead to an abatement effort lower that the accepted quota and they are useless, or they lead to more drastic efforts and they impose other targets than those originally accepted.
9. This expression comes from Jacoby *et al.*, 1998.

10. Criqui *et al.* (1999) find the amount of hot air to be around 300 Mt whereas MIT finds lower but still very significant figures (110 Mt, Ellerman *et al.*, 1998).
11. US dollars, constant 1990 prices.
12. We will not return in this paper to the reasons why a price-co-ordination approach may have been more efficient from a procedural perspective (Hourcade, 1996).
13. We will not enter into the theoretical background of this discussion in this paper; a tentative elaboration can be found in Lecocq *et al.*, 1999.
14. We leave aside in this paper the fact that the choice of these rules should derive from serious consideration of their counterincentive effect if non-Annex B countries have a strong interest in having the highest CO_2 emissions before entering the market.
15. These scenarios were calculated to cover the core of the distribution of IS98 scenarios: emissions in the H.a scenario are comparable to the IS98 A1 (Morita *et al.*, 1998); scenario H.b is similar to the IIASA–WEC highest baseline A2 (Nakicenovic *et al.*, 1998); the L.b scenario is similar to the IS98 B2 (Riahi and Roehrl, 1998).
16. We assumed here that cheap abatement opportunities in developing countries remain unchanged until the middle of the 21st century. This is a very questionable assumption which we will come back to in a future version of this paper. In the same way we are not discussing here the danger of misuse of large financial inflows to developing countries that exceed their absorption capacity over the short term, a danger that was demonstrated by the experience of many oil-exporting countries after the sudden rise in oil revenues in the 1970s.
17. Bollen *et al.* (1999) use a similar step-in procedure (developing countries enter when they reach US$10,000 per capita GDP) in a model with six non-Annex B regions. They also present a very progressive step-in (developing countries enter up to 2065) with an oscillating carbon price which falls each time an important DC enters the market.
18. Note that both entitlement rules lead to inequitable welfare distributions: while the EU and Japan are not too far from the BLS criteria for the US and CNZ in the Convergence case, the US, CNZ and China in the Jacoby case pay far more than they 'should'. If we leave aside the Chinese case because of the dramatic difficulty of representing this country, the result for the US and CNZ is intriguing since citizens from these regions would pay between 32 times and 20 times the Indian level. It is arguable that this does not pose some 'reverted equity' problem (for example in the name of past responsibility); but the resulting welfare distortions will not be accepted unless the overall cost remains moderate.
19. Community strategy on climate change, Council conclusions, European Union, The Council. Brussels, 18 May 1999, 9 pp.
20. The result is the reverse if one considers some simulations circulated within the EU before Kyoto and which provided a very low baseline emissions scenario (plus 8%). But in this case the 'supplementarity condition' no longer holds, since the Kyoto target would mean a far greater abatement in the US (minus 35%) than in the EU (minus 15%). In fact the EU figures do not reflect the studies carried out in each country and the uncertainty of their baseline.
21. Criqui *et al.* (1999) capture the order of magnitude of this extra rent under various assumptions.
22. The market power cases were designed by CIRED, which means the authors are solely responsible for them.
23. The formula used to compute the ceiling in this example is the one proposed by Austria and used as a reference until the meeting of the Council in May 1999. The formula is $C = 0.025*(2*0.948 - K)$, where C is the ceiling (based on 1990 emissions B), K the commitment made at Kyoto by the Party, 0.948 being the mean value of these commitments. This formula is slightly different from the first one of the Council, being more stringent, but this does not change the conclusions (for example for Europe in our example $C = 0.024*B$ whereas in the first Council's formula $C = 0.048B$.

References

Agarwal, A. and Nairin, S. (1998). The atmospheric rights of all people on earth, CSE Statement by Anil Agarwal and Sunita Narain, *http://www.oneworld.org/cse/html/cmp/cmp335.htm*

Agarwal, A. and Nairin S. (1991). Global warming in an unequal world, Centre for science and environment, New Delhi.

Agarwal, A. and Narain, S. (1998). *The atmospheric rights of all people on earth.* CSE Statement.

Baumert, K.A., Bhandari, R. and Kete, N. (1999). What might a developing country climate commitment look like? World Resources Institute, Climate notes. May 1999. 20 pp.

Berk, M., den Elzen, M. and Metz, B. (1999). Global climate protection and equitable burden sharing – an exploration of some options. Paper presented at the EFIEA policy workshop on Integrated climate policies in the European Environment costs and opportunities. Milan, March 4–6. 19 pp.

Bohm, P. (1997). Joint implementation as emission quota trade: an experiment among four nordic countries. Copenhagen, Nordic Council of Ministers, Nordic House of Publishing. 39 pp.

Bohm, P. (1999). International greenhouse gas emissions trading with special reference to the Kyoto Protocol. Copenhagen, Nordic Council of Ministers, Nordic House of Publishing.

Bollen, J., Manders, T. and Timmer, H. (1999). The IPCC SRES Stabilization scenarios. Communication to the *Joint IEA/EMF/IEW Meeting*, Paris. 16–18 June.

Cramton, P. and Kerr, S. (1998). Tradable carbon permit auctions: how and why to auction not grandfather. Discussion Paper 98-34. May 1998. 24 pp.

Criqui, P. *et al.* (1999). Marginal abatement costs of CO_2 emissions reductions, geographical flexibility, and concrete ceilings: an assessment using the POLES model. Energy policy. 27:585–60.

Colombier, M. (1998). Critères de différenciation des engagements en matière d'émissions de gaz à effet de serre, Etude pour la Mission Interministérielle à l'Effet de Serre, International Conseil Energie. December. 88 pp.

Edmonds, J.A. *et al.* (1998). Unfinished business, the economics of the Kyoto Protocol. PNNL report #12021.

Ellerman, D. and Decaux, A. (1998). Analysis of post-Kyoto emissions trading using marginal abatement curves. MIT Joint Program on the Science and Policy of Global Change. September.

Estrada-Oyuela (1998). First approaches and unanswered questions. In: Issues and options: the clean development mechanism. New York, UNDP. 23–29.

Godard, O. and Hourcade, J.-C. (1991). Instruments for a collective management of climatic risk: a price-signal based framework. Climate change and ozone layer protection. Sophia Antipolis, 21–25 January. Paris, UNPD. 401–57.

Godard, O. and Henry, C. (1998). Les instruments des politiques internationales de l'environnement: la prévention du risque climatique et les mécanismes de permis négociables. In: CAE, *Fiscalité de l'environnement.* La documentation française. Paris. 197 pp.

Goulder, L.H. (1995). Environmental taxation and double dividend: a reader's guide. International Tax and Public Finance. 2:155–82.

Grubb, Michaël (1989). The greenhouse effect: negotiating targets. Royal Institute for International Affairs, London.

Ha Duong, M., Hourcade, J.-C. and Lecocq. F. (1999). Dynamic consistency problems behind the Kyoto Protocol. Mimeo CIRED. 24 pp.

Ha Duong, M., M. Grubb and J.-C. Hourcade (1997). Influence of socio-economic inertia and uncertainty on optimal CO_2 emissions abatement. Nature. 390:270–73.

Hahn, R. and Stavins, R. (1999). What has Kyoto wrought? The real architecture of international tradeable permit markets. Resources for the future. Discussion Paper 99-30, 23 p.

Hourcade, J.-C. and Baron, R. (1993). Chapitre 1: Permis négociables. In: OECD *Les instruments économiques internationaux et le changement climatique.* Paris: July: 11–44.

Hourcade, J.-C. (1994). Economic issues and negotiation on global environment. In: Carraro, C. ed., *Trade innovation and environment*, 385–405.

Hourcade, J.-C. (1996). International harmonization of GHG emissions policies. The 1996 Annual Congress of the American Economic Association. San Francisco. 5–7 January.

Hourcade. J.-C. and Le Pesant, T. (1998). Possible impacts of quantitative limits on flexible instruments. Communication to the European–US Conference on Post-Kyoto Strategies. 6–8 September 1998, Semmering.

Hourcade, J.-C. (1998). The thin pathway for reconciling development and climate change mitigation. Environment and Development Economics. 3, part 3; July: 372–76.

IPCC SAR (1996). Climate change 1995, economic and social dimensions of climate change. Contribution of Working Group III to the Second Assessment Report of the Intergovernmental Panel on Climate Change. In: Bruce, P.J., Lee, H. and Haites, E.F., eds. IPCC 1996. Cambridge University Press, Cambridge. 448 pp.

Jaccard, M. (1997). Heterogenous capital stocks and decarbonating the atmosphere: does delay make cents? Draft (submitted to Energy Economics). 40 pp.

Jacoby, H., Prinn, R. and Schmalensee, R. (1998). Kyoto's unfinished business. Foreign Affairs. 77(4):54–66.

Kantorovitch. L. (1976). Essays in optimal planning. White Plains, New York, International Arts and Science Press.

Kopp, R. *et al.* (1999). A proposal for credible early action in US climate policy. RFF feature, 16 February 1999, Washington DC.

Lecocq, F., Hourcade, J.-C. and Ha-Duong, M. (1998). Decision making under uncertainty and inertia constraints: sectoral implications of the when flexibility. *Energy Economics.* 20:539–55.

Lecocq, F., Hourcade, J.-C. and Le Pesant, T. (1999). Equity, uncertainties and robustness of entitlement rules, communication to the *Joint IEA/EMF/IEW Meeting.* Paris, 16–18 June. 23 pp.

Liptak, T. (1965). Two level planning, econometrica. 1965;33(1);141–69.

Mathy, S. and de Gouvello, C. (1999). Mécanisme de développement propre: flexibilité pour les engagements de réductions des émissions des pays de l'annexe 1 ou levier pour le développement des pays en développement. Communication to the PIREE Conference, Strasbourg, 2–3 December. 32 pp.

Michaelowa Axel (1998). Joint implementation the baseline – economic and political aspects. Global Environment Change. 8(1):81–92.

Morita, T. *et al.* (1998). *SRES IS98 A1 marker scenario.* Downloaded from http://www.sres.ciesin.org 15 January 1999.

Nakicenovic, N., Grübler, A. and McDonald, A. (eds) (1998). *Global energy perspectives.* IIASA–WEC, Cambridge University Press, Cambridge. 300 pp.

Newell, R. and Pizer, W. (1998). Regulating stock externalities under uncertainties. Discussion paper 99-10, 15 December. Washington DC, Resources for the future. 29 pp.

Pizer, W. (1997). Prices vs quantities revisited: the case of climate change. Discussion paper 98-02, October 1997. Washington DC, Resources for the future. 48 pp.

Riahi, K. and Roehrl, A. (1998). *SRES IS98 B2 marker scenario.* Downloaded from http://www.sres.ciesin.org 15 January 1999.

Rose, A., Stevens, B., Edmonds, J. and Wise, M. (1998). International equity and differentiation in global warming policy. *Environment and Resource Economics.* 12:25–51.

United Nations (1992). *UN Framework Convention on Climate Change.* Climate Change Secretariat, Palais des Nations, Geneva.

UNFCC (1997). Kyoto Protocol to the United Nations Framework Convention on Climate Change. Conference of the parties, third session Kyoto, 1–10 December.

Weitzman, M. (1974). Prices vs quantities, review of economic studies 41(4):477–91.

Wigley, T.M.L., Richels, R. and Edmonds, J. (1996). Economic and environmental choices in the stabilization of atmospheric CO_2 concentrations. *Nature.* 379.

World Resources Institute (1990). World Resources 1990–91 a guide to the global environment, World Resources Institute in collaboration with the UN Development Program, Oxford University Press.

Glossary

BAU:	Business as usual
CDM:	Clean development mechanism
COP:	Conference of the parties to the UN Framework Convention on Climate Change
ERU:	Emissions reduction unit
FCCC:	Framework Convention on Climate Change
GHGs:	greenhouse gases
GWP:	Global Warming Potential
IET:	International emissions trading
ITEPS:	International tradable emissions permits system
ITPS:	International Tradable Permits System
JI:	Joint Implementation
JUSCANZ:	Japan, United States, Canada, Australia, New Zealand
MAC:	Marginal abatement costs
P&M:	Policies and measures

PART 2

Efficiency of Emission Trading Schemes

CHAPTER 5

International Greenhouse Gas Emissions Trading – With Special Reference to the Kyoto Protocol[1]

Peter Bohm

Introduction

The purpose of this paper is to identify potential benefits of international greenhouse gas (GHG) emissions trading among participating countries committed to emissions limits (Annex B countries in the terminology of the Kyoto Protocol) and comment on various design options for this trading. The principal criterion used for evaluating, in principle, the net benefits of different versions of free as well as constrained emissions trading is the net costs avoided by the individual country when trade is allowed. The discussion is mainly focused on carbon emissions from the combustion of fossil fuel, the dominating source and that likely to be most, or sufficiently, simple to monitor. Other sources or sinks could be added once their monitoring is given a satisfactory solution; this does not require that such monitoring would have to be the same for all the sources and sinks. Although joint implementation (project-related emissions trading among Annex B countries) and the use of the Clean Development Mechanism (project-related emissions trading with non-Annex B countries) also represent a form of trading, they are left out of the discussion below, the main reason being that they do not seem to offer any significant benefits in terms of real cost reductions that are anywhere near those of GHG emissions quota trading among Annex B countries.[2]

The starting point of the discussion is an overview of the likely potential benefits of a fully global system of emissions trading and how the benefits tend to be reduced by a non-global system like that of the Kyoto Protocol (Section 1). This section highlights monitoring issues, carbon leakage and factors promoting flexibility in the design of emissions trading. Section 2 addresses the choice of domestic control systems (taxes, tradable permits, etc.) to support the national emissions quotas *ex post* international quota trade, as to aspects of this choice that seem to influence the potential benefits of international emissions trading to the individual Party.[3] Determinants of the benefits of international emissions trading are discussed in Section 3, focusing on the

C. Carraro (ed.), Efficiency and Equity of Climate Change Policy, 93–119.

potentialities of certain trading rules, the real implications of trading 'hot air' (non-existing emissions) and the supplementarity requirement in Article 17 of the Kyoto Protocol, calling for some minimum of domestic emissions reductions by each Party. The main conclusions are summarized in the final section.

The discussion in what follows is based on literature cited in the reference list. Explicit references to sources will be made only when the argument in question is made solely in one source. The main references for the overview are IPCC, 1996 (Chapter 11); UNCTAD, 1998; two European Commission papers, 1998; Toman and Hourcade, 1998; and Bohm, 1998a,b.[4]

1. The principles of international emissions quota trade

1.1. Fully global emissions trading

It is only with the fully mixed pollutants which delete the ozone layer or risk causing global warming that an 'ideal' case for emissions trading has appeared and, in the last-mentioned case, been made possible by the first step taken in Kyoto 1997. Since the global warming potential of GHGs has been considered to be independent of the location of emissions, efficient policy should treat all emissions in the same way (barring special second-best considerations). Thus, not only should all GHG emissions in a country be subjected to the same deterrents in an efficient climate change policy, but this is true also for emissions regardless of national origin. This forms the basis for a straightforward application of GHG emissions trading on a worldwide basis.

Since the Parties to international agreements to limit GHG emissions are national governments, international emissions trading will, at least formally, involve trade among governments. We will later return to the case where (some) governments may allocate its emissions quota to firms or other legal entities inside their respective countries. The case for *a fully global tradable emissions quota (TQ) system* as a perfectly competitive market for the commodity 'emissions reductions' can be simply stated as the case for international trade of any completely homogeneous commodity.[5] In principle, free trade would equalize prices for this new commodity and implement a global least-cost solution to the target of global emissions given by the agreement. Noteworthy is that this trade is 'ideal' or 'perfect' also in the sense that it will not be constrained by transportation costs as is true for trade in goods and many services. Countries can sell commitments to undertake emissions reductions at prices that will be the same for buyer countries whatever the distance between sellers and buyers; hence, 'consumers' pay the prices the 'producers' get for individual transactions and these prices tend to be fully equalized for emissions trading on a perfect international market.

Furthermore, as we shall see, with certain simple trading rules extensive international emissions trading implies that transaction costs per unit of trade will be small. This is another respect in which global GHG emissions would be an ideal case for emissions trading and in stark contrast to many cases of local emissions trading where transaction costs have constituted a significant drawback (Stavins, 1998).

The lack of relevant experience

There is little experience of emissions trading between nations concerning fully mixed global pollutants. The Montreal Protocol included some provisions for trade of CFC production quotas between (the strictly limited number of) producer countries.[6] The Protocol also allowed CFC consumption quota trade among EU member states, but also in that case trade seems to have been marginal only. Therefore, and especially if future GHG emissions trading among nations would be quite extensive with respect to both the number of traders and the volume of trade, it seems fair to say that no significant international precedents exist that could help evaluate the prospects and consequences of GHG emissions trading among countries.

Domestic emissions trading is well documented by the experience primarily from the US, where this environmental policy instrument was first introduced by the USEPA in 1975 and has been extensively used for controlling local emissions problems. There are many success stories from these applications, but also evidence of unexpected implementation problems, many of which relate to market imperfection problems of thin markets with few traders and, as it seems, unfortunate combinations with concurrent command-and-control policies (Atkinson and Tietenberg, 1991). More recently, emissions trading has been introduced for SO_2 emissions covering mainly those from coal-fired electric power plants in the US. This application seems to have offered significant cost savings, but it should be noted that, since the environmental effects of emissions of this pollutant are location dependent (a non-uniformly mixed pollutant), market emissions permits here are traded so that less harmful emissions may be replaced by more harmful ones or vice versa (e.g., Atkinson and Tietenberg, 1987; UNCTAD, 1998). Hence, the fact that emissions trading up to now has been confined to domestic trading and to pollutants that are analytically different from GHGs makes the existing US experience of limited relevance in the present context.

Market design

To fix ideas for a meaningful discussion of global GHG emissions trading, a few characteristics of the commodity traded and how it can be traded should be specified. First, as in the Kyoto Protocol, the commodity is emissions reductions or units of national emissions quotas that can be used once during a five-year commitment period (or be banked for a subsequent period). Second, it is only at the end of the commitment period that it needs to be checked that the quotas (*ex post* trade) cover the emissions made. It may be helpful to allow Parties some additional time into the next period during which they could purchase additional units from other Parties' first-period excess quota holdings (if borrowing is not permitted) to avoid being in non-compliance, a so-called 'true up' period.

Fully global emissions trading can be carried out, perhaps most easily, on a global exchange similar to a stock exchange. In contrast to present exchanges of this type, an international *emissions exchange* need not close and restart at intervals, since in principle all nations of the world would be operating on this market with, say, three locations of physical exchange branches, one in Eastern Australasia, one in Europe and one in North America with overlapping office hours and an around-the-clock electronic market for transactions (see, e.g., Sandor *et al.*). Important advantages with this design are that trade can be anonymous (although each agent's sales/purchases are registered, of course) and that current prices are made known to all participants. The problem, who should bear the *responsibility for implementation* of the emissions reductions transacted, is then solved by the fact that the seller will be involved in a transaction with a neutral international exchange, the latter of which (or the Protocol secretariat), once the question of sanctions has been settled, holds the seller responsible for its sales of emissions reductions.

Thus, the following discussion proceeds under the assumption of an efficient market design (as if perfectly competitive), which explicitly includes 'seller beware' rules. To avoid confusion about the meaning of emissions trading in a first commitment period and about the consequences for a Party to be in non-compliance at the end of that period, it is taken for granted here that, well before that period starts, stringent emissions quotas for the next period have been negotiated for all countries. We will return to additional aspects of market design later.

Monitoring

While actual delivery of conventional commodities transacted typically can be validated or certified by the buyers in a straightforward fashion, this certification option is absent in the case of emissions reductions traded. When governments directly or indirectly trade in this particular commodity, the delivery of emissions reduction commitments by sellers of quota units can, strictly speaking, be certified on an aggregate national level only and refer to conditions to be ascertained only at the very end of the commitment period, up to some five years later. Against this background, *monitoring of national emissions levels* will be crucial for a reliable and well-functioning TQ system. If *combustion of fossil fuel* were the only source of GHG emissions, monitoring (although not likely, in any case, to be one hundred percent accurate) could be comparatively simple to achieve by estimating the country's fossil fuel use. Disregarding carbon removal for the time being, the carbon (or CO_2) emissions from fossil-fuel combustion would be given by the quantities of the use of all kinds of fossil fuel, multiplied by their different levels of carbon content.

Fossil-fuel use equals production *plus* imports *minus* exports (*plus* inventory changes). In a global TQ system, where all countries are committed to emissions quotas, it is in the interest of every fossil-fuel-exporting country to avoid underestimates of its export volume, and of every importing country to avoid overestimates of its import volume. Thus, the fossil-fuel volumes traded would likely be kept track of by the double-check thus made possible. Fossil-fuel-producing countries may be able to hide quantities produced for domestic use. However, if the quantities smuggled out of the statistics can be expected to be small, especially if pre-announced production plans and unannounced inspection arrangements would be part of the international agreement, the accuracy of this approach is likely to be sufficiently high.

One problem with the monitoring of fossil-fuel use, as compared to estimating carbon emissions directly, should that be feasible, is that initial and end-period fossil-fuel inventories would have to be estimated. This particular estimation task is likely to be of the same character as that of monitoring carbon emissions, but may be approached by using some rule of thumb for the changes in inventories, once a reliable estimate of the initial stocks of fossil fuel has been established at the start of the first commitment period.

Another problem is related to *carbon removal* from emissions.[7] Direct monitoring of carbon emissions would have the advantage of eliminating emissions sources that would be subjected to carbon removal and hence also provide incentives for developing and introducing such techniques. Indirect monitoring via fossil-fuel use would need a separate approach for carbon removal to be taken into account. This approach could be similar to that of the Montreal Protocol (Article 1.5), where it is made acceptable for signatories to subtract CFCs destroyed in the CFC manufacturing process and thus not emitted into the atmosphere, if technologies approved by the signatories are used.

Monitoring obviously belongs to the set of climate-change policy issues that need further elaboration. However, it should be noted that the monitoring problem is nothing specific for the issue of emissions trading. It stems from the demand for international commitments to limit GHG emissions; thus, the problem arises regardless of whether the limits or quotas are tradable or non-tradable. Since both sides of a quota trading market have opposite incentives for deviating from truthful reporting of amounts of fossil fuel traded, the risk for such deviations seems to be small.

1.2. Non-global emissions trading

In one particular sense, the monitoring problem becomes more complicated when we now move to consider the case of *non-global TQ systems*, which is of more immediate concern here. In the Kyoto Protocol, only Annex B countries that have ratified the Protocol may engage in emissions trading. A non-global system implies that one loses the double check on parts of the international fossil-fuel trade; for fossil-fuel imports from (exports to) non-signatory countries one cannot count on any reliable help from the export (import) statistics of these countries. To keep the monitoring of this trade on a high level of accuracy, it may be required to institute measures such as sample unannounced inspections of fossil-fuel transport from non-signatory to signatory countries, given that the rules then also require that the Protocol secretariat is notified of transports when they are contracted.

Without any new additions to the Annex B group, the maximum number of traders would be 36 countries, not counting the EU itself and disregarding Liechtenstein and Monaco. The two largest traders have 35% (the US) and 18% (Russia) of the assigned amounts of all GHGs or of CO_2 for 2008–12. The EU countries account for 23%. Disregarding the possible influence of market power until later, the expectations are that this market for emissions trading will be large enough, and can be given trading rules for it to behave as a perfect market, e.g. according to the exchange design suggested above. However, if trade will be limited in various respects such as in time or that traders meet bilaterally with less room for effective competition among sellers or buyers, the market may not converge to uniform prices and efficiency.[8]

The prospects for having *additional countries joining the Annex B group* and participating in emissions trading during 2008–12 can only be speculated about. Additional countries, if any, would belong to the poorer part of the globe and require large enough quota allocations to keep them fully compensated for the first commitment period. However, there may well be several such countries that might be interested in committing themselves to an emissions quota and participate in emissions trading. In a 1996 study of the acceptability of a global TQ treaty for a sample of countries, where each country was represented by an experienced (Swedish) diplomat (in 24 of 29 cases an ambassador) knowledgeable of the country he/she represented, 9 of 17 non-rich countries were deemed to be willing to join such a treaty (Bohm, 1997b).[9]

Assuming that new trader countries would get at most quotas equal to their BAU emissions levels (but still enough to keep them at least fully compensated),[10] the effect of an expansion of the group of trader countries would mean that (1) new low-cost abatement options would be available, hence reducing quota prices and the costs of the emissions limits for the *original Annex B countries*; this, in turn, might make the pre-existing quota importer countries inclined to accept more stringent emissions limits for the next commitment period. Furthermore, (2) the *new participants* would be better off during the first commitment period as compared to not joining the Annex B group this early, since they would now get a new export commodity – emissions reductions – that would increase their export earnings by selling abatement at prices exceeding abatement costs.[11] However, (3) there would also be a group of losers as a result of the addition of new trader countries significantly lowering the quota prices, *viz.* the *pre-existing sellers* on the quota market. Countries that anticipate their earnings from exporting emissions reductions to fall short of what they expected at the time of the initial quota allocation may require larger emissions quotas in subsequent periods, or they may be expected to try to make it difficult for new sellers to enter market – if so, most likely perhaps by requiring potentially new seller countries to get such small quota offers that they would decline to join the Annex B group.

Carbon leakage
A major difference between global and non-global (but still large) TQ systems with respect to the overall efficiency of such systems relates to the fact that non-signatories to the agreement may increase their carbon emissions as a result of the emissions reductions by the signatories. Such so-called *carbon leakage* occurs partly because carbon-intensive products in signatory countries become more expensive, which causes an increase in imports from, and hence an increase in the carbon emissions of, non-signatory countries. In addition, and especially if certain non-signatories are taken to remain outside international emissions-reducing agreements for a period far into the future, firms using carbon-emitting technologies in signatory countries now have an incentive to move to non-signatory countries where such emissions cost less.

The Kyoto Protocol does not contain any policies or measures to counteract carbon leakage. One reason is that generally acceptable, e.g. WTO compatible, versions of such measures are not easy to come by. But a more relevant reason in the present context for not observing what any such measure might be is that carbon leakage is a problem common to all policies aimed at reducing emissions among a *non-global* set of countries, e.g. also non-tradable quota systems.

The bottom line concerning non-global emissions trading is that, as long as the number of trader countries is large, there is not likely to be any large difference in the principles of trading arrangements and results between global and non-global trading, with the exception of the carbon leakage problem just noted. An important effect of having still more countries joining an agreement like the Kyoto Protocol is that carbon leakage risks will be smaller.

We now turn to address an issue which is equally relevant for global and non-global carbon emissions trading.

1.3. *Flexibility aspects*

According to conventional terminology, emissions trading provides flexibility to commitments by countries to limit their carbon emissions. Additional flexibility to the TQ system itself can be provided by the choice of the length of the base period and the target period and by allowing banking and borrowing of quota units. As the UNFCCC was originally formulated, the *base period* was set to be one year, the base year 1990, in spite of the fact that a number of 'fairness' problems would have been avoided by choosing a longer period of, say, three or five years. Probably for practical reasons, the Kyoto Protocol kept the same one-year base period. The *target period* of five years, 2008–12, in the Kyoto Protocol avoids some of the problems that are related to the variation over time in fossil-fuel use over the business cycle or as a result of extreme weather conditions (e.g. heating needs and precipitation for hydro-electric output).

The Kyoto Protocol allows *banking* of quota units for use during the next commitment period, presumably for 2013–17 (for which quota negotiations will start no later than 2005, but is assumed here to be finalized well before 2008). Banking means that a country is allowed to keep quota units to sell later or to cover future emissions. Thus, the implication of banking is (1) that countries are allowed to speculate in quota prices across periods and (2) that countries can make extra emissions reductions during the first period that are less costly than the quota prices expected for the next period. Furthermore, (3) banking protects the country from the consequences of abating more than required by mistake. The latter two implications explain why banking is not controversial and thus probably why it is included in the Kyoto Protocol.

What is not accepted in the Kyoto Protocol is *borrowing*, i.e. allowing Parties to borrow from their commitments for the next period, thus exceeding their present quotas. This would provide some leeway for a country that, at the end of the commitment period, is hard pressed to stay within its quota – *ex post* whatever quota trade that has occurred up to that point – and finds additional units at that point in time to be 'too expensive' to purchase.[12] Allowing a strictly limited amount of borrowing, say, by 1–2% of the next five-year commitment period's quota, i.e. 5–10% of the first year's share, could make a significant difference for such a country. Borrowing could be subjected to a cost of a fixed 'rate of interest' or an equilibrium rate at which aggregate borrowing equals aggregate banking (to avoid having emissions exceed the target for the first commitment period). The resulting additional flexibility provided by allowing (limited) borrowing may make some countries more willing to accept stringent (or lower) national quotas and hence make the aggregate commitments larger than if borrowing is ruled out. Thus, it might be in the interests of efficient climate change policy to allow limited borrowing.

2. Domestic policies to support carbon emissions trading

At the end of a commitment period, trader countries will be accountable for having, *ex post* quota trade, an emissions quota sufficient for covering the carbon emissions made during the entire period (unless amendments to the Kyoto Protocol are made to allow for limited borrowing). An advantage with tradable quotas as an international climate change policy, as compared to policies such as international carbon taxes or harmonized domestic carbon taxes, is that each country can choose its domestic policy to be in compliance with their commitments in the Kyoto Protocol. Thus, they could use a domestic tradable permit (TP) system, domestic carbon taxes (or, which will not be discussed any further here, less appropriate taxes such as energy taxes) or direct regulation of carbon-emitting activities. These three options will be discussed in turn.

2.1 Tradable permits

In many discussions of the TQ system, it has been more or less taken for granted that this system will be coupled with domestic TP systems. There are several advantages with this combination. First, the initial national quota can be allocated to domestic units, which in turn may trade on the international market either directly or via special trader firms or brokers. This will contribute to an atomization of the market and hence, with special respect to the US, where the preferences seem to be almost unanimously in favour of using a TP system, make those outside the US, who fear a market dominance of that country, more at ease. The government of a country that uses the TP approach is still a Party to the Protocol and only that Party can be held internationally accountable for compliance. Still, by choosing to use a TP system that is integrated into international trade, the government would delegate the everyday quota trade decisions to domestic firms or trading units.

Second, when a TP system is used, the country 'cannot' go wrong in attaining compliance with the Protocol in the sense that rule-abiding domestic units can only use the permits they set out with *plus/minus* permits bought/sold to cover their carbon emissions. This property is clearly absent when taxes or direct regulations are used.[13]

Third, the choice of design of a TP system (and a carbon tax system) is flexible in the sense that there are several options for the level at which permit (tax) liability can be introduced. There are some strong preferences heard for placing permit liability as much *downstream* as possible, hence where the combustion of fossil fuel takes place. This would imply, it is argued, that it is the entity, which knows the technology options and the market for its outputs and inputs, that would trade in permits to cover its emissions. However, there are several problems with this approach; due to (1) there being high transaction costs, (2) small end-users being likely to be excluded from the TP system, and (3) for this and other reasons, the system not having the advantages it is claimed to have. Transaction costs for having individual motorists buying and presenting permits to buy petrol will obviously be very high, and few pro-TP analysts advocate including small users in the TP system, at least not directly.

One way to deal with the problem of a separate treatment of large and small fossil-fuel users is to move permit liability for the group of small emitters *upstream* and place it, by permit-liability regulation, on sellers of petrol to motorists and heating oil, etc. for one-family houses – e.g. on retail sellers, wholesale sellers or importers/producers of petrol/heating oil, etc. This means that the consumer prices of these products will include the current permit (= quota) prices. The remaining question then is whether the individual motorists, house owners or other small-scale fossil-fuel users would make less efficient decisions by adjusting to permit prices *included* in the petrol, etc. price than by adjusting to permit prices that are *excluded* from the petrol price.[14] Since this is unlikely, and equally true for other fossil-fuel users, all fossil-fuel uses can be directly included in the TP system and transaction costs be minimized by placing permit liability as much upstream as possible, i.e. on producers and importers of fossil fuel.[15] Thus, the initial amount of permits, which equals the country's carbon emissions quota, would be distributed to these units. This does not preclude other units from being allowed to hold permits for one reason or another, e.g. as speculators in permit prices changes or as environmentalists who wish to withhold part of the permit volume from emitters (Bohm, 1998a).

An additional advantage of having permit liability placed on the producer (also the exporter) level and the importer level would arise if the country's emissions were monitored simply by its fossil fuel production *minus* exports *plus* imports as mentioned earlier. This presupposes that there is an acceptable solution to the problem of monitoring changes in fossil-fuel inventories. Then, domestic as well as international monitoring would be simplified by having the permit volume distributed to producers and importers of fossil fuel, thus monitoring only the amounts of fossil fuel sold by these entities. Direct monitoring of carbon emissions would not be needed for international or domestic accountability anywhere along the road.

Fourth, the choice of the design of a TP system is flexible also in the sense that the permits can be *auctioned* off by the government to the permit-liable units and/or those who want to buy permits for other reasons or be allocated *for free*, as is the standard procedure in the US.[16] There are two important differences between these two allocation principles.

(1) Auctioning the whole volume of permits (equal to the initial quota) provides government revenue that allows a reduction of pre-existing distortionary taxes, a so-called double dividend. The reason is that auction prices on permits (essentially a carbon tax, see below) are not in themselves distortionary (contrary to claims made in parts of the economic literature). Given that the initial permit volume reflects a global environmental concern filtered through a process of international negotiations, the auction price reflects this environmental concern and emerges as a corrective rather than a distortionary levy just like any other environmental 'tax' reflecting similar concerns would. With *gratis* allocation of permits there would still be permit prices that emerge on the permit market right after the distribution of permits has taken place. These prices would likely be about as high as the auction price, at least in the beginning. They would not reflect any distortion of the economy, any more than the auction prices (and the permit prices that succeed them) would.

(2) *Gratis* allocation such as grandfathering (allocation guided by some historical emissions/fossil-fuel-use data for individual firms, etc.) means that the

wealth (rents) represented by the permits will be given in some proportion to existing legal entities or firms. Strictly speaking the implication is that these firms obtain windfall profits (as compared to not being given the permits *gratis*). This does not mean that it will be less expensive for them to emit carbon. The alternative to emitting a tonne of carbon and having permits cover that amount is always to sell or not to buy permits. Thus, all agents, with or without a free allocation of permits, will be confronted with the same (opportunity) cost for *emitting carbon*.

Still, as a result of making some firms richer than others by giving them (a large volume of) permits, the 'favoured' firms get 'deeper pockets' which on an imperfect capital market enables them (1) to remain in business for some 'non-economic' reason, (2) to have more funds for risky investments like R&D for which borrowed funds are less likely to be available, and (3) to have cheaper access to bank loans and capital markets by simply being richer, making loans less risky for the lenders (Bohm, 1994). Firms that are not favoured in this fashion are definitely the new firms that want to enter the market. Since these firms tend to be more efficient than existing firms, giving away permits for free to existing firms can be expected to slow down productivity growth. It follows from what was said in the preceding paragraph that no such favour is given to incumbent firms when the permits are auctioned.[17]

A major issue in the TQ debate has been that a country which uses grandfathering of permits, say the US, would give its firms (in sectors favoured by the grandfathering rules used, e.g. carbon-intensive sectors) a competitive edge against countries that use carbon taxes or auction permits. However, the two points mentioned above – revenue recycling to reduce tax distortions and neutrality towards new firms – imply that auctioning of permits provides other important efficiency benefits to the economy in these countries. In particular, the recycling of auction revenue to reduce distortionary taxes typically provides financial assistance to existing firms as well and could be designed specifically with that purpose in mind (although that would typically reduce the magnitude of the double dividend).

Finally, it should be noted that if fossil-fuel producers/importers were made permit liable, these firms rather than the end users would be favoured by a free allocation of permits. To avoid such an outcome of unwarranted transfers of wealth, the fossil-fuel producers/importers would have to be taxed and the end-users subsidized. When permits are auctioned, and especially when the government wants to use the revenue to reduce taxes, this cumbersome process would not be needed. This is likely to imply that placing permit liability on fossil-fuel producers/importers is attractive mainly for governments that want to auction the permits.

2.2. *The equivalence of permits and carbon taxes*

For countries that otherwise would use a carbon tax, the revenues of which would be used for tax recycling, an equivalent approach would be to use auctioned tradable permits. Thus, contrary to what is often alleged, tax and TP policies are equivalent as long as political constraints on any of these policy options are disregarded.[18] The price for emissions, the emissions volume and the amount of government revenue would be the same if the permits were auctioned by the government or if the tax had been set at a level equal to the resulting auction price for the permits (disregarding the time difference between tax collection (*ex post*) and the permit auction (*ex ante*)).

If the permits instead were distributed for free in certain proportions to certain firms, the TP option would be the same as a tax that would generate the same emissions volume and where the resulting tax revenue, equal to the wealth value of the 'grandfathered' permits, were redistributed to the same firms in the same proportions as in the case of grandfathered permits. In other words, not only price and quantity but also the distributive outcome can, in principle, be made the same in the two cases of TPs and carbon taxes.

Political reasons may explain why the US does not seem to have access to the same set of options as other countries that do not 'have to' give away the permits for free or 'cannot' use taxes. Thus, for example, a Nordic country that could use carbon taxes and recycle the tax revenue could also accomplish the same thing by auctioning permits, since the tax policy and the permit policy are environmentally – or emission-wise – equivalent in the present context. For reasons given at the end of the preceding subsection, a country like the US may not have any other option than grandfathering permits to *emitters*, i.e. ruling out the use of carbon taxes as well as the cumbersome tax/subsidy scheme, where producers/importers of fossil fuels are made permit liable, as politically unpalatable. This means that the US might be politically 'forced' to use an approach which harms new firms, hence, productivity growth, and misses a chance to reform its tax system through revenue recycling.[19] By contrast, some other countries such as the Nordic countries may use either taxes or permits and have permit liability placed where otherwise tax liability would be placed, e.g. on producers/importers of fossil fuels.

2.3. *Carbon taxes and regulations*

In Section 2.1, one principal argument was given that may support the claim that TPs will be more efficient to use than other options for domestic policies. It was stated that a TP system would lead automatically to fulfilment of a quota commitment, when TPs were issued in the amount of the initial national emissions quota and permit-liable entities (and others who wish to hold permits) trade directly or indirectly to increase or decrease the quota that the country is given. By contrast, options such as direct regulation and carbon taxes would have to be adjusted over time if the target is to meet the government's *ex-post*-quota-trade quota; such adjustments may be more difficult to predict and hence more costly for all those affected by these taxes or regulations than having these entities adjust to changes in market-determined permit prices.[20]

Alternatively, a government would have to adjust its quota trade to meet the outcome of whatever the tax levels or regulations that initially had been decided; this would be costly for the country in that its marginal emissions abatement costs would end up deviating from the 'international market quota price', thus forcing the government to abate too much at home or to buy quota units at prices that exceed abatement costs at home.[21] For the case of regulations, there is, of course, also the well-known argument that regulations are unlikely to be as cost-effective as the market-based instruments, carbon TPs and carbon taxes.

Still, individual countries may not like, or may not find it technically feasible, to use a TP system at all or one that implies that individual permit holders trade on the international quota market. For a government that wishes to use TPs for domestic trading only, prices on the domestic permit market that exceed the international quota price would encourage the government to buy additional quota units and feed them in terms of permits into the domestic market. In the opposite case, when the quota price is higher than the domestic permit price, the government could buy permits on the domestic market and sell them as quota units on the international market.

In the last-mentioned case as well as when carbon taxes or regulations are used, the government will be the trader on the quota market. An argument commonly voiced is that this would be inefficient for the reason that governments are not known to be efficient traders (although they may very well hire professionals to do the job for them)[22] or that they would not be as well informed about the relevant marginal abatement costs as the permit holders would be. However, as discussed in Bohm, 1998a, given that governments will keep negotiating emissions quotas for new commitment periods, there are strong incentives for them to be well informed about their countries' marginal abatement costs. Only then can they evaluate the costs of proposed quota allocations and how internationally fair such proposals are.[23] By contrast, there are many possibly permit-liable firms, which may not be so well informed about their position on their future commodity markets, e.g. those in strong competition with other firms. Given the arguments stated above for having part or all of the permit liability placed upstream, at least some permit-liable firms in such a system would not have any direct information about the relevant market conditions and the carbon substitution technologies downstream. They may even be less well informed than a government, whose perspective for international quota trade is the relevant *aggregate* marginal abatement costs for the country as a whole. Thus, even countries that will not use an integrated TQ/TP system may not be much less efficient than others in trading on the international quota market.

Further work is, of course, needed to identify which of the domestic market-based policy options and their various versions would be best suited to meet the policy targets of an individual country. However, given the efficiency arguments presented above, a likely candidate at this point in time would be TPs auctioned to fossil-fuel producers and importers.

3. Determinants of the benefits of international emissions trading

If emissions trading would arise among countries that act in their own self-interest, the resulting trade is proof enough that the benefits of international emissions trading are large enough. Still, the likely size of these benefits is naturally of great interest to ascertain for countries that need to determine the extent to which they would like to participate and the trade design they would prefer. As a general remark, *aggregate* trade benefits would be larger (and carbon leakage problems smaller) the larger the number of participating countries and the more the quota allocation among a given set of trader countries deviates from the cost-effective distribution of emissions among them. *Individual* country gains from trade would be larger the more distant their initial quotas are from their internationally cost-effective emissions volume, although each country, of course, loses from having a smaller initial quota, other thing equal. Furthermore, aggregate as well as individual country gains from trade would be larger the larger the span between buyers' and sellers' reservation prices (i.e. the span between high and low marginal abatement cost curves) around a given equilibrium quota price.

A more specific indication of what benefits may arise from international carbon emissions trading is given by the Nordic quota trade experiment referred to earlier. Using marginal abatement cost data derived in a fashion that made them unlikely to differ much from what could be expected to be realistic, it turned out that even for these, in many respects, quite similar countries, abatement costs differed a great deal and gave rise to considerable trade gains. The estimated aggregate cost for Denmark, Finland, Norway and Sweden to stay on their 1990 carbon emissions levels by the year 2000 (a Rio 'target') amounted to US$713 million in the absence of trade, but was reduced to US$368 million as a result of (hypothetical) emissions trading, i.e. by almost 50% (for details see Table 1 and Bohm, 1997a).[24]

The outcome of this experiment illustrates another important aspect of the benefits of international emissions trading. The efficient trade volume in the year 2000, i.e. the volume exhausting all potential aggregate gains from trade for these countries, amounted to 6.95 million metric tonnes of CO_2, while negotiated trade actually amounted to only 5.5 million tonnes, i.e. 80% of efficient trade. However, trade achieved 97% of the fully efficient trade *gains* (gains as if under perfect competition). This outcome reflects the fact that, when all potentially valuable transactions are not carried out, it is only the least valuable trade, where buyers' and sellers' reservation prices are very close, that does not materialize.

The design of the experiment also reflected the likely extent of certain aspects

Table 1. Carbon emissions trading among four countries (the fully efficient trade in parentheses)

	Unilateral		Trade	*Ex post* emissions trading		
Country	Em. red. (Mton)	Cost (MUSD)	Exp./Imp. (−) (Mton)	Em. red. (Mton)	Cost (MUSD)	Net gain (MUSD)
Denmark	1.7	61	0.5 (1.19)	2.2 (2.9)	83 (116)	6.7 (5.1)
Finland	6.0	94	5 (5.76)	11 (11.8)	216 (251)	136 (132.2)
Norway	5.4	456	−3.5 (−4.4)	1.9 (1.0)	91 (40)	178 (194.5)
Sweden	1.6	102	−2 (−2.55)	−0.4 (−1.0)	−22 (−51)	24.4 (25.1)
Total	14.7	713	±5.5 (±6.95)	14.7	368 (356)	345 (357)

of *uncertainty* in real-world trading and some of their implications for the gains from emissions quota trade. As already pointed out, countries negotiating quota allocations need to know as much as possible about the cost implications of alternative such allocations for themselves but also about the extent to which different proposed allocations can be regarded as internationally fair; hence, countries will have strong incentives to estimate the marginal abatement costs of their own country as well as those of others. To the extent that the marginal abatement costs are likely to guide emissions trading among Annex B Parties, this information will also allow estimates of quota prices for the upcoming commitment period and thus, the order of magnitude of their expected trade gains. Investigations into the marginal costs have been going on for some time in both international organizations and individual countries and can be expected to make the international community approach a state of near-full common knowledge of what the cost differences are. In the Nordic experiment, the participating countries had been exchanging information for some time about their respective marginal abatement costs as far as the *technical* abatement options were concerned. However, the trade-relevant *social* marginal abatement costs, i.e. after the trader countries have taken income distribution, employment and other 'political' costs into account, which may introduce small or large information asymmetries, will emerge as essentially unknown deviations from the technical costs. The experiment was designed to reflect these information conditions, which can be expected to hold also for the trade among the Annex B countries, although the trade rules may be quite different when the number of countries will be much larger than four.

3.1. *The 'hot air' issue*

Estimating business-as-usual (BAU) emissions ten years from now is obviously difficult. However, it is widely expected that they would fall short of the 'assigned amounts' or quotas in the Kyoto Protocol for some of the Annex B countries, notably Russia and the Ukraine. If so, the outcome is that part of the emissions reductions eventually bought will be 'hot air' only. But the claim that this is a failure of the Protocol may be based on a misunderstanding of what was possible to achieve in the Kyoto negotiations. The resulting Protocol would certainly be a failure if the relevant comparison for the Protocol were that the negotiations could have ended in no country being given a quota with 'hot air', other things equal. But it may well be that other things would not have been equal.

First, Russia and the Ukraine may not have accepted a reduction of their quotas to equal their most likely BAU levels, especially not given the time-pressed conclusion of the Kyoto negotiations. Instead, the risk would seem to have been significant that, confronted by such a quota offer, one or both of these countries would have dropped out of the agreement. Recall that BAU estimates are uncertain, perhaps especially so for countries like Russia and the Ukraine. These countries may also be more optimistic about their GDPs around 2010 than other countries are and predict that their BAU emissions would grow considerably even after an expected increase in their energy efficiency.

Second, if Russia and the Ukraine would have dropped out if offered quotas without any 'hot air' according to some given estimate, other Annex B countries would realize that their commitments then would have become more costly and therefore may have negotiated larger quotas to themselves. Take the case where these countries taken together now would have added the assumed known 'hot air' amount to their quotas, thus keeping aggregate emissions reductions on the level of the Kyoto Protocol. Their costs would then be reduced by the fact that they no longer would have to pay for the 'hot air' bought from Russia and the Ukraine. But they would then also have to replace the amount of *real* emissions reductions that these two countries otherwise would have made at the estimated above-zero quota price. If the approximate marginal abatement cost relationships up to 2012 were known, it would have been possible to estimate the likely quota trade and quota price already now and hence, whether the net costs for the remaining Annex B countries would be smaller or larger as compared to the present Protocol. Since this was not known in December 1997, it can only be said that there is a risk that their costs would be larger for the unchanged aggregate emissions reductions and that they may have wanted to increase their quotas by more than the 'hot air' amount as so far assumed.

In addition, it should be observed that, as a result of Russia and the Ukraine having dropped out, there would be an increased risk for carbon leakage from the remaining Annex B countries and hence, a risk that global emissions reductions then would be smaller than what would follow from the present Protocol. This would add to the risk that the remaining Annex B Parties would have liked to increase their quotas by more than the 'hot air' amount, thus increasing the aggregate emissions level above the level implied by the present Protocol. If so, it follows that the Kyoto Protocol may be the 'best' that could have been achieved in the given circumstances.

More specifically, this means that the 'hot air' component of the Kyoto Protocol could be seen as part of a successful agreement, i.e. as one with possibly larger aggregate emissions reductions than a feasible alternative outcome of the negotiations without 'hot air'. In this alternative case, the argument that now is heard that 'hot air' calls for deviations from free emissions trading – to ensure that the expected quota buyer countries would abate more at home – would not have been made. Then, it hardly makes sense to argue that such deviations are called for in the Protocol that may accomplish larger aggregate emissions reductions than in the alternative unobservable case. (See further the subsection on 'supplementarity' below.)

It should be noted that the introduction of emissions trading may increase the attempts by seller Parties to obtain large quotas, possibly including 'hot air' (which the Russian and Ukrainian behaviour may be an illustration of), as compared to the case of non-tradable quotas, other things being equal. The reason is that making quotas tradable implies that the value of an additional quota unit to a seller country, i.e. a Party with low marginal abatement costs, is increased from the marginal abatement cost (zero if already in the 'hot air' region) to the expected quota price (see Bohm, 1992). Also, if seller countries (countries in transition to market economies and, later on, presumably also developing countries) are countries with expectations to grow at rates above the Annex B average in later periods, they may press for larger quotas already in a first commitment period, given that larger quotas 'today' will help to get larger quotas 'tomorrow' accepted.

The conclusion of this subsection – that the presumed 'hot air' component of the Kyoto Protocol, in fact, may be part of a successful agreement – may be incomplete in one possibly important respect. In Toman and Hourcade, 1998, it is argued that "trading in 'hot air' is seen as a bad precedent in terms of future entry by other countries into Annex B of the Protocol". However, the 'hot air' in the Kyoto Protocol is now a 'fact of life'. The concern that it would work as an effective precedent can only be addressed by Parties being better prepared for future negotiations than they were in Kyoto. In addition, it could be declared up front that the expected 'hot air' component of the Kyoto Protocol should not be regarded as a precedent but as a specific characteristic of this the first round of emissions quota negotiations, where the quota negotiations close to the time limit and a desire to get 'at least something' out of these negotiations called for significant concessions to a set of poorer countries.

3.2. *Trading rules and trade gains*

A possible version of *efficient, free international emissions trading* is, as mentioned earlier, that of an exchange where Annex B countries and, for some of them, private entities, to which a national quota has been distributed, enter bids and asks (offers) in terms of quota units and prices. Such bids or asks made on a market that is worldwide and constantly open would be accepted or rejected in the same way as on an ordinary stock exchange. At the end of the first commitment period, a series of trade/price combinations will have been observed, including possible resales or repurchases as a result of changes in underlying factors determining quota demand and supply, and changes in expectations of new trade opportunities that evolve over time. The gains from these trades will be the result of (1) supply and demand as revealed by the asks and bids and accepted bids and asks made and (2) the dates when these bids and asks will have been made. The latter determinant of the gains made may be referred to as revealing the skill (or 'luck') of individual traders. However, since all participants in this trade are likely to have access to similar trading skills, it would seem that the second determinant of the differing gains by the traders is essentially due to stochastic elements. Although all countries are familiar with, and have experienced, the influence of stochastic elements on trade gains, e.g. when negotiating long-term contracts for fossil-fuel trade, it would seem that the trade rules now suggested may be acceptable to the Parties (see also the subsequent discussion of 'market power').

If trade would take place directly between two parties, outside an exchange or in the absence of an exchange, a strong feeling may arise that some of these *bilateral transactions* are unfair to other traders. The role of such transactions would have to be analysed in detail. However, as a minimum condition for acceptable bilateral transactions, it would seem to be necessary to make all transactions subject to a transparency requirement, where they and their prices are made commonly known to all traders (e.g. via a Protocol secretariat), though not necessarily who trades with whom. For this transparency requirement to be meaningful, it would also seem necessary to require that Parties consent to abide by a rule of no 'side-payments' of any kind, although enforcement of such a rule, of course, would be difficult.

Implementing multilateral transaction systems, as on an exchange market, would make it possible to keep traders anonymous to one another (but not to the market administrator, the Protocol secretariat). If non-anonymous, certain mutually beneficial transactions may not come about for reasons of diplomatic controversy or, when firms trade, for reasons that established firms do not want to sell permits to new competitors. Anonymity would thus seem to make market transactions more efficient.

3.3. Market power

Much attention has been paid to the risks that one or two large trader countries could influence market prices to their own advantage. As a case in point, it has been feared that the US might act as a monopsony and keep prices down, although this would be less likely if the US government delegates its international quota trading to independent firms. The trading rules finally adopted will certainly have been scrutinized from the perspective of market power.

However, it is worth noting that the quota exchange discussed above is likely to make market power inoperative. The primary reason is that one buyer cannot succeed in reducing prices by reducing its demand, when all other traders can keep making transactions as long as there are sellers that can accept prices offered by buyers and vice versa. This means that the aggregate trade volume cannot be made smaller than the perfectly competitive volume and that prices for *final* trades must approach the competitive price level.

Still, it may be possible for a dominant trader to make *early* transactions at prices that are favourable to this trader. If so, part of the conventional market power argument would remain valid. However, since the trader countries, as already pointed out, would have strong incentives to collect information about not only their own but also their trade partners' marginal abatement costs, they would be able to predict fairly well what the competitive price level would be on the international quota market. Then, it becomes less likely that sellers (buyers) would accept prices much below (above) that level, hence reducing this form of market power in early transactions as well (Bohm, 1998; Carlén, 1998).

3.4. Supplementarity: Caps on quota imports or the use of flexibility mechanisms

Article 17 of the Kyoto Protocol calls for emissions trading to be only "supplemental to domestic actions for the purpose of meeting quantified emissions limitation and reduction commitments under Article 3". The principal reason seems to be that it would be unsatisfactory if a Party could meet its commitments only (mainly) by buying emissions reductions from other Parties and thus avoid doing any (significant) reductions at home. It has been suggested that quantitative constraints on imports of emissions reductions be introduced, possibly including not only emissions trading but also JI and CDM activities. Or the constraints may be of a qualitative character implying, e.g. harmonization of policies and measures across countries. Such constraints might increase marginal costs or shadow prices of emissions reductions in the importing countries and be, as has been argued, a more effective stimulus to "technical innovation needed to make future targets more affordable" (Toman and Hourcade, 1998).[25]

Only constraints on emissions trading will be discussed in what follows. The reason is the assumption that, if CDM operations are properly checked for possible distortions of the projects' nation-wide baselines and hence their *de facto* emissions reductions, both strictly speaking unobservable (see e.g. Bohm, 1997a), this flexibility mechanism will most likely play a modest role only.[26] JI among committed countries, also on a project-by-project basis with its demanding estimation problems, can hardly be expected to attract much attention since the simpler and more potent emissions trading mechanism exists for these countries.

Furthermore, no distinction will be made between quantitative and qualitative constraints here. The latter type of constraints may be: changes in average quota prices, levels of subsidies on carbon intensive activities, road pricing, norms and standards for energy efficiency, etc. (UNCTAD, 1998). The reason for discussing only binding quantitative constraints is that the resulting reduction in efficient emissions trading is the crucial point and that the purpose of qualitative constraints in this context is to force countries to do more abatement at home than they otherwise would like to do.

First of all, it should be noted that the opening up of emissions trading implies that a new tradable commodity is introduced in the world economy. As with all other tradable commodities, a case can be made for free trade, which would benefit all trader countries. If all countries eventually would agree to observe negotiated emissions quotas and thus become Annex B countries, global free emissions trading would minimize global costs for a given global emissions target and benefit all countries that participate, especially if we can assume that significant economic distortions do not exist to the extent of making unregulated trade detrimental to some or all countries participating.

Regardless of emissions trading being global or non-global, all participating countries acting in their self-interest would make at least part of their emissions reductions at home, given quota prices non-trivially exceeding zero. The basic reason for this outcome is that the lowest costs for a country to reduce any of its emissions from its BAU level will barely exceed zero, especially when forewarned several years in advance. This presupposes, of course, that there are no absolute political constraints on actions at home, which if they did exist would mean that the country is punishing itself by accepting costly constraints on trade. What this means is that the country would pay more by not undertaking low-cost actions at home – by not producing, as it were, a commodity to the extent the country has a comparative advantage to do so.[27] Moreover, if the country is large, its extra demand for imports of emissions reductions will tend to increase quota prices, thus punishing it even more.

If an individual country expects quota prices in 2008–12 to be much lower than those expected for the subsequent commitment period, it might seem to pay the country to do less – and in an extreme case, nothing – at home now and save, if meaningful, domestic emissions reduction options for the next period. In principle, this may go on for several periods, thus implying that domestic emissions reductions are postponed for quite some time. It is unlikely, however, that this would go on forever, so eventually also such domestic emissions reductions would come about. But the significance of such delays is likely to be small for the following two reasons. First, postponement would not come out in this fashion if many countries tried to act in the same way, thus reducing quota prices in early periods. Second and more important, banking is there to make it possible for countries to do what is worth doing at home now, still import more quota units than needed now, thus banking excess quota units for tomorrow.

It follows from this discussion that regulating quota trade for *all* countries to fall below the level of free trade would not be in the interest of any country (although, as always, it may be to the benefit of some groups inside a country). Forcing countries to produce more of the emissions reduction commodity at home than it wants to is like forcing cold Nordic countries to grow some minimum share of bananas before it is allowed to import bananas from countries that have a comparative advantage in banana production. It would increase the costs of banana consumption in the regulated countries and reduce demand for imports and hence prices to the exporting countries.

Regulating quota imports for *some* but not all countries would, of course, tend to reduce quota prices and hence benefit those importers not affected by the regulations. Likewise, varying the extent to which regulations would limit trade for different countries means that they would be treated differently. This would add to the complexity of ever negotiating the details of a stringent Article 17, a safe prediction for any future ambition to make this article anything else than a paper tiger, it would seem.

Confronted with the argument for putting a cap on emissions trading – to avoid rich industrialized countries 'getting away with' covering 'too much' of their emissions reduction commitments by importing cheap emissions reductions from other countries – we have noted that not only these, but typically all, Parties lose. This has two further important implications. First, by making emissions reductions globally more costly in this way, also emissions reduction commitments in subsequent periods will be made more expensive and therefore less likely to be significant. Second, since trade regulations would reduce the gains from trade for poorer countries, such as Russia and the Ukraine now, it will also reduce the prospects for trade gains for potential entrants from the developing countries in the future. Thus, supplementarity will not only make present Annex B countries less likely to accept more stringent future commitments, it will also make it harder to get new countries to join the set of Annex B countries.

Finally, in the more immediate time frame, the success of the Kyoto Protocol stands and falls with the US ratifying it. It is well known that the likelihood that they would depends to a large extent on emissions trading being unregulated. This, if nothing else, would seem to call for meeting the supplementarity requirement in Article 17 in the least stringent way possible or by replacing this requirement as much as possible by having the rich countries make stringent commitments for the period after 2012 as early as possible.

4. Main conclusions

(1) Although precedents to international emissions trading are almost non-existent, it seems clear that there are enough options available for efficient carbon emissions trading benefiting all participating Annex B countries

and assisting them to make more stringent emissions reductions possible in the future. The aggregate benefits are likely to be large given the expected large international differences in the marginal costs of emissions abatement.

(2) The entry of new Annex B countries, all likely to be sellers at least initially, improves the aggregate trade gains and tends to reduce carbon leakage. Constraints on trade to meet a stringent supplementarity requirement would reduce the willingness of new countries to enter.

(3) The Kyoto Protocol does not allow countries to borrow additional emissions quota units from their quotas in the subsequent commitment period. Strictly limited borrowing, say, up to as little as one or 2% of emissions quota for a subsequent commitment period would still add important flexibility with few drawbacks, if any. Old and new Parties would most likely be more willing to commit themselves to stringent emissions reduction commitments in the future if also this flexibility were available. Thus, it seems worthwhile to have the Protocol amended to allow limited borrowing.

(4) Reliable monitoring of greenhouse gas emissions may be seen as particularly important when international emissions trading is introduced requiring verification that net emissions reductions sold to other countries are actually made. Monitoring would seem more effective the less it relies on direct monitoring of carbon emissions or varying monitoring techniques for different sectors of the economies. Although indirect monitoring via the carbon content of the fossil-fuel use of each country is not without problems – primarily concerning the monitoring of fossil-fuel inventories – it seems to offer the simplest and overall most reliable approach. Checking domestic compliance to a system of tradable permits seems relatively easy to administer if producers and importers of fossil fuel have been made the only permit-liable entities in a country.

(5) Several options exist for the institutional design of international emissions trading. A stock-exchange kind of institution has been in focus here as an example of a design that is likely to be efficient and avoid any significant influence of market power. Bilateral trading is unlikely to be equally internationally efficient and fair.

(6) Among the feasible domestic policies to support international emissions quota trade, a system of tradable permits is likely to be the most efficient, although perhaps not much more than carbon taxes. A tradable-permit system is compatible either with a government in charge of international quota trade or, if the individual country so prefers, by also delegating international trade to the permit-liable firms or their brokers. Given the existing incentives, governments cannot offhand be taken to be less efficient as international traders than firms, contrary to what is often alleged.

(7) Auctioning of permits (= the initial emissions quota volume) is more efficient than *gratis* distribution of permits (grandfathering) and need not

be harmful to existing firms, especially not if part of the revenue is used for compensation when required. Auctioning produces a 'double dividend' in that the (remaining) revenue can be used to reduce pre-existing distortionary taxes. Furthermore, auctioning does not slow down the entry of new and presumably more efficient firms, at least not to the extent grandfathering would.

(8) Tradable carbon permit systems and carbon taxes are equivalent in the sense that the permit prices/taxes, government revenue and firm wealth could be made the same (under certainty), (1) when permits are auctioned or emissions are taxed and the revenue in both cases is kept for revenue recycling by the government, and (2) when permits are distributed *gratis* to a set of firms or the revenue of the alternative tax is redistributed to these firms in the same proportion as the permits were grandfathered.

(9) Making only fossil-fuel producers and importers permit liable is the most comprehensive and transaction-cost-effective design of a system of tradable carbon emissions permits.

(10) The introduction of 'hot air' in some of the emissions quotas of the Kyoto Protocol has been interpreted as a drawback jeopardizing the Protocol's effect on the overall emissions reductions. However, this argument may stem from a misrepresentation of what the alternative to the present Protocol actually would have been. It was shown here that the aggregate emissions reductions may well have turned out to be smaller if 'hot air' had been removed from the countries concerned and they therefore would have chosen to drop out. If so, other Parties may have significantly reduced their emissions reduction commitments. If the relevant 'no-hot-air' alternative would have implied aggregate emissions reductions smaller than in the present Protocol, the reason for even raising a demand for a supplementarity requirement can be questioned.

(11) It is most unlikely to be in the interest of any Annex B Party to avoid making domestic emissions abatement to any significant extent, if quota prices would exceed zero, i.e. if the Protocol's aggregate emissions reductions were non-trivial.

(12) All quota exporting countries stand to lose from the enforcement of a stringent supplementarity requirement, as do all those importing countries that are effectively restrained from reaching the desired import volume. Since this is likely to imply that no importer country will be protected from binding constraints (which otherwise would have changed the cost distribution of the Kyoto Protocol as it stands now), it seems likely that all countries will lose from any binding supplementarity requirement. Hence, benefits of emissions trading would be maximized for all countries if the least possible were done to limit the trading. Particularly important to note is that regulating trade and thus reducing the trade gains is likely to reduce the willingness of existing Annex B countries to commit themselves to lower emissions quotas in the future. This would reduce also the

willingness of new countries to join the Protocol and enter into emissions trading, which would have provided them with a new 'export industry'.

(13) Rather, it would seem better to meet the concerns of those who advocate supplementarity by having Parties make strong commitments for the period after 2012 and leave emissions trading as free as possible. Avoiding the introduction of any (binding) trade constraints would also help in having the US Senate ratify the Kyoto Protocol. Without US participation there may be no significant emissions reductions at all for quite some time.

Notes

1. This study was commissioned by the Nordic Council of Ministers (Copenhagen) and was published, in a somewhat larger version, as TemaNord 1999:506 (*www.norden.org*). The Council has kindly accepted to have the study reprinted here. Helpful comments by Björn Carlén, Carsten Helm, and Patrick Säfvenblad are gratefully acknowledged.
2. The reasons are: JI (project-related emissions reductions in the new Kyoto Protocol sense of that term) is unlikely to play any significant role for Annex B countries already engaged in full-scale emissions trading, which is the topic of this paper. CDM (which is equal to joint implementation in the old sense) refers to project-related emissions reductions in non-Annex B countries that are difficult to monitor given that project baselines are unobservable and incentives exist for both buyer and seller parties to exaggerate the emissions reductions achieved. Hence, stringent verification requirements will hardly allow international acceptance of a great many such projects. However, in a recent paper (Manne and Richels, 1998), it is assumed that CDM may account for as much as 15% of hypothetical full-scale emissions trading among Annex B and non-Annex B parties.
3. The term (national) emissions 'quotas' is used for the assigned amounts of national emissions in the Kyoto Protocol. Hence, the term 'tradable emissions *quotas*' that is used here (as in IPCC, 1996) refers to international emissions trading, carried out by legal entities or by governments. The term tradable *permits* is reserved for domestic emissions trading, i.e. where a national emissions quota has been allocated as permits to legal entities in a country and these entities may or may not be allowed to trade also across national boundaries. For permits traded internationally, they represent a piece of quota trading as well. The primary reason for not using one and the same term for domestic and international emissions trading is that tradable quotas and tradable permits as now defined differ in important respects, for which identical terms are likely to cause confusion.
4. Those interested in the issues of the present paper should probably check UNCTAD, 1998, especially with respect to the detailed account of problems of monitoring, reporting, verification, compliance and enforcement, issues that are not so well covered below. It may also be worthwhile to check the web site *www.weathervane.rff.org* where the Washington-based research institute, Resources for the Future, publishes comments on, as well as analyses of, climate change policies, in particular concerning emissions trading in GHGs.
5. Note that emissions reductions represent a well-defined commodity here since the reductions are counted from the level of the emissions *quota* (or assigned emissions amounts) of an Annex B country in the Kyoto Protocol. 'Emissions reductions' in the sense of reductions from an uncertain future emissions *baseline* (or *business-as-usual* level) are not well defined, of course.
6. A recent overview of historical precedents of emissions trading can be found in UNCTAD, 1998.
7. Although there exist technologies for carbon removal from some forms of fossil-fuel combustion, the costs for using such technologies still seem to be prohibitively high or only now getting close to being economically attractive.

8. The same would be true, if the supplementarity requirement in Article 17 of the Kyoto Protocol led to significant, binding quantitative or qualitative limits on emissions trading. (On 'supplementarity', see further Section 3.4).
9. Of a total of 29 countries, 17 were believed to be in favour of such a treaty; that includes 8 of the 12 rich countries. 'China' and 'India' were not among the non-rich countries saying Yes, as is also in line with what was commonly expected some 18 months before the Kyoto meeting.
10. See the discussion of 'hot air' allocations below.
11. This argument disregards the fact that a country may be offered a better quota allocation for a later budget period if they decline a certain 'profitable' quota offer for joining now.
12. Such problems will essentially remain even if a 'true up' period of a couple of months into the next commitment period is introduced (UNCTAD, 1998) where countries in non-compliance could buy additional units from those who so far have banked quota units from the first period.
13. To be sure, *fraudulent* behaviour is possible in all three cases, in that a permit-liable or tax-liable unit or one subjected to regulations can break the law.
14. To highlight the permit price part of the consumer price, if that is required, this could be done in the same fashion as is often used to make buyers aware of the share of a consumer price that is a sales tax or a VAT, but now by making mandatory a statement of the absolute amount of the permit price.
15. As an illustration, a non-fossil-fuel-producing country, like Sweden that has some 300–400 registered carbon-tax-liable importers of fossil fuel, would in a TP system simply make these tax-liable units permit-liable instead. (See the account of an experiment proposed to make it possible to compare the effects of the carbon tax system in Sweden with those of a TP system, by replacing the carbon tax system for a representative sample of tax-liable importers by permit liability, STEV 1991.)
16. The reason for the US not auctioning permits to any significant extent may be called 'political' and similar to the political resistance to introducing new taxes. A limited amount of auction may exist, such as the 2.8% of the SO_2 permit volume. This amount is aimed at facilitating the access of new firms to permit holdings.
17. This is not to say that there may be *fairness* reasons for transferring some wealth to old firms, which in contrast to new firms have invested in durable equipment guided by relative prices that now are changed by environmental policy.
18. This is correct, strictly speaking, under full information. Under uncertainty, it may not be practically feasible to implement this equivalence, since the permit approach implies a control of quantities and the tax approach a control of prices.
19. However, things may be changing now, at least as evidenced by the June 1998 working paper by two US economists, entitled 'Tradable Carbon Permit Auctions: How and Why to Auction, Not Grandfather' (Cramton and Kerr, 1998).
20. It would seem possible that a country would end up having, possibly as a political compromise, both (low) carbon taxes – say, keeping an already existing such tax – and a TP system, to which also international emissions trading now had been delegated. However, confusion may arise especially if the tax would exceed the international quota price level, hence implying that the domestic permit prices should be negative. Such combinations are not considered anywhere in the text here.
21. In both these forms of adjustment – through changes in the levels of domestic taxes/regulations or the *ex-post*-quota-trade quotas – the cost of deviations from the efficient levels would be reduced if the Parties had access not only to banking but also to (limited) borrowing.
22. It may be noted that the emissions trading experiment with 'government traders' from Denmark, Finland, Norway and Sweden (Bohm, 1997a) performed very close to the theoretical efficiency level of a perfect market.
23. Also for a government that uses carbon taxes or auctioned permits and needs to predict government revenues, it would need information about the aggregate marginal abatement costs in the country.

24. If trade gains to an individual country are considerable, income effects are likely to play a role. General equilibrium models, of which there are now many in use to estimate the net costs of climate change policy, need to take such effects into account.
25. This argument is strange for several reasons (as Toman and Hourcade, no doubt, would agree). First, the argument, as it stands, says that making abatement (artificially) more costly now will make it less expensive later; the first effect is certain whereas the second is not. Second, it may be noted that caps on trade may not always make imported abatement as expensive as domestic abatement. If trading institutions are such that market power can arise, market power on the importer side would tend to reduce prices on quota units traded *below* the free-trade competitive price. The opposite would be true if market power were placed on the seller side of the international market. (For an exercise along these lines, see Manne and Richels, 1998).
26. However, it may be noted that 'limits on Annex B trading could drive more trading to the CDM, which is more difficult to establish and monitor' (Toman and Hourcade, 1998).
27. This argument presupposes that a country cannot successfully hide its low marginal abatement costs by increasing its quota imports (or reducing its exports), hence augmenting its fossil-fuel use, in order to gain international acceptance for being given a much larger quota allocation for the next commitment period, sufficiently larger to cover the extra costs of not using the less costly abatement options at home.

References

Atkinson, S.E. and T. Tietenberg (1987). Economic implications of emissions trading rules or local and regional pollutants. *Canadian Journal of Economics*. 20(2) May 370–86.

Atkinson, S.E. and T. Tietenberg (1991). Market failure in incentive-based regulation: the case of emissions trading. *Journal of Environmental Economics and Management*. 21; July, 17–31.

Bohm, P. (1992). Distributional implications of allowing international trade in CO_2 emission quotas. *The World Economy*. 15(1): January 107–14.

Bohm, P. (1997a). *Joint implementation as emissions quota trade: an experiment among four Nordic countries*. Nord 1997:4. Nordic Council of Ministers, Copenhagen.

Bohm, P. (1997b). *Are tradable carbon emission quotas internationally acceptable? An inquiry with diplomats as country representatives*. Nord 1997:8. Nordic Council of Ministers, Copenhagen.

Bohm, P. (1994). Government revenue implications of carbon taxes and tradeable carbon permits: efficiency aspects. Paper presented at the International Institute of Public Finance 50th Congress, Cambridge, MA. 22–25 August.

Bohm, P. (1998a). Determinants of the benefits of international carbon emissions trading: theory and experimental evidence. In: *Emissions trading – proceedings of the conference on greenhouse gas emissions trading* (Sydney, 21–22 May 1998). ABARE, Canberra 2601.

Bohm, P. (1998b). Public investment issues and efficient climate change policy. In: The Welfare State, Public Investment and Growth. Springer-Verlag Tokyo.

Bohm, P. and B. Larsen (1994). Fairness in a tradeable-permit treaty for carbon emissions reductions in Europe and the Former Soviet Union. *Environment and Resource Economics*. 4:219–39.

Carlén, B. (1999). Large-country effects in international emissions trading: a laboratory experiment. Research Paper 1999:15, Economics Department, Stockholm University.

Cramton, P. and S. Kerr (1998). Tradable carbon permit auctions: how and why to auction, not grandfather. Working Paper, Economics Department, University of Maryland, June (summary on *www.weathervane.rff.org*).

European Commission (1998). An analysis of the Kyoto Protocol. Commission Staff Working Paper. Brussels. 3 March.

European Commission (1998). Flexibility for efficiency in European climate policy: the factual context. DGII, Brussels. 27 April.

IPCC (1996). An economic assessment of policy instruments to combat climate change. Chapter 11

(Brian Fisher, Scott Barrett, Peter Bohm, Rob Stavins *et al.*). In: Climate change 1995: economic and social dimensions of climate change. Intergovernmental Panel on Climate Change (IPCC). Working Group III, Second Assessment Report, Cambridge University Press.

Manne, A. and R. Richels (1998). *The Kyoto Protocol: a cost-effective strategy of meeting environmental objectives.* Working Paper, EPRI, Stanford, CA. July.

Sandor, R., J. Cole and E. Kelly (1994). Model rules and regulations for global CO_2 emissions credit market. In: *Combating global warming: possible rules, regulations and administrative arrangements for a global market in CO_2 emissions entitlements.* Part II, UNCTAD/GID/8. New York.

Statens Energiverk (The Swedish Energy Agency) (1991). Överlåtbara utsläppstillstånd för koldioxid (Tradable carbon emissions permits): Beskrivning av ett möjligt system och förslag till ett försök (A possible system design and the design of a full-scale experiment). Stockholm.

Stavins, R. (1998). Market-based environmental policies. Discussion Paper #98-26. Washington, D.C. Resources for the future.

Toman, M. (1998). Research frontiers in the economics of climate change. Discussion Paper #98-32. Washington, D.C. Resources for the future.

Toman, M. and Hourcade, J.-C. (1998). International workshop addresses emissions trading among 'Annex B' countries. Washington, D.C. RFF (*www.weathervane.rff.org*). August.

UNCTAD (T. Tietenberg *et al.*) (1998). Greenhouse gas emissions trading: defining the principles, modalities, rules and guidelines for verification, reporting and accountability. mimeo. Available at *www.colby.edu/personal/thtieten* August.

CHAPTER 6

Efficiency and Equity of Emissions Trading with Endogenous Environmental Technical Change*

Paolo Buonanno, Carlo Carraro, Efrem Castelnuovo and Marzio Galeotti

1. Introduction

In recent years, particularly the last one, there have been many attempts to quantify the costs of implementing the Kyoto agreement under different policy options, i.e. with or without different degrees of introduction of the so-called 'flexibility mechanisms' (Emissions Trading, Joint Implementation, Clean Development Mechanisms).[1] Despite the high variability, all estimates show that the Kyoto flexibility mechanisms significantly reduce the costs of compliance. Shogren (1999) notes that 'it is estimated that any agreement without the cost flexibility provided by trading will at least double the US costs, ... the key is to distribute emissions internationally so as to minimise the costs of climate policy'. Manne and Richels (1999) state that 'losses in 2010 are two and one-half times higher with the constraint on the purchase of carbon emissions rights; international co-operation through trade is essential if we are to reduce mitigation costs'. These are just two examples of the many studies reaching the same conclusion: emissions trading, and more generally flexibility mechanisms, can achieve a reduction in overall mitigation costs without reducing the effectiveness of the climate policy (see also Rose and Stevens, 1999; Bosello and Roson, 1999; Hourcade, Ha-Duong, and Lecocq, 1999; Tol, 1999). The same conclusion can also be achieved independently of the specific climate model adopted. In fact, Chander, Tulkens, Van Ypersele, and Willems (1999) show that the application of simple economic principles is sufficient to prove that: (1) flexibility mechanisms reduce total compliance costs; (2) the largest cost reduction is achieved when no constraint is imposed on the trading system (e.g. no ceilings);[2] (3) there exists a system of transfers such that this cost reduction benefits all countries.

However, the theoretical conclusions by Chander *et al.* (1999) are achieved with the help of a static model and it is not a priori clear whether they can be generalized to the case in which investment, stock pollution, R&D and technical change are accounted for. In particular, the issue of technical change is very controversial and not yet sufficiently studied in that context.

C. Carraro (ed.), Efficiency and Equity of Climate Change Policy, 121–162.

Some arguments offered in support of the introduction of ceilings on emissions trading, for instance, are indeed based on the role of technical change. It is argued (e.g. Hourcade *et al.*, 1999) that the widespread adoption of flexibility mechanisms reduces the incentives to carry out environmental R&D, thereby reducing the effectiveness and increasing the costs of abatement options in the long run. Moreover, the incentives to R&D induced by the presence of ceilings may spill over onto other sectors, thus speeding up the 'engine of growth' and reducing the impact of climate change control on long-run per-capita income and welfare.

The above argument is not completely convincing. For example, Convery (1999) notes that 'to the extent that constraints are placed on carbon trade, the costs of compliance will be increased ... the negative effects would spill over also to the countries likely to export carbon credits, since the volume of their sales and the price they get will be lower because of lower demand induced by higher costs of mitigation measures taken at home by the EU and the US. All countries lose, and emissions reduction commitments in subsequent periods will be made more expensive and therefore less likely to be significant. Since the gains from trade experienced by Russia and Ukraine will be reduced, it will also reduce the prospects for trade gains from potential entrants from the developing countries in the future'. Bohm (1999) asserts that 'forcing countries to produce more of the emissions reduction quantity commodity at home than it wants to is like forcing cold Nordic countries to grow some minimum share of bananas before it is allowed to import bananas from countries that have a comparative advantage in banana production .... Supplementarity will not only make present Annex B countries less likely to accept more stringent future commitments, it will also make it harder to get new countries to join the set of Annex B countries'. Thus, an additional argument in favour of a free adoption of the flexibility mechanisms is that they will encourage the accession of new countries to the group (coalition) of those who have already decided to accept quantified limits to their GHGs emissions.

In this paper we do not study the validity of this second argument, which is next on our research agenda (see Carraro, 1998, for a theoretical overview of the issue of endogenous formation of climate coalitions). We rather focus here on the link between climate policies and technical change. In particular, we try to provide an answer to the following questions:

- Does endogenous technical change modify the effectiveness and the costs of different ways of implementing the flexibility mechanisms to reduce GHGs emissions in Annex 1 countries?
- Do policy-induced R&D and its impact on productivity and on the emissions/output ratio help to achieve the Kyoto targets? Is R&D a complement or a substitute for emissions trading, i.e. do countries reduce their R&D efforts when trading is allowed for?
- If this is the case, are the benefits from free emissions trading more than compensated for by the losses induced by lower R&D efforts?

– What is the impact on equity of different policy options? Is it true that emissions trading will favour developing countries, thus increasing equity, as argued in Nordhaus and Boyer (1999),[3] or does emissions trading favour mainly Annex 1 countries, because it reduces their abatement costs, thus reducing equity?

In order to answer these questions we take the well-known RICE model of integrated assessment (Nordhaus and Yang, 1996) and incorporate in it a modified version of the endogenous environmental technical change (ETC) model proposed by Goulder and Mathai (1998) (see also Nordhaus, 1997). In this latter model, a representative firm chooses its optimal R&D effort which increases the stock of technological knowledge. This stock enters the production function as one of the inputs and, at the same time, affects the emissions–output ratio. Thus, the idea is that more knowledge will help the firm to increase its productivity and to reduce its negative impact on the environment. Using our 'ETC-RICE' model we then 'solve' the problem set by the six regions into which the world is divided when deciding the optimal level of four instruments: fixed investments, R&D expenditures, rate of emissions control, and the amount of permits which each country wants to buy or sell. There are some regulatory constraints to the solution: with or without ceilings on trading, with the possibility to trade only among Annex 1 countries or under global trade, with or without environmental R&D incentives.[4]

In order to provide our analysis with a benchmark, the ETC-RICE model has been calibrated in such a way as to reproduce the same Business As Usual (BAU) scenario as that of the RICE model of Nordhaus and Yang (1996), where technical change is present, but is exogenous, and has an impact on the emissions output ratio.

The structure of the paper is as follows. In Section 2 we present and discuss the changes introduced in the RICE model as far as modelling of technical change and of emissions trading, implementation of the Nash solution to the international game, and the calibration technique are concerned. We compare the results of two models. In the following two sections we present the results of our simulations whose presentation is organized around a number of propositions concerning the main questions posited above.

In Section 3 we try to assess the impact of endogenous environmental technical change by comparing the results of two models. In the first ETC-RICE model, which we call the model with exogenous environmental technical change, each country optimally sets its R&D level and as a consequence the stock of knowledge, but this choice does not affect the emissions–output ratio, which evolves exogenously in accordance with Nordhaus' and Yang's assumptions. In the second ETC-RICE model, with endogenous environmental technical change, a change of the stock of knowledge also modifies the relationship between emissions and output, which therefore depends on the optimal R&D amount chosen by each country. Hence, in both models technical change is partly endogenous, because R&D is optimally chosen and depends on all relative prices, including the price of permits. But only in the second model does a change of relative prices, e.g. the price of permits, affect the environmental dimension of the technology adopted in each country. Hence, only in the second case can we talk of endogenous environmental R&D or of induced environmental technical change. By comparing the two versions of the model we seek support to the following propositions:

1. The costs of implementing Kyoto are lower when emissions trading is permitted and are even lower when ETC is endogenous;
2. If R&D and emissions trading are substitutes, the price of permits is lower because countries reduce their emissions through R&D and therefore demand fewer permits. Hence, the level of trading is also lower than in the case of exogenous technical change;
3. Permit-selling countries carry out excessive R&D in order to have more permits to sell in the market. As a consequence, more R&D is carried out at the world level, which implies that total emissions are lower because R&D reduces the emissions abatement ratio;
4. When ETC is endogenous, the equity index is lower because emissions-related transfers to developing countries are lower (as a consequence of the lower permit price and demand).

Section 4 of the paper is devoted to analysing the role of ceilings. The RICE model with endogenous environmental technical change enables us to compare the effects of policy scenarios in which increasing restrictions to emissions trading are introduced. In particular, we would like to check whether the following propositions are correct:

5. When ceilings are introduced all countries increase their R&D effort;
6. This increased R&D effort has a positive impact on factor productivity, and for developed countries this impact more than compensates the increased abatement costs induced by the trade restrictions. Hence, for those countries the total costs of implementing Kyoto is lower for stricter ceilings (alternatively, there is an optimal ceiling value which minimizes the total cost);
7. In the long run, technical change induced by stricter ceilings fosters economic growth and yields a higher GNP level;
8. With ceilings, the implementation of the Kyoto Protocol is more equitable because developed countries increase their abatement through domestic policies and do not buy developing countries' 'environmental resources' at a low price.

As shown in the following sections, we find support only for some of the above propositions. In the last section we draw some conclusions and outline directions for further research.

2. Endogenous technical change modelling

Most models used to assess the effects of policies designed to control GHGs emissions adopt an exogenous representation of technical change. Traditional energy economy models include an Autonomous Energy Efficiency Improvement (AEEI) factor, which typically consists of an exogenously set parameter. Of course, this is quite unsatisfactory. The need for a 'new generation' of climate models which endogenize the linkages between economic variables (policy variables, in particular) and technical progress is widely recognized and now strongly advocated (see Weyant, 1997; Carraro, 1998). However, there are currently only a few attempts to introduce these linkages in existing models, especially in models of integrated assessment.

Some early attempts were made in the econometric general equilibrium model for the US economy of Jorgenson and Wilcoxen (1990), where environmental policy decisions affecting relative prices determine an endogenous change in total factor productivity, and in the applied general equilibrium model for Germany of Conrad and Henseler-Unger (1986) and Conrad and Ehrlich (1993), where substitution possibilities are more feasible with the most recent capital vintages and the adoption of new, 'cleaner' vintages depends on climate policies. However, more recently, several models have been developed. One main difficulty faced by modellers when they try to endogenize technical change is the non-observability of this variable. This is why old-fashioned models used a deterministic trend as a proxy of technical change. And this is why that is the starting point of some *ad hoc* attempts to model technical change. For example, in Boone, Hall, and Kemball-Cook (1992), Carraro and Galcotti (1996) and Dowlatabadi and Oravetz (1997), technical progress is still represented by a variable which is added to the main equations of the model. However, this variable is no longer a deterministic function of time. It is rather a stochastic function of time, in which other economic effects are also accounted for.[5]

The problem with this approach is its *ad hoc* nature. There is no explicit solution of firms' optimization problem that determines the optimal amount of R&D and investments in the corresponding types of capital. Therefore, links between these variables are mainly statistical and lack a clear economic interpretation.[6] Similar considerations apply to those energy–environment–economy models that purport to explicitly represent energy–supply–technologies. Here a common modelling device is the use of 'backstop technologies', assumed to be available at a future date in infinite supply at a high price relative to current energy prices (Nordhaus, 1973). While this approach endogenizes the switch to a different technology, and thus the time of its adoption, the price level at which this occurs continues to be exogenously determined.

More recently, some contributions have endogenized the process of technical change with the help of structural models of R&D efforts and innovation. These are typically generated as the solution of a firm's dynamic optimization problem (along with other relevant decision variables), but in which the parameters describing their functional structure are calibrated rather than estimated (see Goulder and Matai, 1998, and Goulder and Schneider, 1999). In this way, it is relatively easy to simulate a model of R&D and innovation, even if the price to be paid is to postulate a set of parameter values that cannot easily be tested. The use of R&D as a policy variable is also a feature of the R&D ICE model of integrated assessment succinctly described in Nordhaus (1997).

In a partial equilibrium model of knowledge accumulation (Goulder and Mathai, 1998), the firm chooses time paths of abatement and R&D efforts in order to minimize the present value of the costs of abating emissions and of R&D expenditures subject to an emissions target. The abatement cost function depends both on abatement and on the stock of knowledge which increases over time via R&D investment.[7] In a similar vein, Nordhaus (1997) lays out a model of induced innovation brought about by R&D efforts. In particular, technological change displays its effects through changes in the emissions–output ratio. This aspect is then embedded in the non-regional version of the author's RICE integrated assessment model for climate change policy analysis (Nordhaus, 1993).

In the present paper, we propose a model of endogenous technical change inspired by the ideas contained in both Nordhaus (1997) and Goulder and Mathai (1998) and accordingly modify the regional RICE model (Nordhaus and Yang, 1996). Doing so requires input of a few new parameter values, some of which we try to estimate using information provided by Coe and Helpman (1995), while the remaining parameters are calibrated so as to reproduce the BAU scenario generated by the RICE model with exogenous technical change. We extend the integrated assessment model thus obtained to allow for trading of emissions permits and analyse several policy options looking at their efficiency and equity implications.

Let us briefly describe our model of endogenous technical change and the estimation/calibration procedure.

The RICE model proposed by Nordhaus and Yang (1996) is one of the most popular and manageable integrated assessment tools for the study of climate change. It is basically a single-sector optimal growth model suitably extended to incorporate the interactions between economic activities and climate. There is one such model for each macro region into which the world is divided (US, Japan, Europe, China, Former Soviet Union, Rest of the World). Within each region a central planner chooses the optimal paths of fixed investment and emissions abatement that maximize the present value of per capita consumption. Output (net of climate change) is used for investment and consumption and is produced according to a constant returns Cobb–Douglas technology that combines the inputs from capital and labour with the level of technology. Population (taken to be equal to full employment) and technology levels grow over time in an exogenous fashion, whereas capital accumulation is governed by the optimal rate of investment. There is a wedge between output gross and net of climate change effects, which depends upon the amount of abatement (rate of emissions reduction) as well as the change in global temperature. The model is completed by three equations respectively representing emissions (which are related to output and abatement), carbon cycle (which relates concentrations to emissions), and climate module (which relates the change in temperature relative to 1990 levels to carbon concentrations). Each country plays a non-cooperative Nash game in a dynamic setting, which results in a Open Loop Nash equilibrium (see Eyckmans and Tulkens, 1999). This is a situation where in each region maximizes its utility subject to the individual resource and capital constraints and the climate module for a given emissions (i.e. abatement) strategy of all the other players.[8]

It is useful to reproduce here the RICE resource and production function constraints as well as the emissions relationship. Letting $n = 1, \ldots, 6$ and $t = 1, \ldots, T$ identify each macro region and time period respectively (in the model each period spans ten years), we have:

$$Q(n, t) = A(n, t)L(n, t)^{1-\gamma}K_F(n, t)^{\gamma} \tag{1}$$

where Q is gross output, A the level of technology, L is the population, and K_F is the stock of fixed capital which accumulates according to the usual recursive law:

$$K_F(n, t+1) = I(n, t) + (1 - \delta_F)K_F(n, t) \tag{2}$$

with I being the rate of investment and δ_F the rate of capital depreciation. Output net of climate change effects, Y, is used for consumption C and investment purposes:

$$Y(n, t) = \Omega(n, t)Q(n, t) = C(n, t) + I(n, t) \tag{3}$$

where Ω is the climate factor which, as said, depends upon the rate of emissions reduction μ and the change in temperature. Finally, carbon emissions E depend upon the rate of abatement μ as well as gross output according to the following relationship:

$$E(n, t) = \sigma(n, t)[1 - \mu(n, t)]Q(n, t) \tag{4}$$

where σ can be loosely defined as the emissions–output ratio. The policy variables of the model are the rates of fixed investment and of emissions abatement. For the other variables the model specifies a time path of exogenously given values. Interestingly, this is also the case of the technology level A and of the emissions–output ratio σ. Thus we can say that the RICE model contains a specification of exogenous environmental and non-environmental technical change. Obviously those time paths need not coincide in general with the optimal ones. In addition, investment fosters economic growth (thereby driving up emissions) while abatement is the only policy variable used for reducing emissions.

We modify the above equations of the RICE model in order to endogenize the process of technical change. In particular we assume that innovation is brought about by R&D spending which contributes to the accumulation of the stock of existing knowledge. Following an approach pioneered by Griliches (1979, 1984) we assume that the stock of knowledge is a factor of production, which therefore enhances the rate of productivity (see also Weyant, 1997). Besides this channel, however, knowledge also serves the purpose of reducing, *ceteris paribus*, the level of carbon emissions. Thus, in our formulation, R&D efforts prompt both environmental and non-environmental technical progress, although with different modes and elasticities.[9] More precisely, we modify equations (1) and (4) as follows:

$$Q(n,t) = A(n,t)K_R(n,t)^{\beta_n}[L(n,t)^{1-\gamma}K_F(n,t)^{\gamma}] \tag{5}$$

and:

$$E(n,t) = [\sigma_n + \chi_n \exp(-\alpha_n K_n(n,t))][1-\mu(n,t)]Q(n,t) \tag{6}$$

In (5) the stock of knowledge has a region-specific elasticity equal to β. Note that to the extent that this coefficient is positive, the output production process is characterized by increasing returns to scale, in line with current theories of endogenous growth. Also, note that while we allow for R&D-driven technological progress, we maintain the possibility that technical improvements can also be determined exogenously (the path of A is the same as in RICE). In (6) knowledge reduces the emissions–output ratio with an elasticity of α, which also is region-specific; the parameter χ is a scaling coefficient, whereas σ_n is the value to which the emissions–output ratio tends asymptotically as the stock of knowledge increases without limit. The stock accumulates in the usual fashion:

$$K_R(n,t+1) = R\&D(n,t) + (1-\delta_R)K_R(n,t) \tag{7}$$

where $R\&D$ are the expenditures in research and development and δ_R is the rate of knowledge depreciation. We finally recognize that some resources are absorbed by R&D spending. That is:

$$Y(n,t) = C(n,t) + I(n,t) + R\&D(n,t) \tag{8}$$

In summary, our formulation introduces R&D as a further policy variable of the model which on the one hand contributes to output productivity and, on the other hand, affects the emissions–output ratio, and therefore the overall level of pollution emissions.[10]

As for parameter calibration and data requirements for the newly introduced variables, we proceed as follows. Firstly, coefficients already present in the original RICE model are left unchanged. Next, for each region we calibrate the coefficient β in the production function (5) so as to obtain in the year 2000 a value of the R&D–output ratio equal to that of 1990. R&D figures for 1990 were taken from Coe and Helpman (1995), while the 1990 stock of knowledge for the US, Japan and Europe comes from Helpman's Web page.[11] Here for the other three macro-regions, 1990 values of the knowledge stock have been

Table 1. Coefficients of the ETC-RICE Model

	α_n	β_n	δ_R	χ_n	σ_n	$K_R(n, 1990)$
US	0.195440	0.04355	0.05	0.019369	0.00971	1.24200
Japan	0.522430	0.04550	0.05	0.005270	0.00600	0.27773
Europe	0.296490	0.03180	0.05	0.007659	0.00699	0.75520
China	0.618650	0.01080	0.05	0.112771	0.00904	0.03145
FSU	1.197400	0.01660	0.05	0.095579	0.00935	0.07269
ROW	0.072926	0.00927	0.05	0.022409	0.00845	0.39343

Note: The stock of knowledge is expressed in trillions of constant 1990 US dollars.

constructed by taking the ratio knowledge/physical capital of the three industrialized regions and multiplying it by the 1990 physical capital stock of the other regions as given in the RICE model. The regional parameters α and χ in equation (6) are OLS estimated using time series of the emissions–output ratio and of the stock of knowledge (the sample runs from years 1990 to 1999, i.e. it consists of ten years of data). The data for the former variable are those used by Nordhaus and Yang (1996), while those for the latter variable are recovered from a BAU simulation conducted using an emissions–output ratio $\sigma(n, t)$ as in the original RICE model.[12] The asymptotic values σ_n are computed by simulating the pattern of the exogenous emissions–output ratio considered by Nordhaus and Yang (1996) for 1000 periods: the values of the last period are then taken as asymptotes. Finally, the rate of knowledge depreciation is set at 5%, following a suggestion contained in Griliches (1979). The relevant quantitative information is presented in Table 1.[13]

For each of the policy options described below we consider model simulations specifying both an exogenous and an endogenous environmental technological change process. In other words, for each policy option two versions of the model are simulated using either equation (4) or (6) respectively.[14] This allows an assessment of the role and impact of endogenous environmental technical change.

Simulations of the model under the two aforementioned hypotheses on the nature of technical change represent our BAU (Business As Usual) scenario, where no constraint on emissions is imposed. As said, the calibration is such that a unique BAU scenario is yielded by the two versions of the model. The principal use of our ETC-RICE model in the present paper is to study the main implications of the Kyoto Protocol, and in particular the role of the flexibility mechanisms contained therein. To this end, we further extend the RICE model in order to incorporate the possibility for the world's regions to trade pollution permits among themselves. This calls for additional modifications of the equations considered above as well as a redefinition of the process leading to an Open Loop Nash Equilibrium.

We first introduce in our ETC-RICE model a constraint that sets a limit to the amount of emissions each Annex 1 macro-region is allowed to produce under the provisions of the Kyoto Protocol. We represent this as follows:

$$E(n,t) = Kyoto(n) \tag{9}$$

Note that (9) holds only for the signatory countries. In particular the US, Japan, Europe, and Former Soviet Union (FSU) must reduce their emissions, over the period 2008–2012, to 7%, 6%, 8%, and 0% below their 1990 level respectively. We take 2010 as an approximation of the above-mentioned five-year period.

Simulations of the model with constraint (9) effective represent a first policy option that we consider, which we term simply as 'Kyoto'. Note that China and Rest of the World are not subject to the restriction in (9) while the FSU is subject to a zero percent reduction target. In the BAU and Kyoto scenarios, policy variables are the amount of investment and of R&D expenditures, as well as the rate of pollution abatement.

In order to reduce the costs of complying with the limits to emissions, the Protocol introduces the possibility for Annex 1 countries to exchange pollution permits: countries with lower abatement costs will sell permits in an international market, while countries with high costs will buy permits. In equilibrium, net purchases are zero at the price of permits which equalizes marginal abatement costs across regions. In order to introduce this flexibility mechanism in our model, we modify equations (8) and (9) as follows:

$$Y(n,t) = C(n,t) + I(n,t) + R\&D(n,t) + p(t)NIP(n,t) \tag{10}$$

$$E(n,t) = Kyoto(n) + NIP(n,t) \tag{11}$$

According to (10), resources produced by the economy must be devoted, in addition to consumption, investment, and research and development, to net purchases of emissions permits. Equation (11) states that a region's emissions may exceed the limit set in Kyoto if permits are bought, and vice versa in the case of sales of permits. Note that $p(t)$ is the price of a unit of tradable emissions expressed in terms of the numéraire output price. Moreover, there is an additional policy variable to be considered in this case, i.e. net demands for permits *NIP*.

In accord with the possibility of emissions trading, the sequence whereby a Nash equilibrium is reached must be revised as follows. Each region maximizes its utility subject to the individual resource and capital constraints, now including the Kyoto constraint, and the climate module for a given emissions (i.e. abatement) strategy of all the other players and a given price of permits $p(0)$ (in the first round this is set at an arbitrary level). When all regions have made their optimal choices, the overall net demand for permits is computed at the given price. If the sum of net demands in each period is approximately zero, a Nash equilibrium is obtained; otherwise the price is revised in proportion to the market disequilibrium and each region's decision process begins again.

The above modification of the model enables us to analyse different ways of

Table 2. Policy options

Coalition structure	Flexibility (ceilings) 0	15%	33%	100%
No coalitions	×			
⋮				
Annex 1 regions	×	×	×	×
⋮				
All regions		×	×	×

implementing emissions trading. Hence, a third policy option, in addition to the BAU and Kyoto scenarios, that we consider is when there is an active market of emissions permits limited to Annex 1 countries. In this case, a country can affect its level of emissions by exchanging permits in the market. We call this scenario Et-A1, for Emissions Trading among Annex 1 countries, in order to distinguish it from a further possibility where all countries, including those with no institutional limits to emissions, participate in the market for permits. We call this latter option Et-All. In this latter policy option, countries which do not have quantity commitments within the Kyoto Protocol are allowed to sell emissions if their actual emissions levels are below their BAU level.

Some have argued that full access to a market of emissions permits provides no incentive to individual countries to undertake domestic action to cope with the environmental problem. This leads to the possibility of imposing a ceiling on the emissions right that can be imported. We model and study this aspect by simulating the ETC-RICE model under the additional 'ceilings' constraint:

$$NIP(n, t) = CEIL[E_{BAU}(n, t) - Kyoto(n)] \qquad (12)$$

where E_{BAU} is the level of regional emissions obtained from the BAU simulation of the model and $CEIL$ is the percentage ceiling to participation in emissions trading. In the present paper we consider four additional ET (Emissions Trading) policy options, setting the ceiling to either 15% or 33% and having only Annex 1 or All countries exchanging polluting rights.[15] The eight policy options analysed in this paper are summarized in Table 2.

A final important feature of this paper is its focus both on efficiency and on equity of the different policy options described above. Efficiency will be measured in terms of abatement costs and total costs of complying with Kyoto, the latter including both the costs of abatement and the costs (benefits) from buying (selling) permits. We will also consider impacts on GNP growth, R&D efforts, emissions control, and emissions price. Equity will be measured by an equity index IE which, following Bosello and Roson (1999), compares an 'equally distributed level of consumption' *EINC* with the actual average consumption per head. More precisely, the equity index is:

$$IE(t) = EINC(t) \Big/ \left[\sum_n s(n,t) PVC(n,t) \right] \quad (13)$$

where:

$$EINC(t) = \exp\left[\sum_n s(n,t) \ln PVC(n,t) \right] \quad (14)$$

and where s is the region's share in the world population and *PVC* is the present discounted value of regional consumption. This is the maximized value of the objective functions generated by the model simulations. The index *EI* ranges between zero and one: the closer to unity, the more equitable the distribution.

3. Efficiency and equity effects of induced technical change

With the help of the ETC-RICE model just described we analyse the eight policy options summarized in Table 2. From the optimization runs we derive the optimal time paths of the control variables and their impacts on the endogenous variables, for the period 2000–2100. In particular, we focus on those variables that enable us to verify the propositions spelled out in the Introduction.

We start by analysing the role of the endogenous technical change. Our first proposition was:

Proposition 1. The costs of implementing the Kyoto Protocol are lower when emissions trading is permitted and are even lower when ETC is endogenous.

In what follows we use two definitions of costs. In the first one, we consider only domestic abatement costs, whereas in the second one we add the costs (gains) arising from the exchange of tradable permits. Assuming exogenous environmental technical change, Figures 1 and 2 show (for all world regions) the total costs according to the two definitions just given as we move from a situation in which there is no trade of permits (Kyoto hereafter) to one in which only Annex 1 countries trade (Et-A1 hereafter) and then to the case in which all countries are allowed to trade (Et-All hereafter). Figures 3 and 4 shows the differences between the two notions of costs when environmental technical change is endogenous and when it is exogenous respectively. It is easy to see that: (1) both total abatement costs and overall costs decrease as more countries are allowed to trade. The lowest cost is achieved in the case of

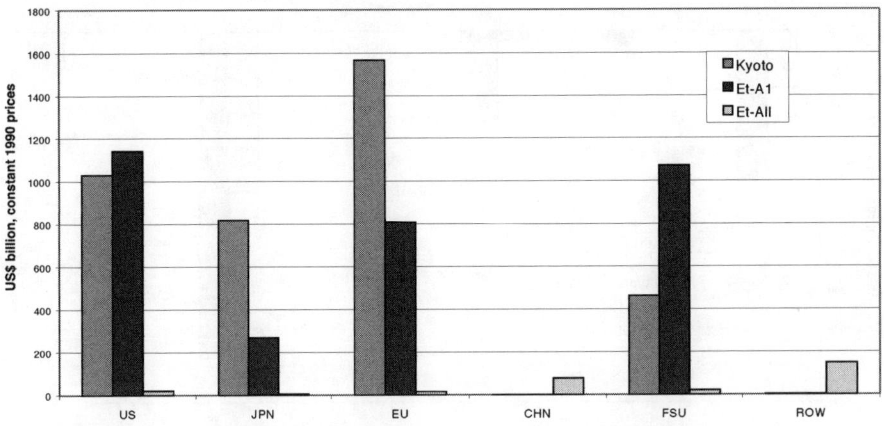

Figure 1. Total abatement cost with exogenous environmental technical change.

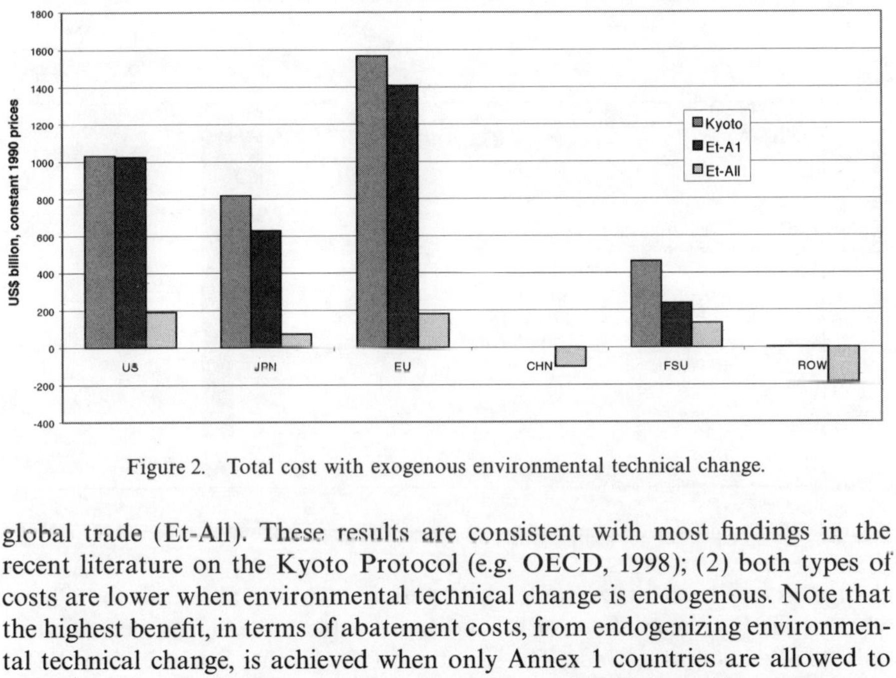

Figure 2. Total cost with exogenous environmental technical change.

global trade (Et-All). These results are consistent with most findings in the recent literature on the Kyoto Protocol (e.g. OECD, 1998); (2) both types of costs are lower when environmental technical change is endogenous. Note that the highest benefit, in terms of abatement costs, from endogenizing environmental technical change, is achieved when only Annex 1 countries are allowed to trade. This suggests that countries can actually exploit the incentive to innovate when their firms are not excessively penalized by high compliance costs. Indeed, one might think that the impact of environmental technical change is highest when emissions trading is not allowed for. This intuition is not supported by our analyses, as will be further shown later on.

When total cost is considered, the situation is less clear. For example, the US decreasingly benefits from the incentive to carry out environmental technical

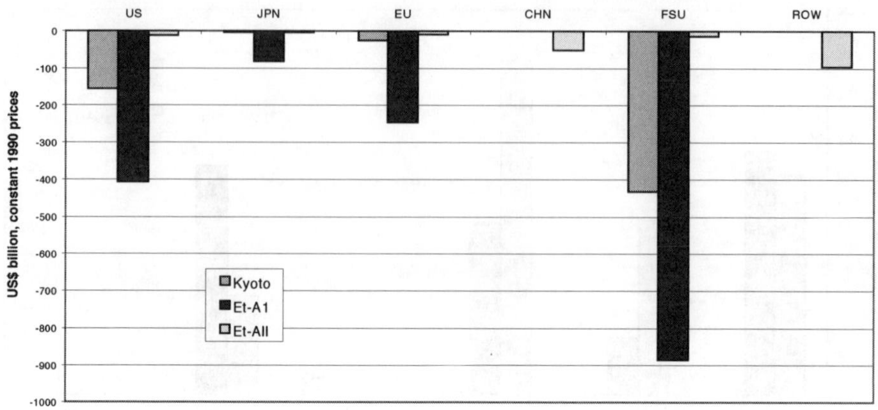

Figure 3. Endogenous vs. exogenous technical change: differences in total abatement cost (period 2010–2100).

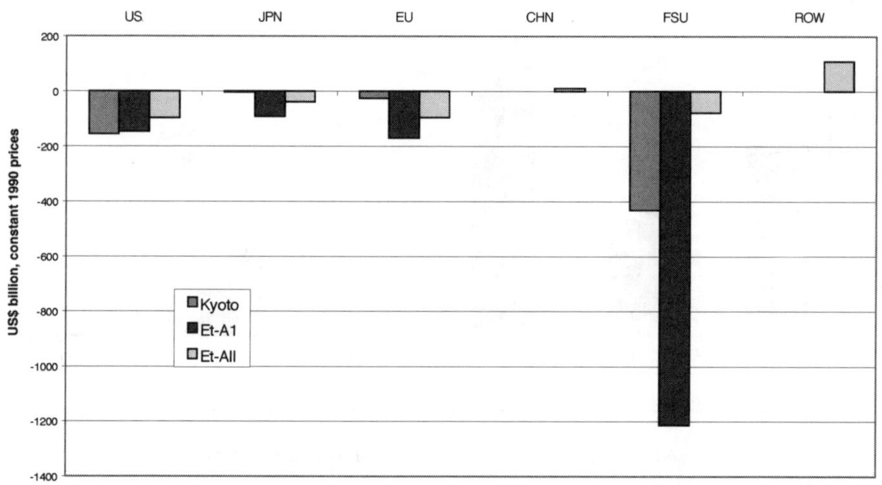

Figure 4. Endogenous vs. exogenous technical change: differences in total cost (period 2010–2100).

change as more countries trade their permits. This is also true for China, whereas the European Union (EU) and Japan achieve the highest benefit when only Annex 1 countries trade. The situation of the Former Soviet Union (FSU) is peculiar. This is the region which derives the highest benefit from the endogenization of environmental technical change when permits are traded only among Annex 1 countries. As we will see below, one reason is that countries of the FSU are the only sellers of permits in the Et-A1 case. Hence, they have a strong incentive to further reduce their emissions to sell permits in the market. Another reason is that they have lower marginal abatement costs: their own R&D has therefore a large impact on abatement costs.

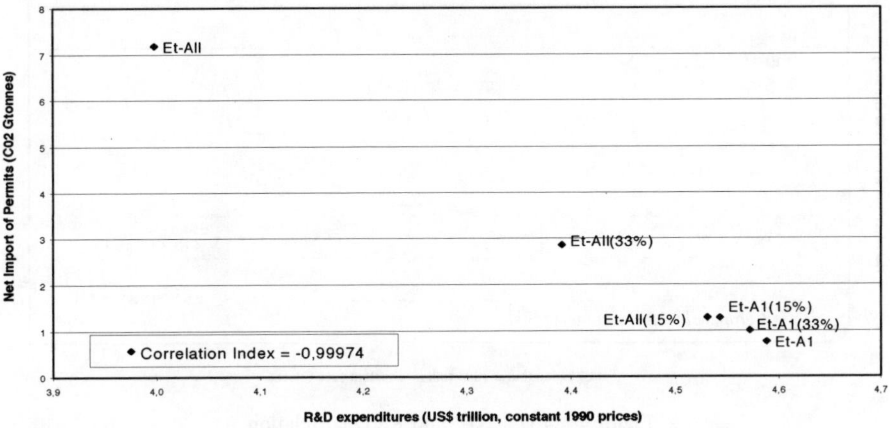

Figure 5. US – NIP and R&D correlation.

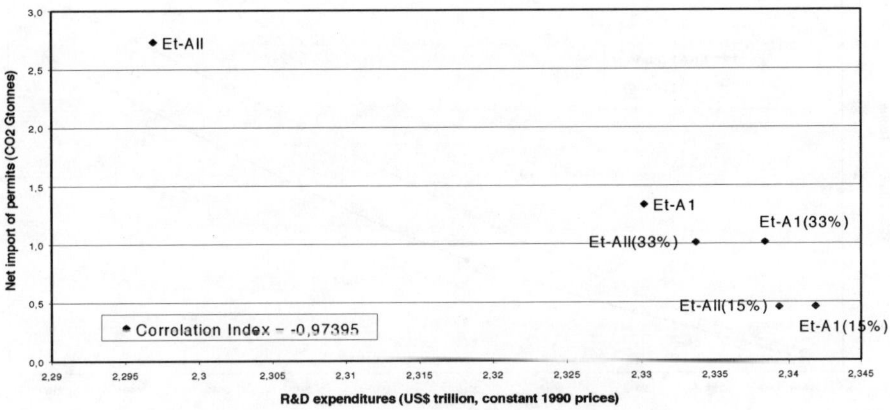

Figure 6. JPN – NIP and R&D correlation.

We can therefore conclude that the first proposition is supported by our policy exercises. The second proposition was:

Proposition 2. If R&D and emissions trading are substitutes, the price of permits is lower because countries reduce their emissions through R&D and thus demand fewer permits. Hence, in the case with endogenous environmental technical change, the level of trading is also lower.

This proposition is only partly true. The fact that emissions trading and R&D are substitutes, at least for countries which have to choose between buying permits and carrying out R&D, can easily be seen from Figures 5, 6 and 7. They show the amount of R&D and import of permits in the different policy

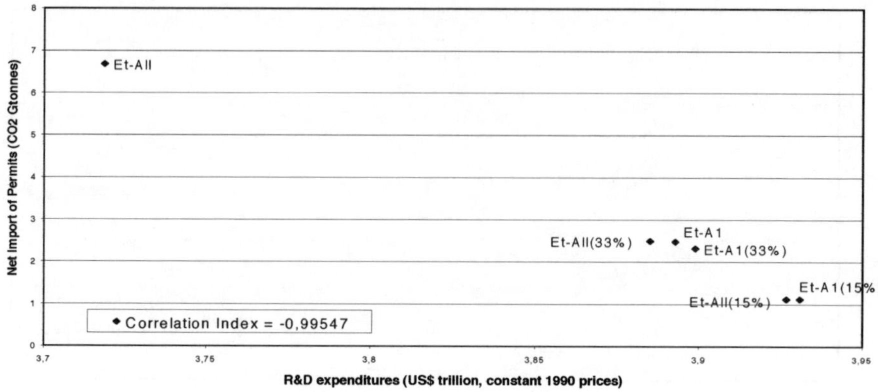

Figure 7. EU – NIP and R&D correlation.

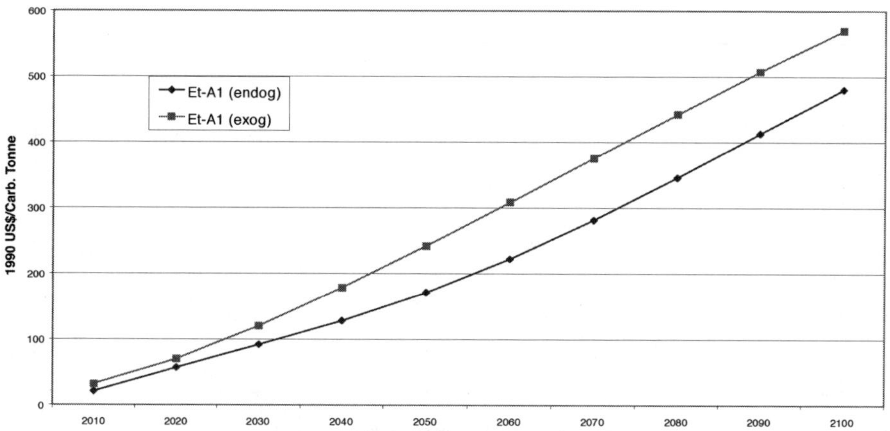

Figure 8. Et-A1 price of permits: exogenous vs. endogenous.

scenarios. Note the strong negative correlation between import of permits and R&D, which supports the conclusion that these two control variables are substitutes.

The consequence is that the price of permits is indeed lower when ETC is endogenous, both when only Annex 1 countries trade (Figure 8) and when all countries trade (Figure 9). The reasons, however, are not only those identified in the proposition: R&D actually increases, in particular when no trade is allowed (Figure 10). However, the demand for permits does not decrease for the EU and Japan, and decreases for the US only when all countries trade permits (Figure 11). A possible explanation for this fact is that the supply of permits also increases when ETC is endogenous (this is what the next proposition states and our results support). As a consequence, the price of permits is

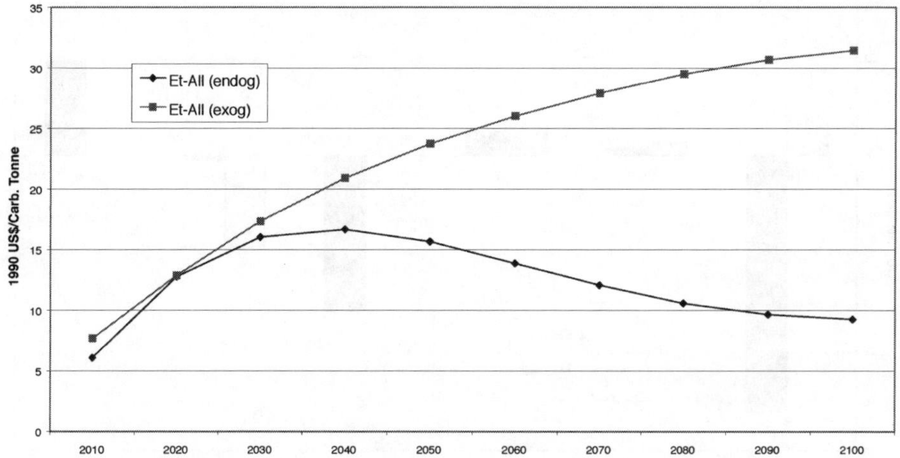

Figure 9. Et-All price of permits: exogenous vs. endogenous.

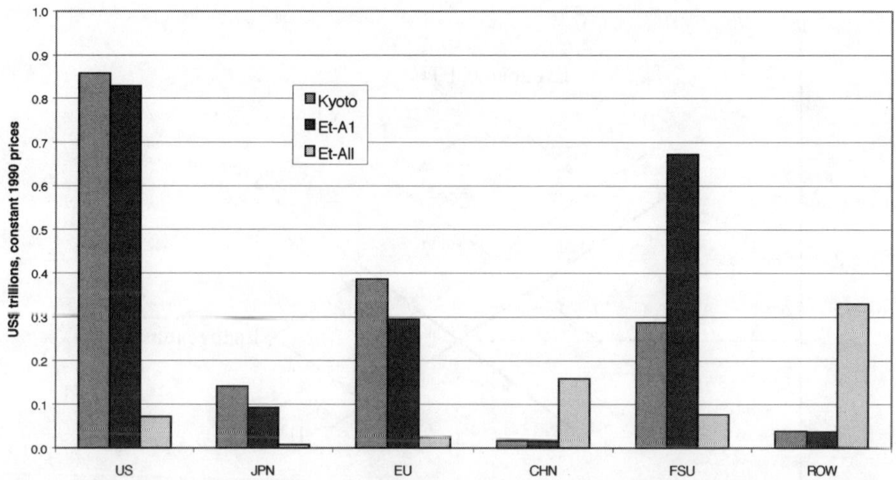

Figure 10. Endogenous vs. exogenous technical change: differences in total R&D expenditure (period 2010–2100).

lower, but the exchanged quantity slightly increases (Figure 12). The proposition should therefore read:

Proposition 2 (revised). If R&D and emissions trading are substitutes, the price of permits is lower in the case of endogenous environmental technical change, both because countries reduce their emissions through R&D, and thus demand fewer permits, and because the supply of permits increases.

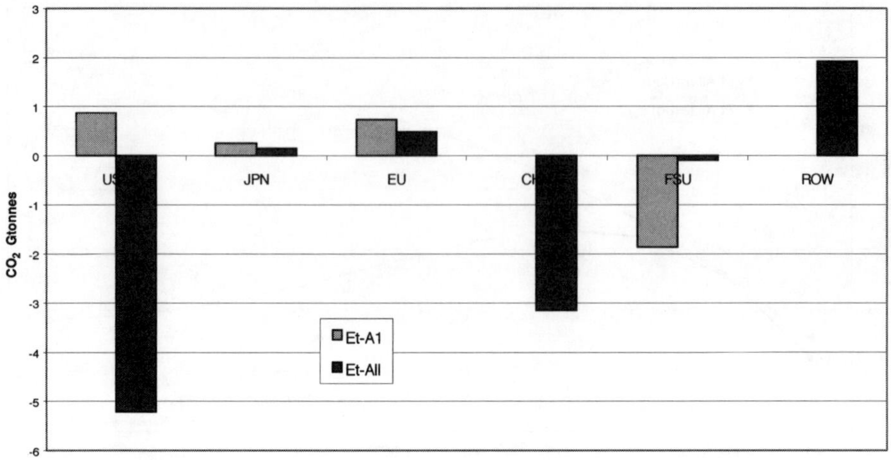

Figure 11. Endogenous vs. exogenous technical change: differences in total net import of permits (period 2010–2100).

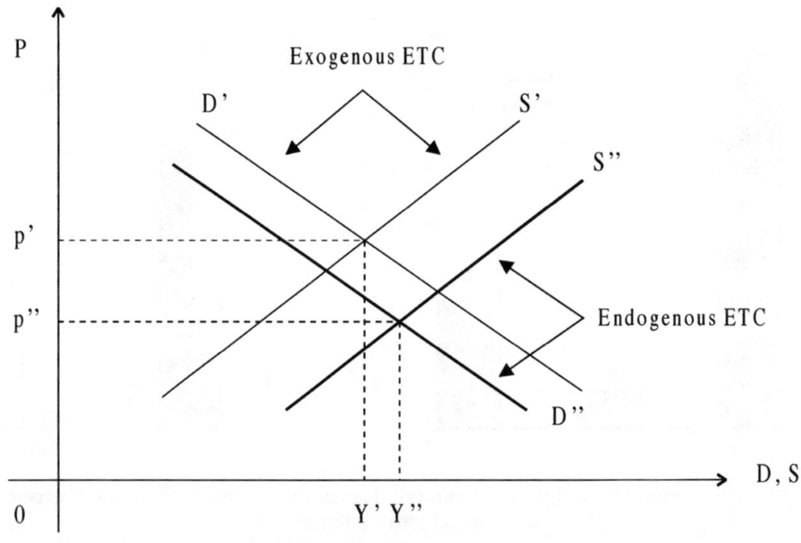

Figure 12. Permit market – demand and supply shifts.

The above situation is represented in Figure 12. The supply of permits indeed increases because:

Proposition 3. Permit-selling countries carry out excessive R&D in order to have more permits to sell in the market. As a consequence, more R&D is carried out at the world level, which implies that total emissions are lower because R&D reduces the emissions–output ratio.

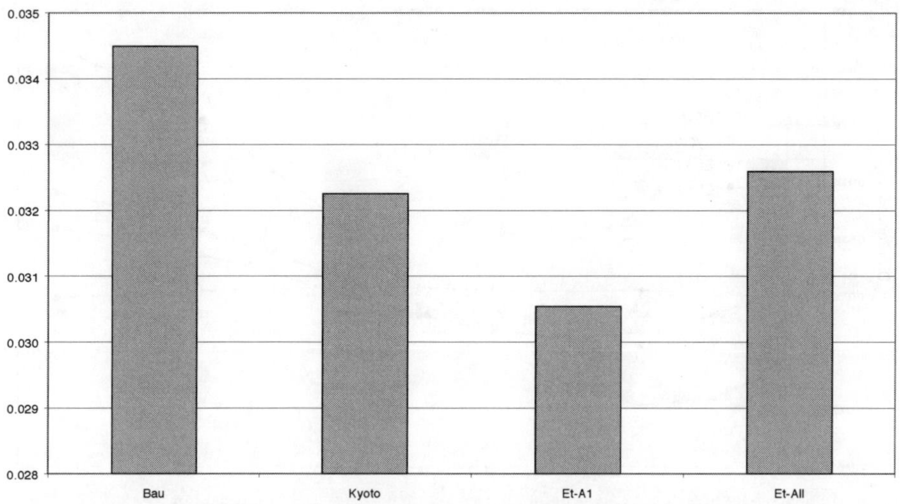

Figure 13. World emission-output ratio (average 2000–2100).

This proposition is confirmed by Figure 11 which shows that sellers of permits increase their R&D expenditures. When only Annex 1 countries trade, the FSU sharply increases its R&D expenditure and sells more permits. When all countries trade, both China and ROW strongly increase their R&D expenditure. The impact is a lower emissions–output ratio at the world level (see Figure 13).

Finally, let us consider the impact on equity of endogenous environmental technical change. Here our proposition was:

Proposition 4. When ETC is endogenous, the equity index is lower because emissions-related transfers to developing countries are lower (as a consequence of the lower permit price and demand).

We have already seen that the price is lower, but the exchanged quantity is usually larger. Hence the proposition should be modified as follows:

Proposition 4 (revised). When ETC is endogenous, the equity index is lower because the total value of emissions-related transfers to developing countries is lower (the effect of the lower price dominates the effect of the increased quantity).

This proposition is actually supported by our simulation runs only in the short term, as Figure 14 shows. In the long term, instead, the equity index is larger with endogenous technical change. This implies that the adoption of flexibility mechanisms can actually increase equity, at least in the long run. When R&D is exogenous, developing countries increase their supply of permits to exploit

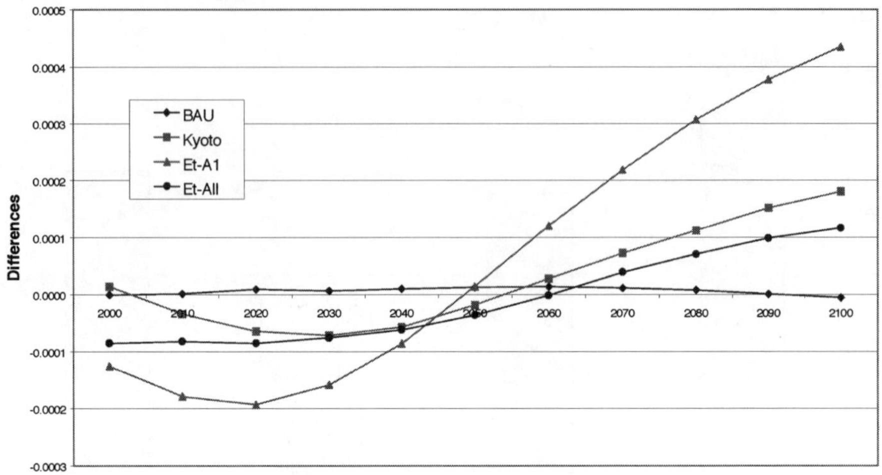

Figure 14. Equity index: endogenous vs. exogenous (difference).

the benefits arising from selling their permits. It appears that this effect becomes relevant only after a few decades, i.e. in the short run the price effect dominates the quantity effect, but in the long run the importance of the two effects is reversed.

4. Efficiency and equity of ceilings on emissions trading

This section of the paper is devoted to analysing the role of ceilings. The type of ceilings whose effects are investigated here results in a supply of emissions permits becoming perfectly inelastic when the constraint binds. Our ETC-RICE model enables us to compare the effects of alternative policy scenarios in which increasing restrictions on emissions trading are introduced. Our first proposition was:

Proposition 5. When ceilings are introduced, all countries increase their R&D effort.

This proposition is in fact confirmed by our numerical analysis as far as developed countries are concerned, but the behaviour of the other world regions is quite different (see Figures 15–20). In the US, Japan and the EU, the lowest R&D effort, as measured by the percentage R&D expenditures to GNP ratio, is realised in the Business-As-Usual scenario and when all countries are allowed to trade without any restrictions. The ratio then increases when Annex 1 countries are allowed to trade without ceilings, and further increases in the presence of ceilings. The R&D effort is highest in the Kyoto scenario, which

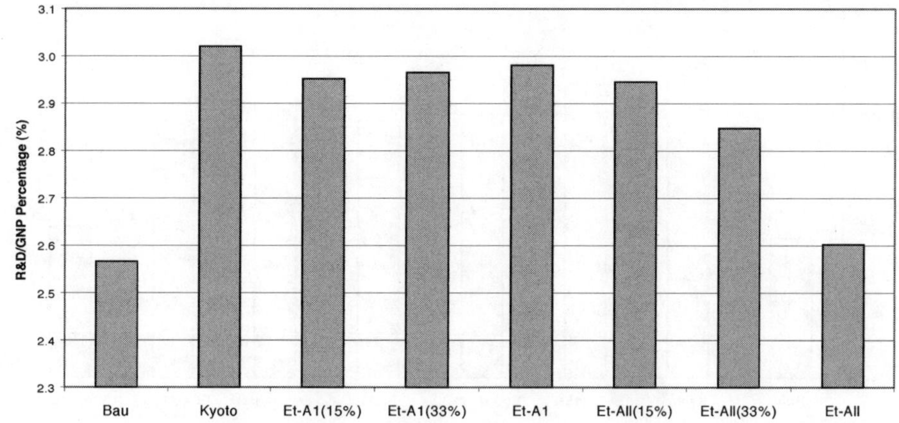

Figure 15. US R&D/GNP percentage: average (period 2010–2100).

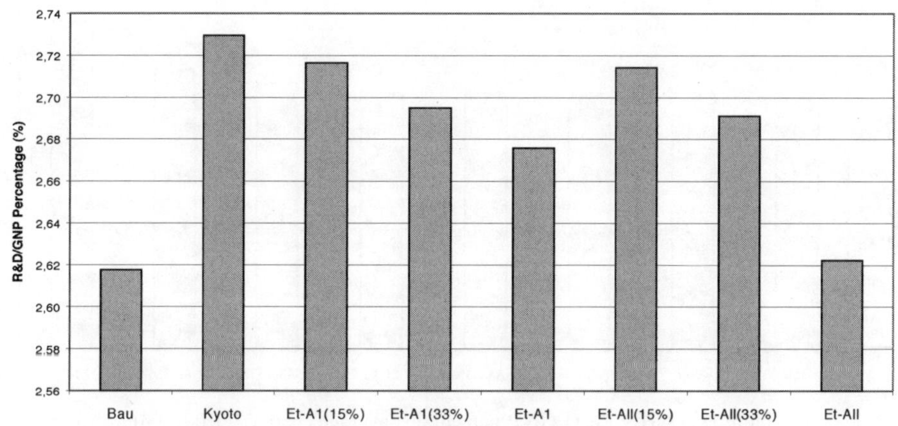

Figure 16. JPN R&D/GNP percentage: average (period 2010–2100).

corresponds to a zero percent ceiling (all abatement is carried out through domestic measures). Hence, these results support the conjecture of those who are in favour of ceilings, namely that those restrictions would stimulate R&D expenditures (but the effects on costs and growth will be presented below).[16]

In the other world regions, the R&D effort depends on their role in the permit market. For example, the FSU carries out the largest R&D effort in the case in which Annex 1 countries trade without ceilings. The reason is that this is the most favourable situation for the FSU, because the demand for permits is high and the FSU is the only seller. Hence, it increases R&D to reduce their own emissions levels, thus increasing the amount of permits that can be sold in the market. When ceilings are introduced, however, the demand for permits is lower and the FSU's R&D effort is consequently reduced. It becomes even

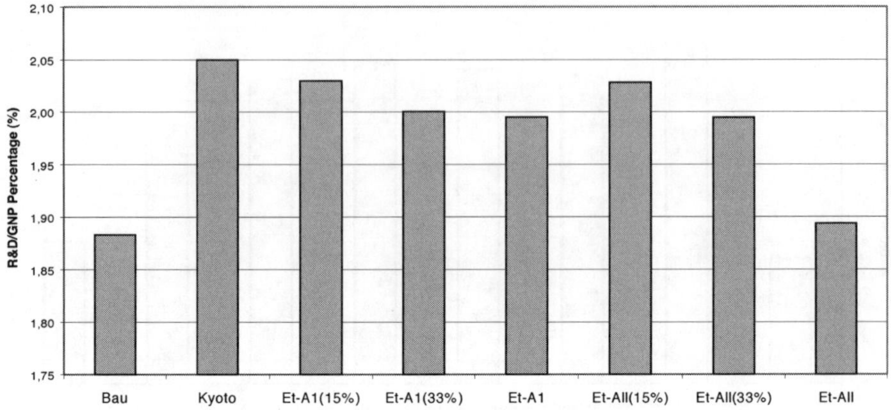

Figure 17. EU R&D/GNP percentage: average (period 2010–2100).

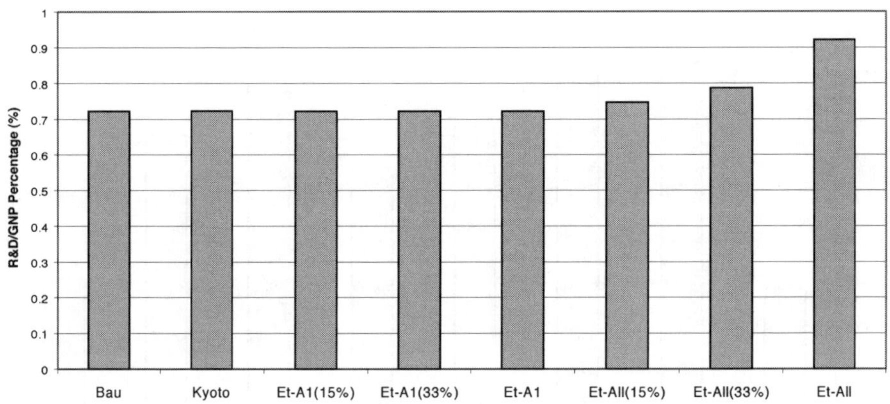

Figure 18. CHN R&D/GNP percentage: average (period 2010–2100).

lower when all countries are allowed to trade. In this case, the FSU is no longer the unique seller, nor the one with the lowest abatement costs. Hence, China and the Rest of the World sharply increase their R&D effort to exploit the benefits arising from selling permits. This R&D effort is obviously reduced in the presence of ceilings. The conclusion is that in general, ceilings are likely to bring about higher R&D expenditures (relative to GNP) in OECD countries, but they induce a reduction in R&D in FSU, China, and other developing countries, where the greatest stimulus to carry out R&D comes from the possibility of trading emissions permits without restrictions.

The next proposition was:

Proposition 6. This increased R&D effort has a positive impact on factor productivity and this impact more than compensates the increased abatement costs for developed countries induced by the trade restrictions.

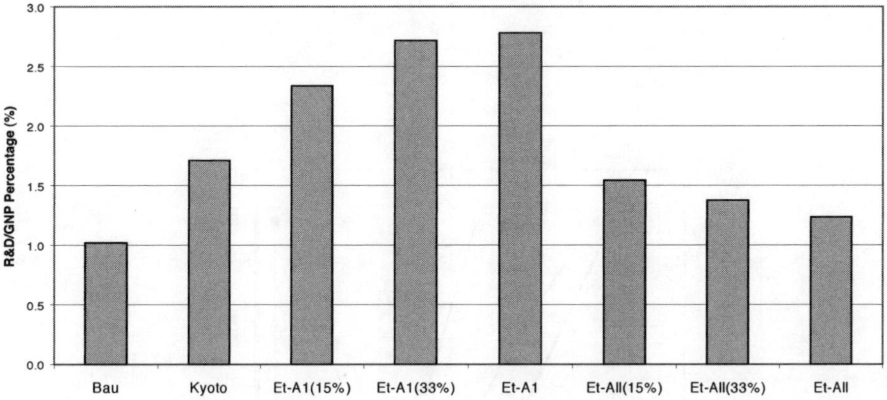

Figure 19. FSU R&D/GNP percentage: average (period 2010–2100).

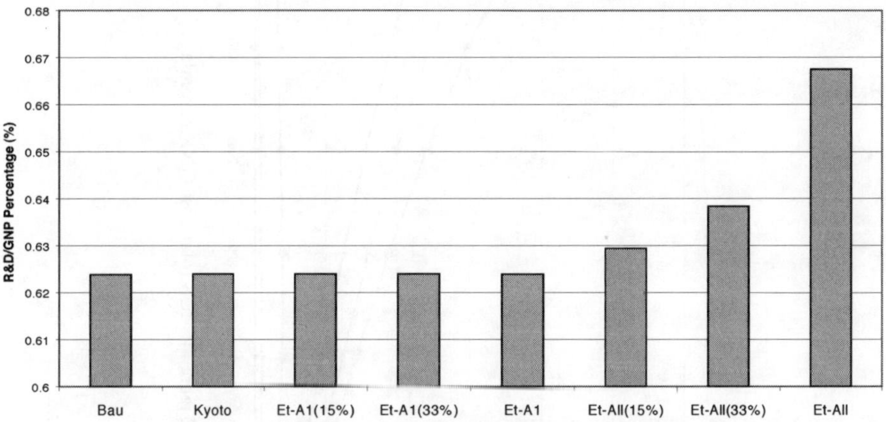

Figure 20. ROW R&D/GNP percentage: average (period 2010–2100).

> *Hence, for these countries the total cost of implementing Kyoto is lower for stricter ceilings (alternatively, there is an optimal ceiling value which minimises the total cost).*

Looking at Figures 21–26 we find that the proposition is not supported by our results.[17] In all OECD regions the highest cost is achieved in the case where all abatement is carried out domestically and there is no trade (the Kyoto scenario), whereas the lowest cost is achieved when all world countries are allowed to trade without restrictions (the US is a partial exception). The presence of ceilings generally increases costs, with the exception of the US where a 33% ceiling in the case of global trade may be slightly beneficial (but the ceiling is binding only in the first decade). The situation of the US is

Figure 21. US total cost/GNP percentage.

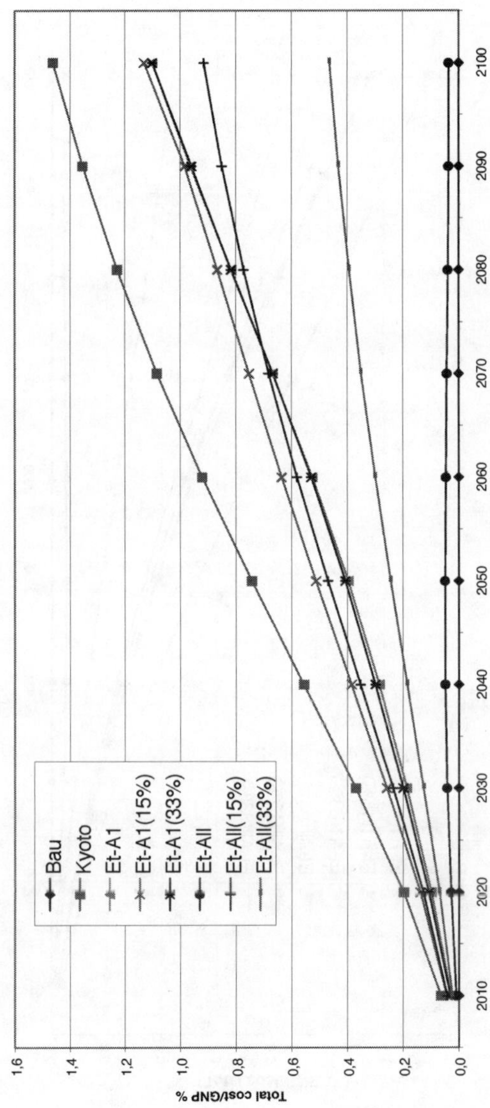

Figure 22. JPN total cost/GNP percentage.

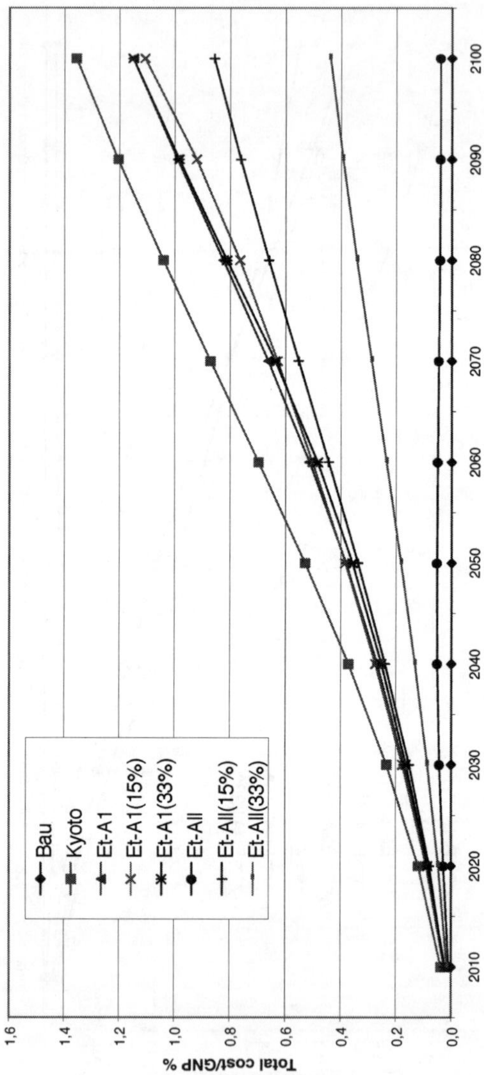

Figure 23. EU total cost/GNP percentage.

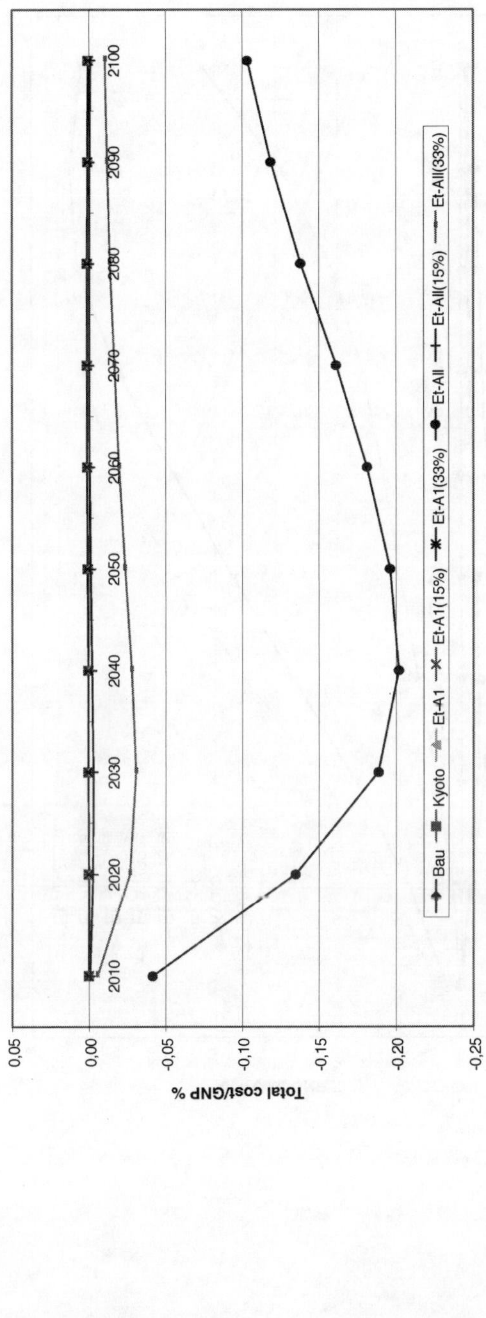

Figure 24. CHN total cost/GNP percentage.

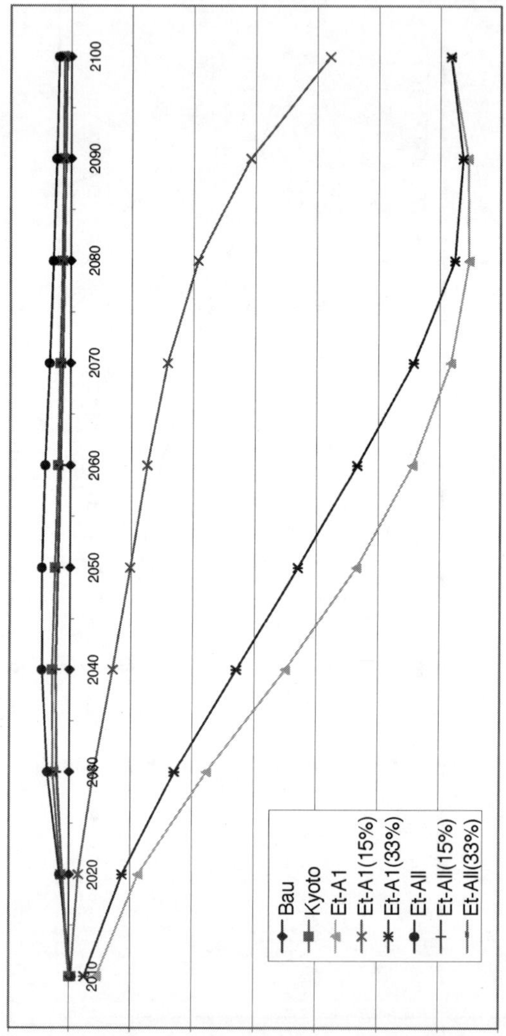

Figure 25. FSU total cost/GNP percentage.

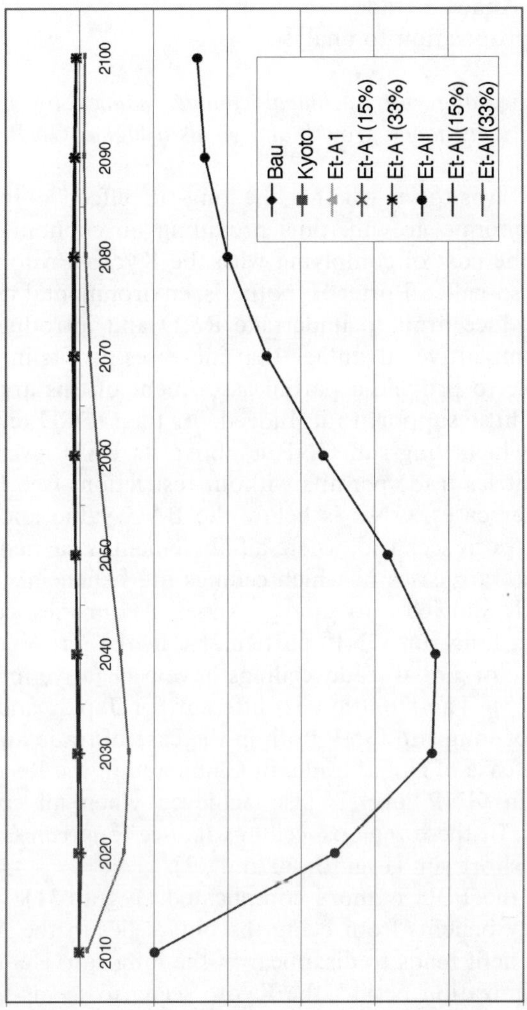

Figure 26. ROW total cost/GNP percentage.

explained again by the fact that, for countries where the ceiling is not binding, its introduction is beneficial because it reduces the equilibrium prices without affecting the quantity demanded in that country. In the developing countries the cost is lowest when they can fully exploit the benefits of emissions trading: for China and the Rest of the World in the case of global trade, and for the FSU in the case of Annex 1 trade.

The subsequent proposition to analyse was:

Proposition 7. In the long run, technical change induced by stricter ceilings fosters economic growth and yields a higher GNP level.

In this case, the goal was to see whether the long-run effects of increased R&D could stimulate economic growth, thus providing an economic benefit that could compensate the cost of complying with the Kyoto Protocol. The argument is close to the so-called Porter Hypothesis: environmental regulation, and the related costs, induce firms to undertake R&D and introduce innovation, thus achieving a competitive advantage that increases profits in the long run.

Our model is able to provide a partial assessment of this argument, but it unfortunately gives little support to it. Indeed, for the OECD regions the only policy scenario in which long-run GNP is above its BAU level is the Et-All case, where all countries trade permits without restrictions (see Figures 27, 28 and 29). In all other cases, GNP is below the BAU value and achieves the lowest level in the Kyoto scenario, where all abatement is carried out domestically. However, there are cases in which ceilings are beneficial. Take the case of the US: if trade is allowed only among Annex 1 countries, ceilings have a beneficial impact on long-run GNP, particularly in the stricter 15% version. However, in the case of global trade, ceilings have a negative impact on long-run GNP (Figure 27). The situation is different for Japan and the EU, for which ceilings reduce long-run GNP both in the case of trade among Annex 1 countries and in the case of global trade. In China and in the Rest of the World the highest long-run GNP level is also achieved when all countries trade without restrictions. In these regions, ceilings reduce long-run GNP, but may be beneficial in the short run (Figures 30 and 32).

The situation of the FSU is more complicated (Figure 31). The reason is that the FSU largely benefits from being the only seller in the Annex 1 trade scenario, but this benefit tends to disappear in the long run. Hence, in last two decades of our optimization period, the Kyoto scenario seems to provide the largest difference between GNP and its BAU value. The second- and third-best long-run options are those in which a 15% ceiling is introduced, with global trade and Annex 1 trade respectively.

We finally focus again on equity. One of the arguments in favour of ceilings is that they would yield a more equitable outcome by preventing developed countries from exploiting developing countries' natural resources. This argument is summarized by the proposition that follows.

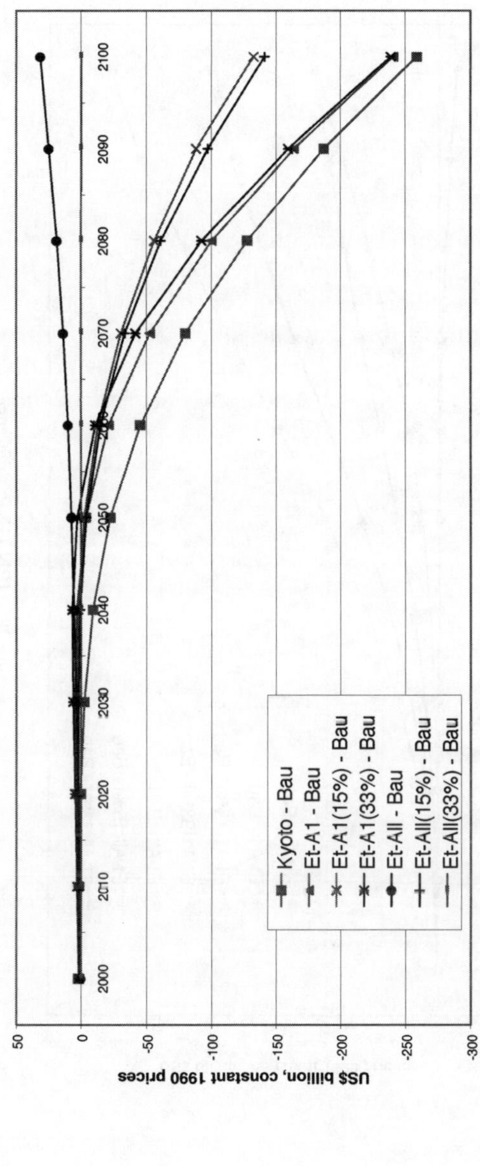

Figure 27. US GNP: difference.

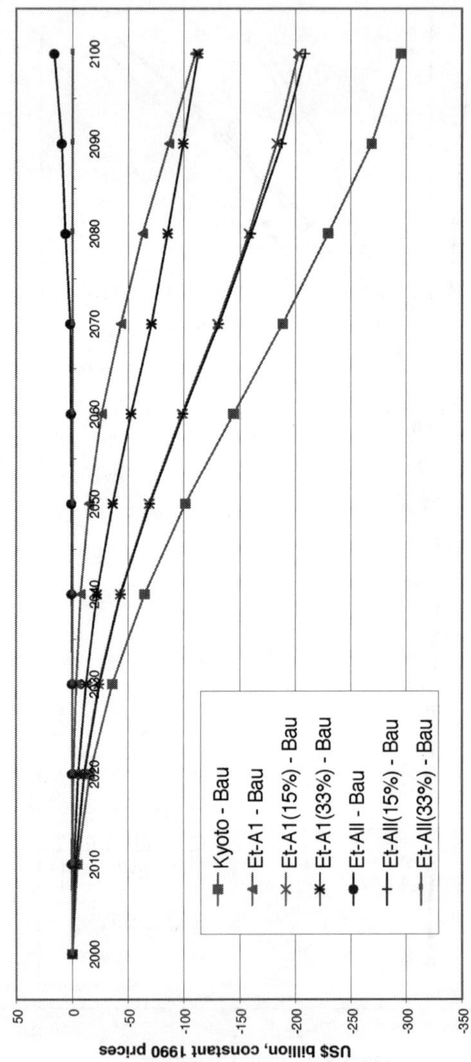

Figure 28. JPN GNP: difference.

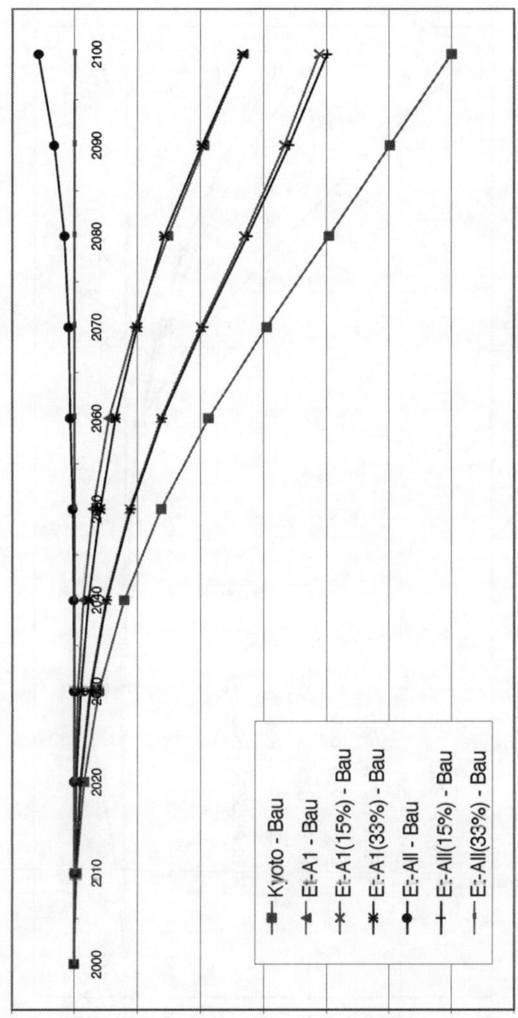

Figure 29. EU GNP: difference.

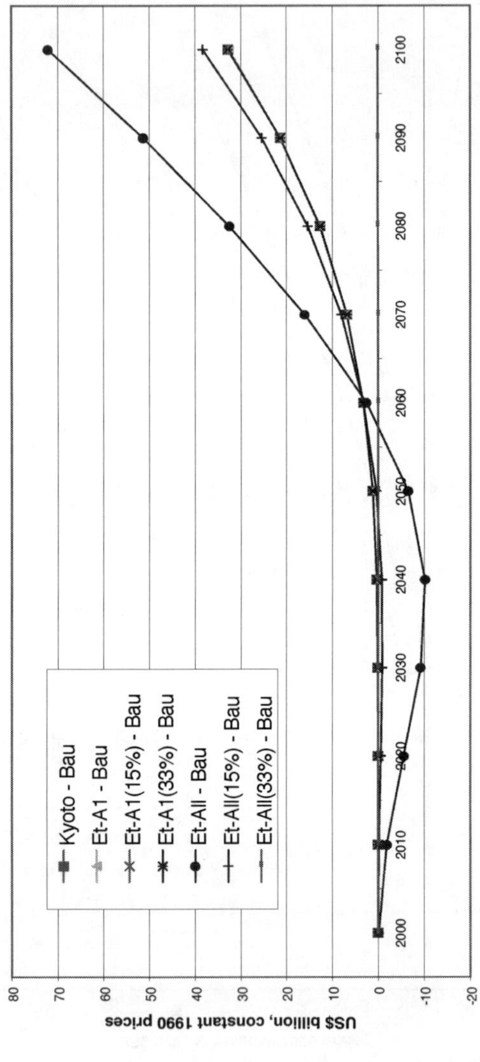

Figure 30. CHN GNP: difference.

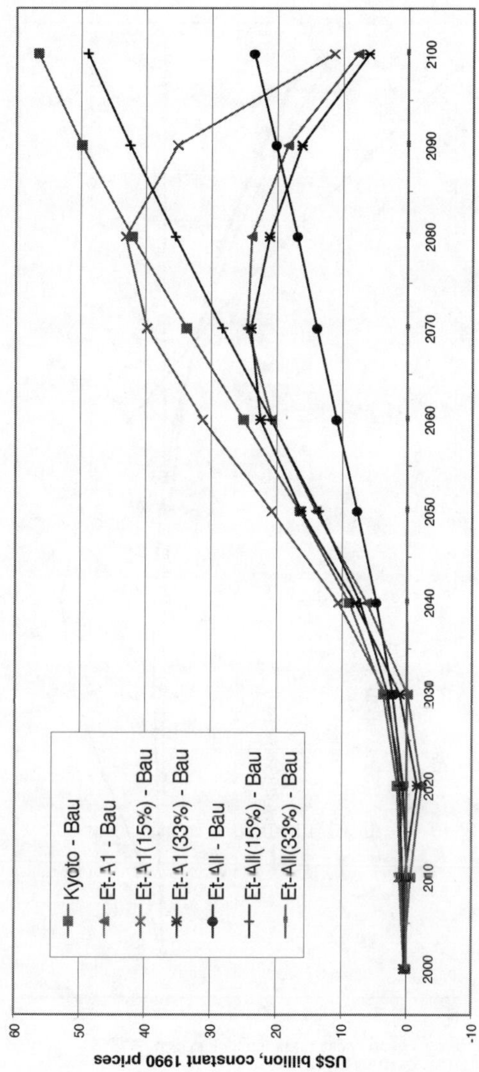

Figure 31. FSU GNP: difference.

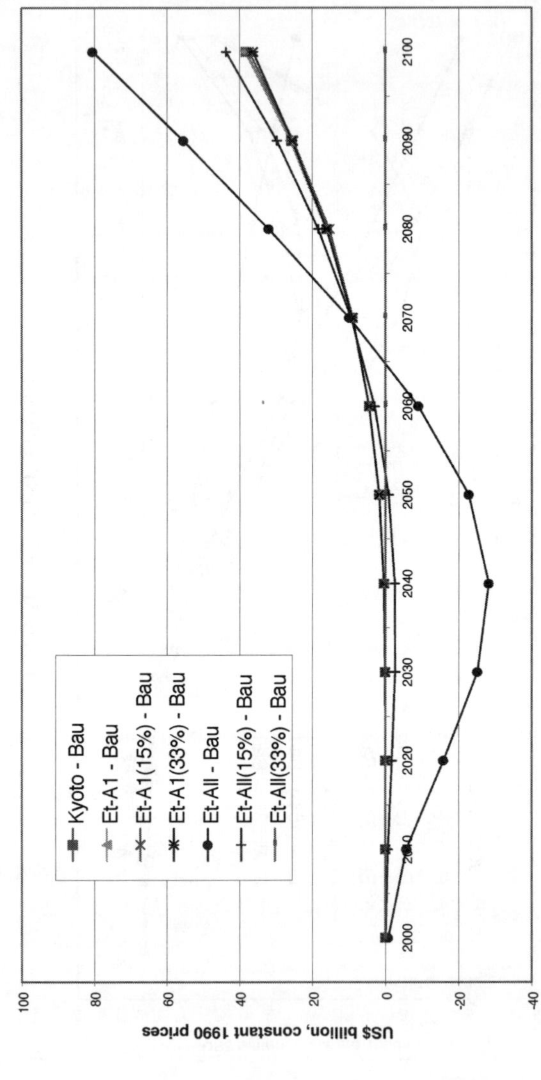

Figure 32. ROW GNP: difference.

Proposition 8. With ceilings, the implementation of the Kyoto Protocol is more equitable, because developed countries increase their abatement through domestic policies and do not buy developing countries' 'environmental resources' at low prices.

Also here we must distinguish short-run and long-run effects on equity. As shown by Figure 33, the long-run most equitable policy scenario is the one in which trade is allowed among Annex 1 countries without ceilings. Next is the case with a 33% ceiling (almost never binding), and then comes the case with global trade without ceilings. Hence we can conclude that our results do not support the above proposition, because in the long run the highest level of equity is achieved when no ceilings are introduced (or, equivalently, ceilings are not binding). This is true also in the short run, where the most equitable policy option is the one in which all countries trade without ceilings. We conclude that, as far as equity is concerned, there seems to be no support for assertions emphasizing the potential benefits of quantitative restrictions to emissions trading.

5. Conclusions

In this paper we have used a simple climate model with endogenous environmental technical change, obtained by integrating the RICE model of Nordhaus and Yang (1996) with the model proposed in Goulder and Mathai (1998), in order to analyse the efficiency and equity of different climate policy options. In particular, we had three goals:

- to identify the effects arising from the endogenization of environmental technical change;
- to assess the pros and cons of the introduction of ceilings to emissions trading;
- to quantify the distributional effects of climate policy scenarios.

As far as the first goal is concerned, we have highlighted some significant new effects. In particular, we have shown that the cost of complying with the Kyoto Protocol is lower when environmental technical change is endogenous, even when emissions trading is permitted. Moreover, the introduction of endogenous technical change leads to an equilibrium in which the demand for permits is higher, but supply is also higher, because developing countries invest in R&D to increase the amount of permits they can sell in the market. As a consequence of these two effects, the price of permits is lower.

As for ceilings, our analysis provides little support for quantitative restrictions on emissions trading. Even if the introduction of ceilings increases the R&D efforts of buyer countries and fosters technological innovation, the overall effect on abatement costs and economic growth is negative. The reason is that the

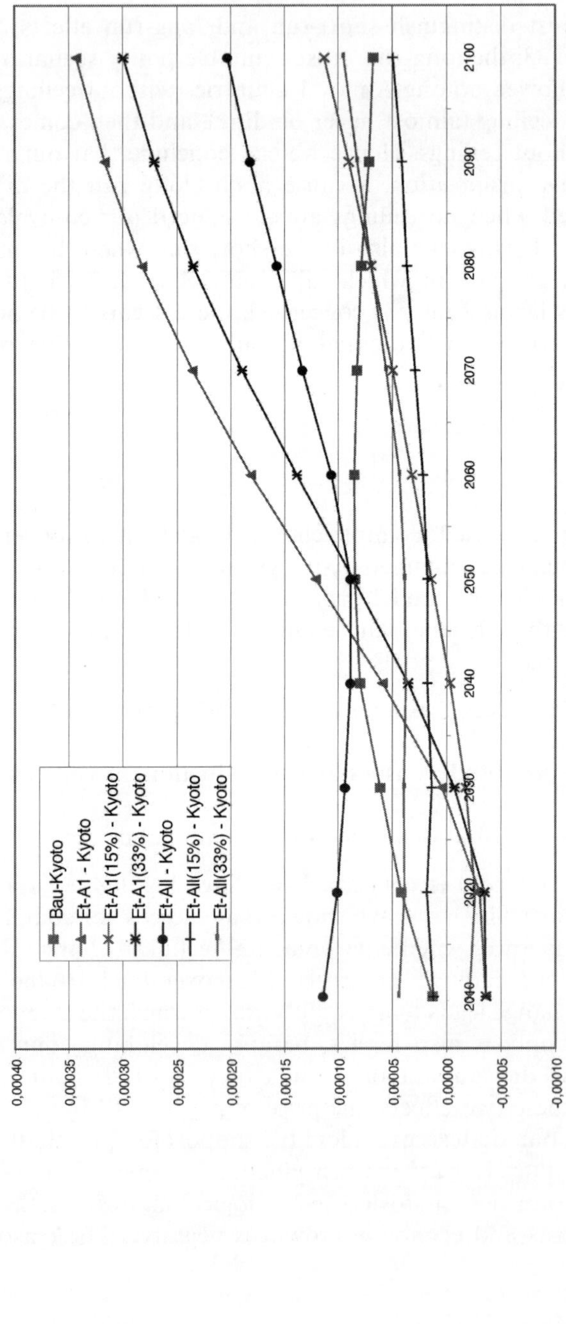

Figure 33. Equity index: difference.

benefits from technological innovation are lower, even in the long run, than the costs of adopting a more costly approach to climate change control. In other words, firms benefit more from a low impact of climate policies on their costs than from the stimulus to innovation that these policies induce.

Finally, equity is not positively affected by ceilings. We have shown that flexibility mechanisms in the presence of endogenous technical change increase equity, and that the highest equity levels are achieved without ceilings, both in the short term and in the long term.

Albeit useful as a first attempt to compare policy options in which technical change plays a crucial role, these results must nevertheless be considered as preliminary. Indeed, further research work is under way in order to extend the analysis in two main directions:

- the generalization of the model of endogenous technical change to account for international spillovers, decreasing returns of R&D, public incentives to R&D, and related issues;
- the implementation of the model of endogenous technical change proposed in this paper in other climate models (MERGE, ICAM, IAM, etc.) in order to verify the overall robustness of the results and to allow for other relevant economic aspects such as the role of energy inputs and of international trade.

Notes

* Paper presented at the FEEM/IDEI/INRA 3rd Toulouse Conference on Environment and Resource Economics, Toulouse, June 14–16, 1999, and at the IEA/EMF/IEW International Energy Modelling Meeting, Paris, June 16–18, 1999. This paper is part of the research work being carried out by the Climate Change Modelling and Policy Unit at Fondazione ENI Enrico Mattei. The authors are grateful to Francesco Bosello, Alessandra Goria, Michele Moretto and Roberto Roson for helpful discussions, and to Alain Bousquet, Hadi Dowlatabadi, William Nordhaus and Leo Schrattenholzer for their comments. We are grateful to Z. Yang who kindly provided us with the RICE model software.

1. Many of these papers are gathered in OECD (1998) and in Carraro (1999).
2. Article 17 of the Kyoto Protocol calls for emissions trading to be only 'supplemental to domestic actions for the purpose of meeting quantified emissions limitation and reduction commitments under Article 3'. To make it operational, it has been suggested that quantitative constraints (ceilings) on imports of emissions permits be introduced.
3. Nordhaus and Boyer (1999) actually claim that the Kyoto Protocol, even if implemented through emissions trading, will be excessively costly to the US and extremely beneficial to developing countries.
4. This last case is not yet implemented in the present version of the paper.
5. Another recent interesting development of the literature on induced technical change is constituted by the attempt to represent it as a learning-by-doing process. The model by Grübler and Messner (1996) is based on the concept of learning curves and describes future costs and performance improvements of new technologies as a function of accumulated R&D, and learning and experience gained in the diffusion of new technologies.
6. An exception is the proposal by Newell, Jaffe and Stavins (1998). Here the model is more sophisticated and represents the economic structure of innovation decisions. However, the

problem lies in the necessity of defining a statistical *ad hoc* model to generate the time series to be used in estimating the structural model of technological innovation. Hence, the model is still somewhat ad hoc, even if the problem is confined to the estimation procedure.

7. A second problem studied by Goulder and Mathai (1998) assumes that the rate of change of the knowledge stock is governed by abatement efforts themselves. This form of technological change is termed learning by doing. The analysis we conduct in the present paper can be easily adapted to this case as well, although we have selected R&D-driven technological change as it appears to be more popular in the literature.
8. As there is no international trade in the model, regions are interdependent through climate variables.
9. Obviously, we could have introduced two different types of R&D efforts, respectively contributing to the growth of an environmental knowledge stock and a production knowledge stock. Such an undertaking, however, is made difficult by the need for specifying variables and calibrating parameters for which there is no immediately available and firm information in the literature.
10. A relevant extension to the notion of an R&D-driven endogenous technical change is to allow for the possibility of international spill-overs. This is an issue whose study has been receiving increasing attention in recent years (see Coe and Helpman, 1995, for example). It is also of special importance to the question of substitutability between R&D efforts and emissions trading and for the ceilings debate. The next version of our ETC-RICE model will incorporate international R&D spill-overs.
11. Helpman's Web page is at the URL http://www.economics.harvard.edu/faculty/helpman/data.html.
12. More specifically, for each region we regress $\ln[\sigma(n, t) - \sigma_n]$ against an intercept and $-K_R(n, t)$. The antilog of the intercept provides an estimate of χ_n while the slope coefficient produces an estimate of α_n.
13. We are currently conducting sensitivity analyses concerning the most crucial parameters. Preliminary results show that none of the conclusions contained in the following sections are altered. All the coefficients not explicitly mentioned in this section are those of the original RICE model.
14. Note that we maintain the specification of the production process (5) where the state of the technology evolves partly exogenously and partly endogenously. Thus we never replicate the working and results of the original RICE model.
15. For this policy option we simulate the model only under endogenous environmental technical change.
16. However, those countries for which the ceiling is not binding purchase more permits at a cheaper price and undertake less domestic abatement as well as less R&D effort. This is the case for instance of the US in the Et-A1 (33%) policy option. On the other hand, Japan and Europe have to make more domestic efforts and R&D investment when ceilings are imposed as these are always binding. These results are consistent with those presented in Ellerman, Jacoby, and Decaux (1998).
17. We would like to stress that this result, like the others in the paper, depend crucially upon parameters whose value has been calibrated on the basis of limited (sometimes absent) empirical evidence. Hence, the results are to be considered cautiously, although, as said, preliminary sensitivity analysis suggests their robustness so far.

References

Bohm, P. (1999). International greenhouse gas emissions trading – with special reference to the Kyoto Protocol. Forthcoming in: C. Carraro (ed.) *Efficiency and Equity of Climate Change Policy*. Dordrecht: Kluwer Academic Publishers.

Boone, L., S. Hall and D. Kemball-Cook (1992). Endogenous technical progress in fossil fuel demand. Mimeo. Centre for Economic Forecasting, London Business School.

Bosello, F. and R. Roson, (1999). Distributional consequences of alternative emissions trading schemes. Forthcoming in: C. Carraro (ed.) *Efficiency and Equity of Climate Change Policy.* Dordrecht: Kluwer Academic Publishers.

Carraro, C. (1998). Beyond Kyoto: a game theoretic perspective. Proceedings of the OECD Workshop 'Climate change and economic modelling. Background analysis for the Kyoto Protocol'. Paris, 17–18 September.

Carraro, C. (ed.) (1999). *International environmental agreements on climate change.* Dordrecht: Kluwer Academic Publishers.

Carraro, C. and M. Galeotti (1996). WARM: a European model for energy and environmental analysis. *Environmental Modeling and Assessment.* 1:171–89.

Chander, P., H. Tulkens, J.Van Ypersele, and S. Willems (1999). The Kyoto Protocol: an economic and game theoretic interpretation. Université Catholique de Louvain, CORE Discussion Paper N. 9925.

Coe, D.T. and E. Helpman (1995). International R&D spillovers. *European Economic Review.* 39:859–87.

Conrad, K. and I. Henseler-Unger (1986). Applied general equilibrium modeling for long-term energy policy in Germany. *Journal of Policy Modeling.* 8:531–49.

Conrad, K. and M. Ehrlich (1993). The impact of embodied and disembodied technical progress on productivity gaps – an applied general equilibrium analysis for Germany and Spain. *Journal of Productivity Analysis.* 4:317–35.

Convery F. (1999). Insights for climate policy in Europe. Forthcoming in: *Integrated Assessment.*

Dowlatabadi H. and M. Oravetz (1997). Is there autonomous energy efficiency improvement? Department of Engineering and Public Policy, Carnegie Mellon University.

Ellerman, A.D., H.D. Jacoby and A. Decaux (1998). The effects on developing countries of the Kyoto Protocol and CO_2 emissions trading. World Bank Policy Research Working Paper N. 2019.

Eyckmans, J. and H. Tulkens (1999). Simulating with RICE coalitionally stable burden sharing agreements for the climate change problem. Université Catholique de Louvain, CORE Discussion Paper N. 9926.

Goulder, L.H. and K. Mathai (1998). Optimal CO_2 abatement in the presence of induced technological change. National Bureau of Economic Research Working Paper N. 6494.

Goulder, L.H. and S. Schneider (1999). Induced technological change and the attractiveness of CO_2 abatement policies. *Resource and Energy Economics.* 21:211–53.

Griliches, Z. (1979). Issues in assessing the contribution of R&D to productivity growth. *Bell Journal of Economics.* 10:92–116.

Griliches, Z. (1984). *R&D, patents, and productivity.* Chicago: University of Chicago Press.

Grübler, A. and S. Messner (1996). Technological uncertainty. In: N. Nakicenovic, W.D. Nordhaus, R. Richels and F.L. Toth (eds.) *Climate change: integrating science, economics, and policy.* CP-96-1, International Institute for Applied Systems Analysis. Laxenburg, Austria, 295–314.

Hourcade, J.C., M. Ha-Duong and F. Lecocq (1999). Decision making under uncertainty and inertia constraints: sectoral implications of the when flexibility. *Energy Economics.* 20: 539–56.

Jorgenson, D.W. and P.J. Wilcoxen (1990). Intertemporal general equilibrium modelling of US environmental regulation. *Journal of Policy Modeling.* 12:715–44.

Manne, A.S. and R.G. Richels (1999). The Kyoto Protocol: a cost-effective strategy for meeting environmental objectives? Forthcoming in: C. Carraro (ed.), *Efficiency and Equity of Climate Change Policy.* Dordrecht: Kluwer Academic Publishers.

Newell, R.G., A.B. Jaffe and R.N. Stavins (1998). The induced innovation hypothesis and energy-saving technological change. Resources for the Future Discussion Paper N. 98 12.

Nordhaus, W.D. (1973). The allocation of energy resources. *Brookings Papers on Economic Activity.* 3:529–76.

Nordhaus, W.D. (1993). Rolling the 'DICE': an optimal transition path for controlling greenhouse gases. *Resource and Energy Economics*. 15:27–50.

Nordhaus, W.D. (1997). Modeling induced innovation in climate-change policy. Paper presented at the IIASA Workshop on Induced Technological Change and the Environment. Laxenburg, 26–27 June.

Nordhaus, W.D. and J.G. Boyer (1999). Requiem for Kyoto: an economic analysis of the Kyoto Protocol. Mimeo. Yale University.

Nordhaus, W.D. and Z. Yang (1996). A regional dynamic general-equilibrium model of alternative climate-change strategies. *American Economic Review*. 4:741–65.

OECD (1998). Climate change and economic modelling. Background analysis for the Kyoto Protocol. Proceedings of the Workshop held in Paris, 17–18 September.

Rose, A.Z. and B. Stevens (1999). A dynamic analysis of the efficiency and equity of tradeable greenhouse gas emissions permits. Forthcoming in: C. Carraro (ed.), *Efficiency and Equity of Climate Change Policy*. Dordrecht: Kluwer Academic Publishers.

Shogren, J. (1999). Benefits and costs of Kyoto. FEEM Note di lavoro N. 44.99, forthcoming in: C. Carraro (ed.), *Efficiency and Equity of Climate Change Policy*. Dordrecht: Kluwer Academic Publishers.

Tol, R. (1999). Equitable cost–benefit analysis of climate change. FEEM Note di lavoro N. 41.99. Forthcoming in: C. Carraro (ed.), *Efficiency and Equity of Climate Change Policy*. Dordrecht: Kluwer Academic Publishers.

Weyant, J.P. (1997). Technological change and climate policy modeling. Paper presented at the IIASA Workshop on Induced Technological Change and the Environment. Laxenburg, 26–27 June.

CHAPTER 7

Making CO_2 Emission Trading More Effective: Integrating Cross-Sectoral Energy Efficiency Opportunities*

Johan Albrecht

1. Introduction

Reducing greenhouse gas emissions will require a strategy that combines various policy initiatives and economic instruments. Next to traditional instruments like taxes on energy or the reduction of subsidies to energy-related sectors, some relatively new instruments have been launched at various international negotiations. Systems of tradable permits for greenhouse gases (GHG), Joint Implementation (JI) and the Clean Development Mechanism (CDM) are currently being considered or are already in an experimental phase. Inspired by the potential role of international emission trading, new instruments with a national scope followed. In late 1998, the Credit for Early Voluntary Action Act of 1998 was submitted to US Congress (EDF, 1998). The Early Credit was part of President Clinton's Climate Change Proposal of October 1997 and can provide emission reduction credits for early voluntary action (pre-2008) to US industries that reduce greenhouse gas emissions.

New initiatives also came from business. At the end of 1998, BP Amoco voluntarily started with an internal system of carbon dioxide emission trading. In 1999, the European electricity industry experimented with an international emission trading simulation exercise in which 19 electricity companies from 14 countries took part to test the effectiveness of parallel electronic trading in electric power and CO_2 (carbon dioxide) emission permits (Unipede-Eurelectric, 1999).

It is known that, compared to national and less flexible climate policy measures, international emission trading will lower average abatement costs for developed countries as the trading will involve the participation of developing countries and regions with lower abatement costs. The estimated savings through emissions trading with developing countries, compared to the GDP cost of unilateral emission stabilization policies, vary from 50% for the group of Annex I countries to 75% for specific countries like Japan (Mullins and

C. Carraro (ed.), Efficiency and Equity of Climate Change Policy, 163–184.

Baron, 1997). Even among developed countries, there could be cost savings up to 50% by implementing carbon dioxide emission trading (Bohm, 1998). There are, however, still many uncertainties and discussions on implementation issues. An important issue will be the share allocated to (international) emission trading in total greenhouse gas reduction policies.

The question we try to answer in this paper is the following: although emission trading leads by definition to the desired level of emissions – and is therefore an effective policy instrument – are there opportunities to reduce the total cost of this emission reduction by some modifications in emission trading designs?

In the next sections, we will elaborate on the need and opportunity to stimulate cross-sectoral energy efficiency investments, using data from the production and use of cars and trucks. We start with a section that develops a basic allocation scheme for emission reductions in an economy with only two sectors. We then use the conclusions of this section to introduce in Section 4 a specific type of tradable allowance that makes it possible to integrate cross-sectoral efficiency investments in existing designs of emission trading. In Sections 5 and 6, we discuss the advantages of the new allowance and define some performance standards that can be used for it. In Section 7, we present the output of basic simulations for the car and truck fleet that we have developed to estimate the impact of our instrument on CO_2 emissions over a period of 15 years. We focus on carbon dioxide in this paper since it is the most important of the six greenhouse gases that need to be reduced as a result of the Kyoto Protocol.

2. International emission trading and national reduction priorities

Most proposals for designs of systems of tradable emission rights start from an initial distribution of permits based on the production of carbon: 'The first step is to measure emissions of carbon dioxide into the atmosphere in terms of the fuels that consumers and industry actually buy' (EC, 1998a). This initial entitlement of rights to emit is of course a crucial element for the political acceptability and the effectiveness of the system. Every year there will be an allocation of emission rights and, by reducing the annual entitlement, total emissions can be reduced.

After the allocation of emission rights to a few sectors like energy producers and importers, price implications for the other sectors need to be considered. If we assume that administrative costs need to be limited and that permits are allocated or 'grandfathered' based on the carbon *produced* by the heavy industries or the importation of energy, the most important consequence of emission trading will be that energy prices increase for the energy-consuming sectors in the economy; households, light industries and the tertiary sector. These sectors will pay more for their electricity needs and for fuels used in transport and for internal power generation. The main difference with an energy tax is that permit trading – if effectively monitored and enforced – will always lead to the desired level of total emissions. Another difference with energy taxes is that grandfathered allowances provide an implicit subsidy to existing energy-intensive and energy-producing or importing industries. One could argue that other sectors are probably more entitled to receive climate policy subsidies, e.g. for the development of cleaner technologies or cheap renewable energy.

The increase in the energy price will depend on the permit or allowance price that reflects the imbalances between demand and supply on the emission allowance market. The chance that emissions from households and from the transport sector will strongly decrease is limited, due to low negative energy price elasticities and, especially, to positive income elasticities that are currently being experienced in most developed countries (EC, 1997a; Bürgenmeier *et al.*, 1997; Albrecht, 1998; Cobb, 1999). It is a striking reality that in every developed country, even in those with high energy taxes, transport CO_2 emissions *continue* to increase. With international emission trading, mostly sectors other than households and transport will reduce their emissions. Only with very high permit prices will transport CO_2 emissions also be reduced.

Although the economic rationale can be discussed, it is a political priority to reduce emissions from households and transport in most developed countries (see the recent EU agreement with ACEA – the European Automobile Manufacturers Association – to strongly reduce the average CO_2 emissions of new cars by 2008). Compared to low-cost abatement opportunities in the poorest developing countries, most emission reduction strategies in OECD countries are not cost-efficient from an economic perspective. The Kyoto Protocol does, however, impose binding national emission reduction targets. As a result, national measures need to be taken even when these are not economically efficient from an international perspective. In this paper, we try to reconcile these political priorities with the efficiency outcomes of unlimited international emission trading.

3. The allocation of emission reductions

Emission trading will allocate emission reduction efforts to the sources with the lowest abatement costs. Marginal abatement costs need to be compared to find the most efficient emission reduction strategy. It is, however, very difficult to calculate marginal abatement costs for all sources in the international economy and it is even more complicated to estimate their long-term development (e.g. for the period 2000–2010 or longer periods).

In Figure 1, which represents a country with only two sectors, we assume that the marginal abatement costs are known for the sectors of electricity generation (origin 0^{eg}) and transport (origin 0^{tr}). We further assume that they initially emit the same amount of CO_2. The overall target for national emission

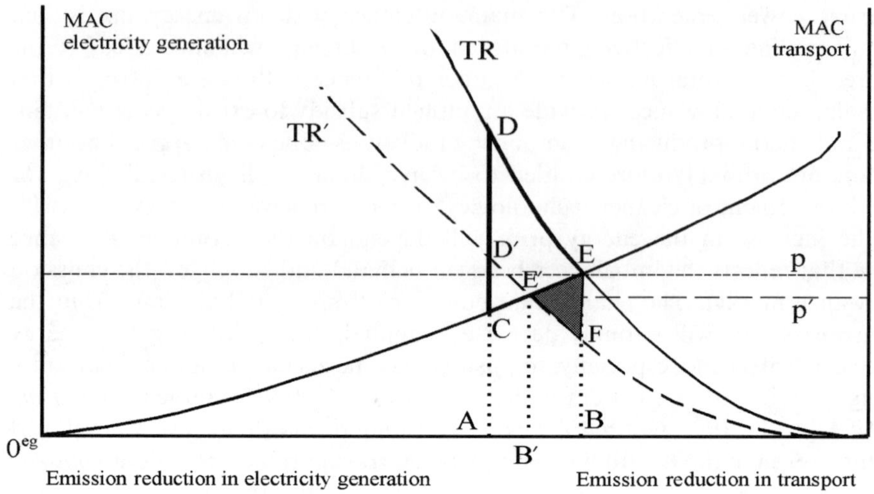

Figure 1. The allocation of emission reduction in an economy with two sectors.

reduction over a specific period is $0^{eg}0^{tr}$. No international emission trading is allowed. We first analyse the optimal allocation of national reduction efforts in this economy. The marginal abatement costs for emission reduction in the transport sector are denoted as TR and start in origin 0^{tr}.

In the most efficient distribution of reduction efforts, emissions from electricity generation are reduced by 0^{eg}B and the reduction of emissions in the transport sector is 0^{tr}B. The uniform price of this emission reduction that establishes this allocation in Figure 1 is p. When the regulator would however decide that emissions in transport need to be reduced by 0^{tr}A, this would lead to an additional welfare loss of CDE since these abatement costs could be saved with the efficient solution.

These long-term abatement curves depend on various factors and assumptions. From Figure 1, it becomes obvious that in this scheme with only two sectors, specific regulatory measures with respect to one of these sectors can influence the emission reduction price level, the allocation of reductions and total costs of emission reduction. Considering the rather direct relationship between fuel efficiency and CO_2 emissions, the long-term marginal abatement costs in transport depend on specific assumptions like possible intermodal shifts, the accelerated introduction of highly fuel-efficient vehicles and the subsequently accelerated scrapping of old and polluting vehicles. An example of such regulation is a ban on the use of cars and trucks older than 20 years. This measure was taken in India in 1998. If policy measures to introduce very efficient vehicles at an acceptable cost for manufacturers – i.e., comparable to normal costs when introducing new engines – are a great success and dirty vehicles are eliminated from traffic flows and replaced by clean vehicles in a relatively short period, the long-term marginal abatement cost curve shifts from

TR to TR′. When dirty cars are scrapped, their CO_2 emissions follow. In the next section, we will propose a mechanism that could realize this shift.

With TR′, the optimal allocation shifts from B to B′. The emission reduction price shifts from p to p′ and ensures the diffusion of lower-cost emission reduction opportunities over the whole economy. The share of tranport emission reductions in total reductions increases as a result of the lower TR′. The electricity sector reduces its emissions by $0^{eg}B'$ while transport emissions are reduced by $0^{tr}B'$. Total abatement costs for the economy are reduced by the area $EE'0^{tr}$. As a result, total welfare increases when the diffusion of fuel-efficient vehicles is accelerated and when producers do not have to bear excessive development costs.

The electricity producers benefit most from this new situation: their total abatement costs are reduced by EE′B′B. These costs are now partially borne by the transport sector (for the area BB′E′F) and partially eliminated (the dark area EE′F). The eventual net reduction of total abatement costs for transport depends on the difference between BB′E′F and $EE'0^{tr}$.

It is of course also possible that the long-term marginal abatement costs of the electricity sector shifts to a lower curve since there exist interesting fuel-switching possibilities for countries that use coal and oil for electricity generation. Estimates of long-term cost developments depend strongly on assumptions like no capital destruction and it is therefore not evident that abatement efforts will be less expensive in the electricity sector. Furthermore, sector-wide empirical studies made for Belgium indicate that cleaner vehicle technologies deliver the least-cost strategy for national climate protection (Electrabel, 1999). The second option in terms of cost-efficiency was the closing down of old coal-based electricity generation. These results were found when the development cost of fuel-efficient engines was assumed to be comparable to the development cost of a new reaction engine (for aviation) and that car use does not increase as a result of more fuel-efficient engines (i.e., no rebound effect).

In the next section, we focus on the role of incentives to stimulate the development and diffusion of fuel-efficient and clean vehicles. A central element is that we need to create incentives that make it worthwhile (not to say profitable) for industry and the electricity sector to reduce the GHG emissions of other sectors (transport, households and the tertiary sector).

4. Car and truck manufacturers and the tradable certificate

While there are many other sources, a significant amount of GHG emissions result from the production and the use of cars and trucks in transport. In most OECD countries, transport accounts for more than 25% of all GHG emissions. The relative share of transport is estimated to increase further (EC, 1997b). The CO_2 emitted in the production phase is, however, identical to the CO_2 emitted during usage of the car or truck. As a result of emission trading, energy

Table 1. CO_2 emissions for the production of one car in 1997

Operations	Volvo Plant	Unit	Tonne CO_2/Unit
Pressing of car body components	Olofström, Sweden	1 tonne of sheet steel	0.125
Production of gearboxes, rear axles and drives	Köping, Sweden	1 set of components	0.013
Foundry; engine manufacturing	Skövde, Sweden	1 car engine	0.219
Assembly and painting	Ghent, Belgium	1 car	0.281
Total			0.638

Source: own calculations based on Volvo (1998a), environmental data for Volvo production plants 1997

prices will increase and the already efficient car and truck *manufacturing* industry will invest further in abatement at the production phase or will buy emission rights.

With CO_2 emission trading, the households that buy the cars will pay higher energy prices. Considering the very low energy price elasticities for transport purposes, the diffusion of energy-efficient cars will not be especially stimulated because industry is not *directly* rewarded for investing in reducing transport GHG emissions. Industry is only rewarded for reducing its *own* GHG emissions. We think that this should not be the major environmental priority for car and truck manufacturers. Therefore, policy measures should stimulate cross-sectoral efficiency investments.

We argue that emission trading systems should integrate cross-sectoral transactions like permits allocated to car manufacturers because they did sell very energy efficient cars to households. Therefore a new type of permits should be created for efforts that lead to reductions of emissions in other sectors. If the electricity sector, or another industry, provides a technology to a firm or industry that can reduce, just by the implementation of this technology, its own GHG emission by x tonnes, the provider of the technology should receive a number of tradable allowances – in terms of GHG reduction units – for its sold product or technology. We will use the term 'certificate' for this type of allowance.

If other manufacturers, like the car industry or producers of heating systems or refrigerators, present a new energy-efficient type, they could also receive a similar allowance or permit for the realized reductions. These new certificates are grandfathered to sectors not because of their mere existence or their energy-intensity, but because they make it possible for other sectors to reduce their emissions.

These allowances can be sold on the GHG Permit Market. The benefits from selling will reward industries for investing in GHG reductions realised by other sectors.

In Table 1, we calculated total CO_2 emissions for producing one car in 1997.

Box 1 – CO_2 emissions during production and use of the product

	Car	Truck
Emissions during production	1–2 tonnes CO_2	3–5 tonnes CO_2
Emissions during lifetime	37.5 tonnes (150 000 km)	1500 tonnes (1.25 million km)
Relative importance of consumption phase	37.5/1.5 = 25	1500/4 = 375

→ Conclusion: instruments should target emissions during product use

The data are collected for European Volvo plants. For other manufacturers, there could be significant differences. In our example, we limit the in-house production phase of cars to the four operations in Table 1. The total CO_2 emissions in this table are relatively high compared to data for some other Volvo plants. We also included one tonne of steel per car in the example. This is probably too much; the 1995 US family car weighed 1470 kg (RMI, 1998). Another consideration is that CO_2 emissions depend on the fuel mix used during production. A good illustration is the Volvo plant in Floby (Sweden) that has *no* CO_2 emissions because it uses district heating based on biofuel.

For the other supplies (seats, glass, electrical components, etc.) that are used during the assembly, we assume that the resulting CO_2 emissions amount to half of the in-house generated emissions. This results in total CO_2 emissions around one tonne per Volvo passenger car produced. Adding emissions during transportation and the end-of-life phases, we used in our further calculations (Box 1) a figure between one and two tonnes carbon dioxide emissions per car produced. As a result of future emission caps and allocations to car manufacturers in permit systems, this figure could decrease over time if the producer (here Volvo) does not prefer to buy permits on the emission market but will opt for internal emission reduction. The reduction will probably not be spectacular, at most a reduction of 0.5 tonne CO_2 per car. The cost of this emission reduction could be considerable.

The finished car is then sold and used in traffic. The average fuel efficiency of this car is around 10 litres for 100 kilometres (Volvo, 1998b). We estimate that one kilometre in traffic results in an average emission of 25 g CO_2 per litre fuel consumed (calculation based on EC, 1998b). For the car in our analysis, CO_2 emissions will be around 250 g/km with a consumption of 10 litres fuel for 100 kilometres. Assuming that the car will be used for 150 000 kilometres and that the average fuel efficiency will be constant as a result of good maintenance, total emissions of CO_2 during the lifetime will be 37.5 tonnes. Emissions during car use clearly outweigh emissions during production.

On the issue of total CO_2 emissions of popular cars, some organizations even publish lists with the worst and the best cars in terms of environmental damage. In 1990, it was calculated that the BMW 750iL with an average fuel efficiency around 17 l/100 km produced 66 tonnes of CO_2 during its lifetime (Public Citizen, 1990).

If industry should make a choice between investing in reducing emissions during manufacturing or investing in cars that need less fuel during their lifetime, the best environmental results will be achieved by the latter option, probably at the lowest cost.

This is even more clear in the similar case of trucks and buses. We define trucks as heavy-duty vehicles, starting from 16 tonnes. Trucks that are used for long-distance transport have an average lifetime of 1–1.5 million kilometres, depending on the quality of the truck and the maintenance. For trucks that are used for short distances (e.g. ten journeys of 30 km each day), the lifetime ranges from 750 000 to 1 million kilometres.

If we start from average emissions of CO_2 around 30 g/km per litre fuel needed – truck engines mostly burn diesel – and an average fuel efficiency[1] of 40 l/100 km, a lifetime of 1.25 million kilometres leads to total emissions of 1500 tonnes CO_2. If the truck were very efficient and consumed only 30 l/100 km with a shorter lifetime of only 750 000 kilometres, total emissions of CO_2 would still be 675 tonnes. Compared to emissions during the production of trucks, estimated around 3–5 tonnes, the difference is extreme. Box 1 summarizes our two examples.

As shown in the examples, we think that policy instruments should focus on the emissions during the use of the product. The problem is that the car or truck manufacturer does not yet receive a tradable permit or allowance for his investment in clean vehicles. But if the new product of this manufacturer emits 10 tonnes CO_2 less over its lifetime, why not allocate allowances for this reduction to the manufacturer?

The manufacturer of cars and trucks would then be able to sell these certificates or allowances on the permits market and use the benefits for lowering the price of the new and highly efficient product or for financing further R&D in ecodesigns. This lowered price, that of course includes a normal profit ratio, will stimulate the market penetration of the highly efficient vehicles.

5. Advantages of the tradable certificate

Next to the stimulation of intensive research in energy efficiency, the main advantage of the tradable certificate is the broadening or acceleration of industry involvement. Every firm with products that need energy can be rewarded for energy savings during the lifetime of its product. If the permit or certificate price is high, the price of these highly efficient products can be significantly lowered and this will stimulate their market penetration. For some producers, a competitive advantage can be created when cross-sectoral reductions of CO_2 can be sold on allowance markets. Without this mechanism, they would only have an incentive to develop cleaner products when energy prices increased dramatically.

Another advantage is that we create a permits market with more activity and less dependency on some economic sectors. Not only the electricity providers or energy importers will be in the emission market – probably buying permits from Russia and increasing the energy price for consumers – but every firm that realized cross-sectoral energy savings can participate. The supply of energy efficiency will come from many parties. The stricter the caps imposed on the main emitting sectors, the more attractive the cross-sectoral efficiency gains.

Some other issues need to be considered. In the case of manufacturers of cars or refrigerators being able to improve the energy efficiency of their products by making use of light materials or insulation solutions provided by the chemical or other industries, the allocation of the certificate is partly made possible by the inputs from other industries. These supply industries could claim a proportion of the certificates for themselves. It would, however, be very complicated to calculate the specific contribution of each input or material in total energy savings. Another option is for car or refrigerator manufacturers just to pay a higher price for these specific inputs. On the other hand, the increased interest in energy-saving technologies and inputs creates new and important markets for these supplying industries. They can increase their sales and this is partly made possible by the allocation of the certificates to the car and refrigerator industries.

If we introduce certificates for car manufacturers or other sectors to increase their CO_2 emissions during production, it is clear that the allocation of other – probably grandfathered – emission rights needs to be reduced. Otherwise, total emissions would increase. In our later simulation, we will assume that each year only part of the total received certificates – calculations are based on complete product life cycles – will be available on the market for emission rights. The certificates that are not available in the first trading period will be banked. We assume that this annual inflow of certificates on the emission market will be deducted from the annual allocation that is planned for the country or industries in the analysis.

Compared to the Early Credit, the main difference is the cross-sectoral incentive to improve energy efficiency. Many aspects associated with the Early Credit are also valid for the tradable certificate: reward real reductions and not gaming, no predetermination of the eventual domestic regulatory programme to control domestic GHG emissions, and focus on domestic action.

An important element is of course the development cost of cleaner technologies. In the context of automobile production, it is important to stress that every 8–10 years completely new engines are developed. We should also consider the recent expansion of safety features that required enormous investments and did increase average car weight. Furthermore, an indication that the development cost of fuel-efficient cars is not that excessive can be found in the

Box 2 – Calculations of emission reduction based on performance standards

	Refrigerator	Car	Truck
Baseline	2 kWh/day	250 g CO_2/km	1200 g CO_2/km
Expected lifetime	15 years	150 000 km	1 500 000 km
Efficiency new type	0.6 kWh	150 g	900 g
Savings (lifetime)	3.8 tonnes CO_2	15 tonnes CO_2	450 tonnes CO_2
Sales (reference period)	10 000	40 000	5000
Total savings			
– lifetime (15 year)	30 000 tonnes	600 000 tonnes	2 250 000 tonnes
– per year	2533 tonnes	60 000 tonnes	150 000 tonnes

impressive recent elimination of needed maintenance services. While only 10 years ago it was normal to change the oil every 5000 to 10 000 kilometres, some recent cars require now their first maintenance service only after 20 000 to 40 000 kilometres. Even in countries with high fuel prices, car maintenance costs exceed total fuel costs and this explains the efforts of manufacturers in this cost aspect. For the Belgian car fleet, total maintenance costs for the period 1988–1997 exceeded total fuel costs by some 25% (Febiac, 1999).

6. Performance standards

If we want to allocate certificates based on improved energy efficiency, a baseline to measure the efficiency gains is needed. The measurement of the energy efficiency improvement is quite easy to establish. For cars and trucks, detailed information on CO_2 emissions per kilometre is available for most models. In our example, we calculated the efficiency gain as the reduction of emissions per kilometre.

Electronic devices need electricity, and their electricity, consumption is expressed in kilowatt-hours (kWh). In Europe, the most efficient refrigerators are already differentiated from the least efficient by energy labels. The price difference was in some countries reduced by a subsidy for the most efficient or A-types (Albrecht, 1998). If we want to allocate a certificate to the manufacturer, we can use an average CO_2 emission rate per kWh. Most figures fall inside the category for 400–800 g CO_2 per kWh used, depending on the input mix of the national electricity providers (EC, 1998b). In the calculations in Box 2, we assumed 500 g CO_2/kWh. For countries that depend strongly on renewable or nuclear energy, lower values should be used. If we assume that all refrigerators are equal in volume and quality, the difference in energy consumption per sold refrigerator should be multiplied by the number of sold products on the relevant market.

In Box 2 we present three examples with a sales volume typical for an important manufacturer selling in an average European country. We assume a lifetime of 15 years for the three products in the example and find that selling energy-efficient products can reduce annual emissions of CO_2 by thousands of tonnes.

On the permits markets, the certificates allocated to the manufacturers of only three products would lead to an annual inflow of 212 533 tonnes CO_2 equivalents during a period of 15 years. This inflow of certificates needs to be deducted from the total annual allocation or grandfathering of CO_2 emissions. If the baselines in Box 2 are set more strictly, the inflow will be lower.

7. The impact of the certificate on the car and truck market

In this section we present two simulations to estimate the potential reductions of CO_2 emissions that could be realized if the tradable certificate were integrated in permits systems. We present two separate markets: the car market and the truck market.

The truck market receives little attention compared to the many policy initiatives for cars that have been taken in many countries. Heavy-duty vehicles are of course fewer in number but with a mileage ten times the car mileage and an average fuel efficiency one-fifth of the car fuel efficiency, a small truck fleet is responsible for the same level of CO_2 emissions as a car fleet that is 50 times bigger.

7.1. *The car market*

We start with a car fleet of 5 million vehicles and annual sales of 500 000 cars. These assumptions are close to the actual situation of the Belgian car market.

We introduce three stock variables: existing car fleet, ecocars and other new cars. There are only two types of new cars sold: an ecocar or a 'normal' new car. The difference between the two categories is based on average fuel efficiency. Over a period of 15 years, we assume that the fuel efficiency of ecocars will improve from 5.5 l/100 km to 3.5 l/100 km. This is a realistic assumption (Von Weizsäcker, 1997; Lovins, 1996). A good example is the recent commercialization of Toyota's Prius, a hybrid electric–gasoline car. The Prius consumes only 3 to 4 l/100 km with CO_2 emissions that are half those of a conventional car. Emissions of toxic gases are reduced by 90%. Since the launch in December 1997, 3500 hybrid cars have been sold each month in Japan, and Toyota employees need to work overtime (Hinrichs, 1998). The reason for this 'success' is the competitive price of the Prius in Japan (US$16 500), which is only 10–15% higher than the price of comparable but less fuel-efficient cars. Some market analysts suggest that Toyota is not making profits on the Prius but wants to build up experience with the coming generation of ecocars. A mechanism that rewards Toyota for the reduced emissions could lower the financial burden of the introduction of the Prius.

The experience with the Prius indicates that there could be a relatively high cross-price elasticity for the demand for eco and other cars. Unfortunately, there are not enough ecocars on the market to generalize this conclusion. Therefore, we work with only two types of cars in our simulation – ecocars and normal cars, both rather expensive – that have strongly comparable characteristics with the exception of fuel efficiency. This assumption takes away the need to have information on cross-price elasticity values.

Most surveys on the costs of ecocars estimate that hybrid and fuel-cell vehicles would cost US$4000 to US$7000 more than comparable cars with traditional internal-combustion engines. Some manufacturers, like Ford and Mercedes – both corporations have invested heavily in the application of fuel cell technology – predict that the difference could be even smaller (Leslie, 1998).

The fuel efficiency of the other new cars in our model will start at 8 l/100 km and will remain more or less constant in the first five years. Then the fuel efficiency will also improve and converge to the level of the ecocars. This assumption is made because we believe that when major corporations develop new engines and car bodies for their ecocars (35–75 kW), these new technologies will also be used – in a later phase – in their other types (more than 75 kW). Manufacturers will not develop car types based on two completely different technological trajectories. This would be too expensive.

For each of the stock variables, we defined a scrapping rate. Evidently, we used in the first phase a scrapping rate for the existing car fleet that is higher than the scrapping rates for new cars. During the subsequent periods, the scrapping rates converge. We also assumed the scrapping rate of the existing car fleet to be dependent on the declining price of the energy-efficient cars. Once these ecocars overcome their initial price disadvantage as a result of scale economies in the production, their inflow in the car market will increase and the scrapping of older cars will accelerate.

In our model, we introduced tradable certificates based on a fuel efficiency baseline of 9 l/100 km. The setting of this baseline is arbitrary. In the car example in Box 2 we took a baseline of 10 l/100 km. We calculated that the use of the more efficient new car will make it possible to avoid the emission of 15 tonnes CO_2. This improvement will result in an allocation of 15 certificates to the manufacturer, each with an intrinsic value of abating 1 tonne of CO_2 emissions. These 15 certificates cannot be sold in the first year because they are based on the use during the complete lifetime of the vehicle. We assumed that manufacturers can only sell 20% in the first year. Of the total certificates, 80% will be banked and sold in the following years.

We assumed that increasing the average fuel efficiency by 1 litre (for 100 kilometres) results in receiving four allowances, each representing 1 tonne CO_2. The efficient car in Box 2 consumes 4 litres less fuel than the baseline type. As a result, emissions will be reduced by 100 g/km. As already mentioned, over the lifetime of the car, 15 tonnes less CO_2 is emitted compared to the less efficient type.

The value of these earned certificates is linked to price developments in the permits market. Prices will depend on abatement costs for carbon-producing industries, major political developments and the functioning of the market. In order to attract developing countries, it is obvious that permit prices need to be relatively stable, preferably at a high level. It will be hard to convince developing countries to join emission trading schemes if the price of the permits is very unstable and crashes frequently. We therefore assumed for our simulations a mechanism of market intervention to keep permit prices inside a range. Using annual emission allocation quotas that depend on average price developments and on the number of introduced certificates, could be an approach for this market intervention. If the allowance or permit price is falling, the allocation to energy-intensive industries could be lowered to increase demand for permits and support the price level.

If average abatement costs turn out to be around US$30–40/tonne CO_2, we assume that the permit price will be around US$40–50. In our simulation we will test for some widely accepted price levels, but not exceeding US$100–150 (Bohm, 1998).

To reduce emissions, the market share of the efficient ecocars needs to increase. The commercial success of the ecocars will depend on the price difference compared to the other new cars. The price of the Prius is comparable – i.e. some 15% higher – to the price of its direct competitors and this is a crucial part of its success.

It would be too optimistic to assume that consumer preferences will shift to light microcars that are of course very efficient. We opted therefore only to work in our model with rather large cars with high comfort levels. There are no other quality differences than fuel efficiency. Of these cars with a high price of US$30 000, only two types are sold on the market. We clearly do not limit the category of ecocars to small cars because we think it is necessary to compare levels of comfort offered by cars. Many very efficient cars are small to reduce weight.

We assume in our model that the ecotype of the average family car is 15% more expensive compared to the other type. It is probable that the difference is smaller, especially for average cars with turbo diesel engines. In most countries, cars with the very efficient turbo diesel engines produced by Volkswagen and Seat cost some 5–10% more than the models of their competitors. The assumptions in our model are therefore rather conservative. If price differences were smaller, many stories like that of the Prius could follow. The same conclusion will hold when the market share of small cars increases.

Production prices of ecocars will fall as a result of economies of scale. For the new engines and materials, the economies of scale will be more important than for existing models.

We further assumed that prices to consumers will be reduced by the total actual value of the certificates that will be allocated to the manufacturer when

Box 3 – Determining the market share of ecocars

Fuel efficiency baseline (9 l/100 km): *b*
Fuel efficiency ecocars: $f_{eco}(< b)$ Fuel efficiency other cars: $f_{other}(< b)$
Production cost eco: P_{eco} Production cost other cars: P_{other}
Value tradeable certificate (4 certificates/saved litre): VTC

If $(P_{eco} - 4*(b - f_{eco})*\text{VTC}) < (P_{other} - 4*(b - f_{other})*\text{VTC})$
THEN market share ecocars increases, or:

If $(P_{eco} - P_{other}) < ((f_{other} - f_{eco})*4*\text{VTC})$ THEN market share ecocars increases

selling an ecocar. We assumed that the banked certificates will not lose or gain value.

It is clear that buyers will opt for the ecocar only if the price is good and the comfort level comparable. In our model, 500 000 new cars will be sold each year. If the price of the very efficient ecocar equals the price of the other car, we assume that the market share of the ecocars will be 75% because comfort levels are identical. This is a modest assumption because with a population of 100% rational consumers, the market share of ecocars would then have to be 100%. But there will always be consumers that do not care about fuel efficiency and consider other characteristics more important. If the price of the ecocars is only 5% higher than the price of the other cars, we assume a market share of 30% for the ecocars. These buyers include the discounted energy savings in their decision to buy. If the price of the ecocar is 10% less than that of the other cars, we assume a market share of 95%.

The distribution of the ecocars and the reduction of CO_2 emissions depend mainly on the difference in production cost and the value of the certificates that the producers of ecocars receive *more* than the producers of the other cars. If the baseline is set at 9 l/100 km, for *every* new car sold certificates will be received.

Box 3 summarizes the main interactions that determine the market share of ecocars in our simulation.

When the baseline is higher than the two levels of fuel efficiency, the total difference $(f_{other} - f_{eco})$ will result in receiving certificates. If the baseline is between the two levels of fuel efficiency, the instrument of certificates is less powerful because the manufacturer of the ecocars does not receive certificates for the total reduction in fuel needs he offers to his customers. He will receive certificates only for the difference $(b - f_{eco})$. We conclude therefore that relatively high baselines could indeed lead to high inflows of certificates but they guarantee that the total difference in fuel efficiency is valuated on the market.

7.2. *Results of the car model*

To estimate the potential of the permits system with tradable certificates, we first calculated the future development of CO_2 emissions without the tradable

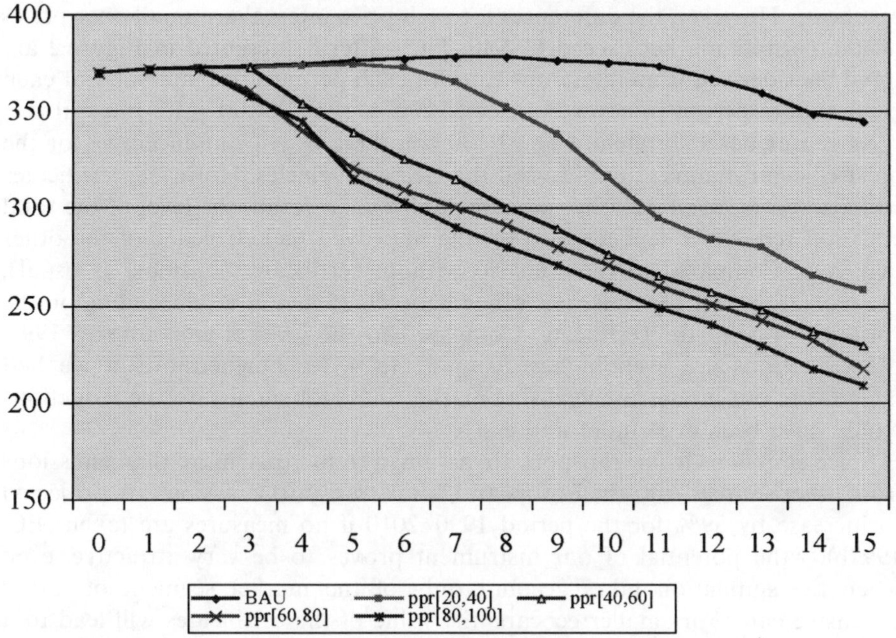

Figure 2. CO_2 emissions of the car fleet (million tonnes, 15 years).

certificates. This is the business-as-usual scenario (BAU in Figure 2). The start of our simulation is the year 1999. This should be kept in mind when comparing our simulation output to other models that start in 1990. We find that without certificates, the market share of the ecocars will remain very low.

In our simulation, the total car fleet increases by 15% but duc to the scrapping of the oldest cars first, CO_2 emissions do not rise similarly. Cars sold before 1985 might have an average fuel efficiency that is 50 to 75% lower compared to the cars sold in 2000. This improvement in efficiency can outweigh increases in the fleet and car use (measured in km/year). As a result, emissions rise slightly only during the first eight years, and when the most inefficient and polluting cars – cars sold during the 1980s – are scrapped, the improved average fuel efficiency of the car fleet will lead to a stabilization and modest reduction of CO_2 emissions starting from 2010. This (positive) development depends of course on the ability and goodwill of manufacturers to produce efficient cars without strong incentives as in the case with the certificates.

We then introduced tradable certificates into our simulation for the average car of the upper segment of the market (price: US\$30 000). We took four price levels (in US\$) for the certificates or permits on the emission market: 20–40, 40–60, 60–80 and 80–100.

From Figure 2 – permit price ranges are indicated as ppr[,] – it is clear that the reduction in CO_2 emissions depends strongly on the introduction of the certificate. However, the differences for each price interval are small once prices on the permits market exceed US$40. The patterns presented in Figure 2 are trend lines derived from numerous runs for each permit price interval. For each run, random permit prices were selected out of the relevant price interval.

Since our baseline remains at 9 l/100 km, there is a strong incentive for the most efficient manufacturers to sell the cleanest vehicles. However, certificates will also be received by other manufacturers. As a result, the level of the total emission reduction depends also on the improved fuel efficiency of the other new cars. Compared to the scenario without certificates (business as usual), the reductions in CO_2 emissions range between 25 and 38%, depending on the value or price of the certificate. Compared to the level of emissions in 1999, the reductions are even higher: from 32 to 43%. Furthermore, if we had included in our model an increasing market share for small cars, the reductions would have been even more impressive.

Since emissions from transport are assumed to increase more than emissions from other sources – for the European Union, transport emissions are projected to increase by 39% for the period 1990–2010 if no measures are taken (EC, 1997b) – the potential of our instrument proves to be very attractive, even when the simulation takes account only of the market segment of rather expensive cars. For smaller ecocars, the value of the certificates will lead to a more marked price advantage.

7.3. The truck market

We developed a similar simulation for heavy-duty trucks, starting from 16 tonnes. The success of CO_2 emission reduction policies will also depend on an increase in fuel efficiency for these vehicles. There are a few fundamental differences between developments for the car and for the truck market.

First of all, the truck market is more competitive than the car market. Buyers are very interested in the energy consumption of the truck that they will use for many kilometres. The strong competition in the transportation market guarantees that transportation firms want to lower their energy costs.

Another difference from the car market is related to the link between fuel efficiency and payload. The fuel efficiency of cars can be improved by reducing the weight of the cars. Trucks are developed to transport a heavy load. The use of weight-saving materials in the production of trucks will not yield similar results as in the car industry because the load remains heavy. The heavy load also has other implications. The average fuel consumption of heavy trucks ranges between 30 and 40 litres for 100 km. In very congested traffic, the fuel consumption can even double. The two most important determinants of the fuel efficiency are not engine performance aspects but very basic elements: aerodynamics of the truck and load and tyre pressure (the latter as a result of the weight of the load). Surprisingly, driving an empty or full container results in a difference in fuel consumption of only 5 l/100 km. More surprisingly,

Table 2. Development of emission requirements for trucks, g/kWh, 1982–1999

Standard	Year	NO_X	HC	CO	Particulates
ECE R 49	1982	18.0	3.5	14.0	–
Germany	1986	14.4	2.8	11.2	–
EEC	1990	14.4	2.4	11.2	–
Euro I	1992–development	8.0	1.1	4.5	0.36
	1992–commercial	9.0	1.25	5.0	0.40
Euro II	1995/1996	7.0	1.1	4.0	0.15
Euro III	1999	under disc.	u.d.	u.d	u.d.
SCANIA	1989	14.0	0.5	1.0	–
SCANIA	1991	7.8	0.5	1.0	0.25

Source: SCANIA, 1998

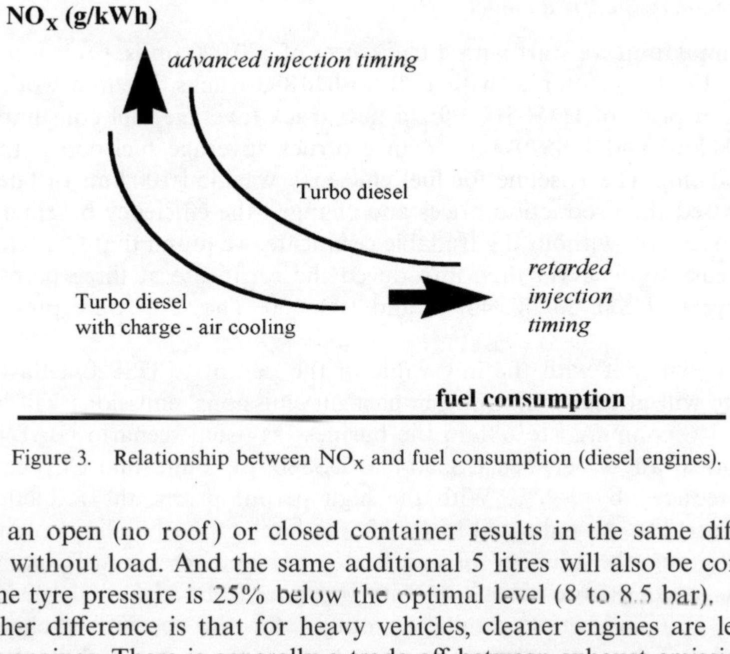

Figure 3. Relationship between NO_X and fuel consumption (diesel engines).

driving an open (no roof) or closed container results in the same difference, with or without load. And the same additional 5 litres will also be consumed when the tyre pressure is 25% below the optimal level (8 to 8.5 bar).

Another difference is that for heavy vehicles, cleaner engines are less fuel-efficient engines. There is generally a trade-off between exhaust emissions and fuel consumption (Samuelsson, 1998). Reducing emissions of pollutants resulted in increased fuel consumption by turbo diesel engines. This depends to a large extent on the link beteen NO_X (nitrogen oxides) emissions and fuel efficiency. From Table 2, it is evident that the focus of emission reduction requirements by the EC has been on reducing NO_X.

The relationship between NO_X and fuel consumption is shown in Figure 3. The emissions of turbo diesel engines depend on the timing of the injection. When the injection is retarded, the emissions of NO_X (g/kWh) will be reduced but the fuel consumption will increase. The challenge in most engineering departments of truck manufacturers is to anticipate the stricter Euro III standard while keeping fuel consumption stable. For SCANIA, it is expected that fuel consumption will increase by 0.5–1 litre per 100 kilometres as a result of the Euro III standard. For other manufacturers that are less advanced in the reduction in NO_X, the increase in fuel consumption is expected to be up to 5 litres.

As a result of this trade-off, we cannot expect the average fuel efficiency of the truck fleet to be significantly reduced over a short period of time. It will take more time. Furthermore, it is not speculative to state that the firms with the cleanest engines will have a significant advantage over their competitors who invested later in the development of clean engines. This difference could lead to lobbying activities to influence the implementation of new standards.

7.4. *Results of the truck model*

In our simulation we start with a truck fleet of 570 000 units. Over a period of 15 years, the fleet increases with 12% to 638 883 trucks. We first work with a production price of US$65 000 for a new truck (average fuel consumption of 37.5 l/100 km) and US$70 000 for an ecotruck (average fuel consumption of 32.5 l/100 km). The baseline for fuel efficiency was 45 l/100 km. In later runs, we increased the production prices and changed the efficiency baseline.

In the scenario without the tradable certificate, we found that CO_2 emissions will increase by 6%. We then introduced the certificate at three permit price levels (ppr in US$): 20–40, 40–80 and 100–150. The results are presented in Figure 4.

It is shown that with the low value of the permits ([US$20, US$40]), the certificate will not have a strong impact on emissions: emissions will increase by only 3% compared to 6% in the business-as-usual scenario (BAU). If the permit price ranges between US$40 and US$80, we found that CO_2 emissions will be reduced by 3–6%. With the high permit prices, the reductions are significant: 12%. As indicated in Box 3, the level of the baseline is important in determining the number of certificates a manufacturer will receive. We therefore ran the truck simulation with baselines from 35 to 50 l/100 km. The results – best and worst simulation output for the specific baseline – are presented in Table 3. As already suggested, too low and too high levels for the baseline reduce the attractiveness and environmental impact of the tradable certificates. The best baseline in the emission trading design is a fuel efficiency that slightly exceeds the fuel efficiency of the other new vehicles sold on the market. For the truck market, the differences resulting from the changes in the baseline are not that spectacular. From Table 3, it is shown that we reach the best results with a baseline of 40 l/100 km.

For all the runs used for Table 3, the price of the permits was selected at random from the interval [US$20, US$100]. The percentage reductions are always calculated compared to the start of the model.

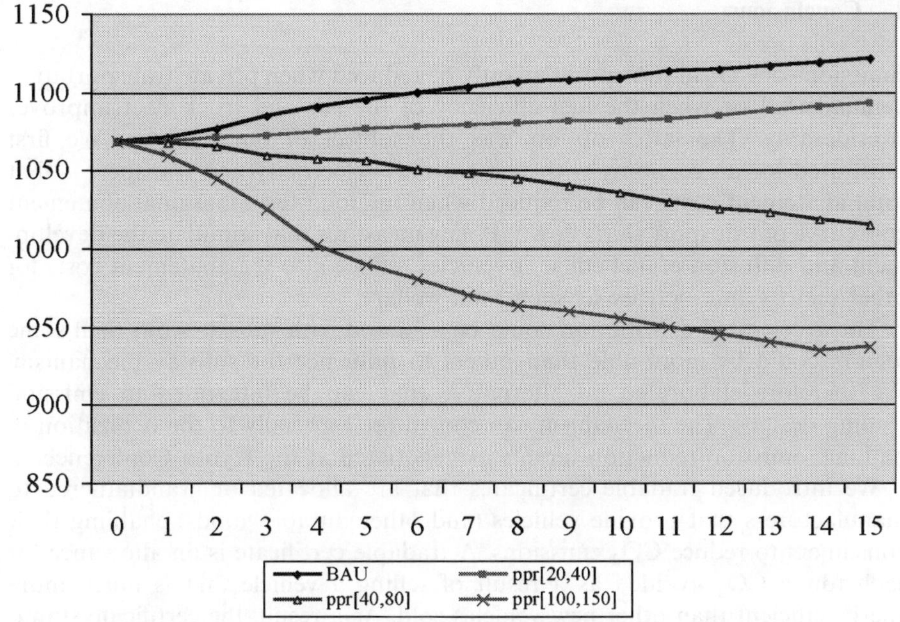

Figure 4. CO_2 emissions of the truck fleet (million tonnes, 15 years).

Table 3. Fuel efficiency baselines for trucks and CO_2 emissions

Fuel efficiency baseline for the certificate	Change of CO_2 emissions *worst case (% change)*	Change of CO_2 emissions *best case (% change)*
35 l/100 km	+2	−7
37.5 l/100 km	+1	−5
40 l/100 km	+0.5	−8
45 l/100 km	+1	−6
50 l/100 km	+1.5	−5

Finally, we present some results where the average production costs in the truck model were increased from US$65 000–US$70 000 to US$80 000–US$85 000 and to US$90 000–US$95 000. The price difference of US$5000 is relatively smaller for expensive trucks. The output of the model is similar for the two cases with higher prices. The reduction in fuel costs and the value of the certificates will in both cases lead to an earlier shift to ecotrucks if the certificate is introduced. We found that with prices for trucks between US$80 000 and US$85 000 and permit prices taken from [US$20, US$100], CO_2 reductions will decrease by 4–10% over 15 years. With production prices of US$90 000 and US$95 000, the projected reductions range between 3 and 9%.

8. Conclusions

Emissions of CO_2 in transport can only be reduced when private transportation demands fall or when the fuel efficiency of the car and truck fleet improves considerably. The latter option was the subject of our analysis. We first illustrated for an economy with two sectors – electricity and transport – that total abatement costs can be reduced when the long-term marginal abatement cost curve of transport shifts down. Policy measures that stimulate the development and diffusion of fuel efficient vehicles reduce also the abatement costs for other sectors and increase as such total welfare.

The accelerated distribution could be achieved with subsidies but then some sectors could be more able than others to influence the subsidy mechanism. We therefore elaborated an alternative that can be integrated in emission trading designs. The mechanism can contribute especially to the realization of national emission reduction targets as negotiated at the Kyoto Conference.

We introduced tradable certificates that are allocated or grandfathered to manufacturers that provide vehicles (and other durable goods) enabling their consumers to reduce CO_2 emissions. A tradable certificate is an allowance for each tonne CO_2 avoided as a result of selling a vehicle that is much more energy efficient than other new vehicles sold. As a result, the certificate stimulates cross-sectoral energy efficiency investments. Starting from reliable data on CO_2 emissions during the production and the use of cars and trucks, we found that opportunities for reducing emissions are impressive in the consumption phase. Emissions during the production phase are almost negligible compared to emissions during the use of the vehicles. These results indicate the opportunities for the certificate.

When the newly introduced allowances are taken from the existing pool of grandfathered emissions, total emissions do remain constant but the total abatement costs could fall, especially when other national emission reductions are expensive. This reduction of the average abatement cost could lead to lower levels of other grandfathered emissions and hence lower total emissions.

Manufacturers can sell the certificates on the emission market and use the revenues to lower the price of their cleanest vehicles. This mechanism should partially overcome the price difference with less efficient cars. This linking of the value of these certificates to the distribution of the cleanest vehicles has been used in two simulations that are based on three stock variables (the existing fleet, new ecovehicles and other new vehicles).

In the simulations, all new cars have a comparable quality: the only difference is fuel efficiency. We found that the introduction of the certificate in tradable permit systems can lead to very significant reductions in CO_2 emissions. Next to the price implications of the certificate, the accelerated scrapping of old and energy-inefficient cars is very important in this process. We found that emissions resulting from the car fleet will be reduced by 25–38% over a period of 15 years (starting in 1999). For the truck fleet, the reduction potential is more limited but still very interesting. However, if the value of the certificate is high enough, emissions resulting from the truck fleet could be reduced by 12% over the same period.

The integration of the certificates in emission trading design can contribute to the realization of national emission reduction targets as established in the Kyoto Protocol. Although unlimited international emission trading will lead to the lowest total climate protection costs, many developed countries are seeking low-cost national reductions strategies. Cross-sectoral energy efficiency strategies can be part of these strategies and they will also reduce total abatement costs for other sectors in the economy.

Notes

* Funded by the OSTC (Belgian Federal Office for Scientific, Technological and Cultural Affairs, Programme on Sustainable Development). The views presented are those of the author.
I want to thank Bart Ameels, Niko Gobbin, Freddy Heylen and Tom Verbeke for their helpful comments. Remaining errors are mine.

1. In this paper, all assumptions on fuel efficiency and lifetime or mileage of trucks are based on interviews with experienced maintenance engineers at SCANIA Belgium. We would especially like to thank Mr. Roger Lauwers. Data from Volvo (1998) and interviews at four transportation firms confirm our data.

References

Albrecht, J. (1998). Environmental consumer subsidies and potential reductions of CO_2 emissions. Working Paper Nr. 98/59 Universiteit Gent, Faculteit Economie en Bedrijfskunde.

Bohm, P. (1998). Benefits of international carbon emissions trading: theory and experimental evidence. Stockholm University, Department of Economics, mimeo.

Bürgenmeier, Beat, Harayama, Yuko and Wallart, Nicolas (1997). Théorie et pratique des taxes environnementales. Economica, Paris.

Cobb, Clifford (1999). The roads aren't free. *Challenge*. 42(3), 63–83.

Electrabel (1999). CO_2–Project Electrabel en SPE. Vergadering Synthese en Communicatie. 18 January 1999. Electrabel, Brussels.

Environmental Defense Fund (1998). Credit for voluntary early action act. Submitted to Congress. (<http://www.edf.org/issues/newbill.html>, visited 19/11/1998).

European Commission (1997a). Tax provisions with a potential impact on environmental protection. Office for Official Publications of the European Communities, Luxembourg.

European Commission (1997b). Climate Change COM(97)481. European Commission, Brussels.

European Commission (1998a). Domestic tradable quotas as an instrument to reduce carbon dioxide emissions. Proceedings of the domestic tradable quotas workshop. Brussels. 1–2 July 1998.

European Commission (1998b). Press Release, ip/98/234.

Febiac (1999). Boekjaar 1998. Verslag ter gelegenheid van de Algemene Vergadering van 23 juni 1999. Febiac. Brussels.

Hinrichs, D. (1998). Green cars: start your engines. Enviroworld. February 1998. (<http://enviroworld.com/February98/031298.HTML>, visited 20/6/98).

Leslie, J. (1997). Dawn of the hydrogen age. *Wired* October 5, 1997.

Lovins A., Brylawski, M., Cramer, D. and Moore, T. (1996). Hypercars: materials, manufacturing and policy implications. The Hypercar Center, Rocky Mountain Institute, Colorado.

Mullins, F. and Baron, R. (1997). International GHG emission trading. Annex I expert group on the UNFCCC. Policies and Measures for Common Action Working Paper 9.

Nylin, S. (1991). Emission regulations today and tomorrow. In: SCANIA (1991). The diesel engine and the environment.

Public Citizen (1990). Driving up the heat: a buyer's guide to automobiles, fuel efficiency and global warming. Washington DC: Public Citizen.

Rocky Mountains Institute (1998). A typical 1995 U.S. family car. (<http://www.rmi;org>).

Samuelsson, H. (1998). Scania launches low-emission engines for city environments. Scania press release, 1 September 1998. <http://www.scania.com/news/archive/>.

SCANIA (1998). Efficiency of diesel engines. Slides provided by SCANIA Training Belgium.

Unipede-Eurelectric (1999). Simulation exercise for trading in electricity and CO_2 emissions permits: the players draw conclusions. Press Release 20/10/99 (available at <http://www.eurelectric.org>).

Volvo (1998a). Environmental data for Volvo production plants, 1997.

Volvo (1998b). Environmental report 1997.

Von Weizsäcker, E., Lovins, A. and Lovins L. (1997). Factor four. Doubling wealth – halving resource use. The new report to the Club of Rome. London: Earthscan Publication Ltd.

CHAPTER 8

Early crediting of emissions reductions – a panacea or Pandora's box?

Axel Michaelowa and Marcus Stronzik

1. Introduction

After long international negotiations, the Kyoto Protocol of 1997 – if it enters into force – sets legally binding emissions targets for a basket of six greenhouse gases. These targets apply to most OECD countries and countries with economies in transition. A novel feature is the use of a commitment period that runs from 2008 to 2012 instead of a single target year. Moreover, the Protocol allows for the use of so-called flexible mechanisms: emissions trading (Art. 17), Joint Implementation (JI) (Art. 6) and projects of the 'Clean Development Mechanism' (CDM) with countries without emissions targets (Art. 12). All these instruments must be 'supplemental' to domestic measures.[1] Supplementarity has not been defined in the Protocol, however. If average emissions in the commitment period are lower than the emissions target, the difference can be banked for the next commitment period. In the case of higher emissions the country will be in non-compliance.

Since the Kyoto Conference a growing number of interest groups and politicians have criticized the fact that pre-2008 emissions reductions are not accounted for in the Protocol. They argue for ways of crediting such reductions either nationally or internationally. This is called 'early crediting'. One mechanism of the Kyoto Protocol already allows early crediting: emissions credits from the CDM will accrue from 2000, not only for the commitment period 2008–2012.

After an explanation of three different possibilities for crediting early action – two on the international and one on the national level – we will consider the macroeconomic as well as the microeconomic implications of these concepts. The main focus will be on the domestic case.

2. Ways of early crediting

As the Kyoto emissions targets apply only from the year 2008 on, there are no strong incentives for countries to introduce regulations that would cover the time before the commitment period begins. This is due to the competitive disadvantage such unilateral regulation is thought to bring. In such a case,

C. Carraro (ed.), Efficiency and Equity of Climate Change Policy, 185–197.

national actors, e.g. companies, have no direct incentives to reduce their emissions before 2008. However, even in the absence of a binding regulation before 2008, if companies face emissions reduction objectives from 2008 on they will anticipate their future commitments in current investment decisions. Therefore, companies will adjust their emissions according to a range of parameters, such as abatement costs, time preference, etc. The resulting optimal emissions path will likely deviate from the business-as-usual one. However, this inherent incentive for early action in an ideal world is counteracted by two effects in reality:

- It is by no means certain that the Kyoto Protocol will enter into force. Therefore, companies will stay on the business-as-usual path until they have more certainty about implementation of Kyoto.
- Early movers could be disadvantaged by the distribution of reduction commitments at a later date, e.g. if the allocation of allowances will be fixed on the basis of emissions in some future year.

To encourage reductions prior to the first commitment period without risking a competitive disadvantage one could think of granting credits which can be used against future obligations. Within the Kyoto framework three possible starting-points can be envisaged. As mentioned above, the CDM already allows early crediting. There exist two further ways: domestically and via JI. While the former would not violate the Protocol, the latter would be contrary to the Protocol in its current form but is also discussed here. At first, a macroeconomic perspective is taken and no cost aspects will be considered.

CDM

Certified emissions reductions (CERs) from CDM projects can be banked by the project participants for use in the commitment period.

CDM projects create certified reductions from the business-as-usual emissions path of the non-Annex B country (area A in Figure 1) that enhance the Kyoto budget of the investing country if all credits accrue to the latter. Any form of credit sharing ($b < 100\%$) will reduce the budget increment (Dutschke/Michaelowa 1998).

A general formula in the case of linear business-as-usual growth would be:

$$y = bx\frac{z}{5}$$

with:

y: enhancement of the Kyoto budget (GHG tonnes)
b: share of credits accruing to the investing country
x: reduction in business-as-usual growth (GHG tonnes per year)
z: number of years in which the CDM projects are active

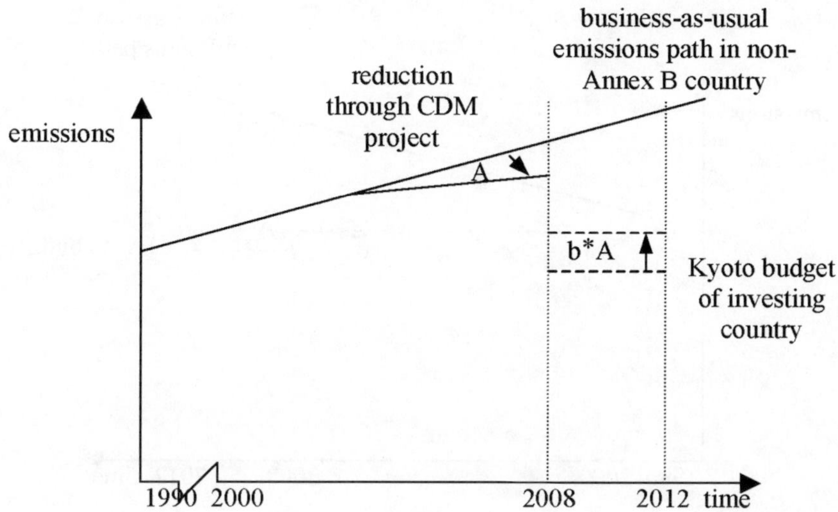

Figure 1. CDM early crediting.

Parkinson *et al.* (1999) have criticized early crediting in the CDM context as this would lead to an implicit overvaluation of the CDM projects. This, however, is exactly the intent as CDM projects have to bear an adaptation and administration tax that shall be offset by the early crediting rule.

Joint implementation

Early crediting through projects between Annex I countries would be quite similar to the just-introduced case of CDM. Investor countries would also profit from early credit JI as it enhances their options (see Figure 2; Center for Clean Air Policy 1998).

The difference exists in how the host country will be treated as it faces – in contrast to CDM – a reduction target. If early JI credits are handled in the same way as regular JI credits host countries will be obliged to subtract them from their budget (see Figure 3). Otherwise host countries would have an incentive to maximize early JI that is not additional and thus the Annex B budget would be inflated. Therefore, without a corresponding subtraction, early JI would be similar to the CDM and face the same problem of additionality (Michaelowa 1998).

If JI leads to a reduction from business-as-usual by the amount A, the same amount has to be deducted from the Kyoto budget to avoid non-compliance. The same calculations apply as above.

The situation does not change in principle if the host country has 'hot air', i.e. its baseline is beneath the Kyoto budget (see Figure 4).

In this case, one is led to think that JI reduces the amount of 'hot air' by

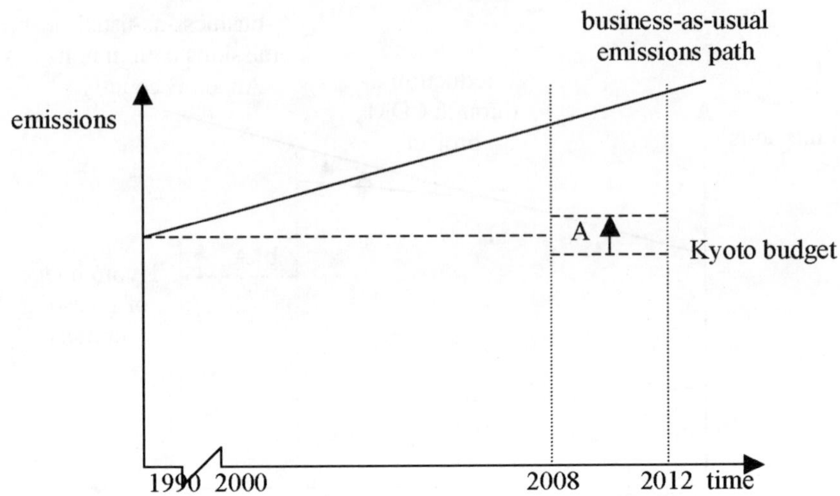

Figure 2. Early JI credits in the investor country.

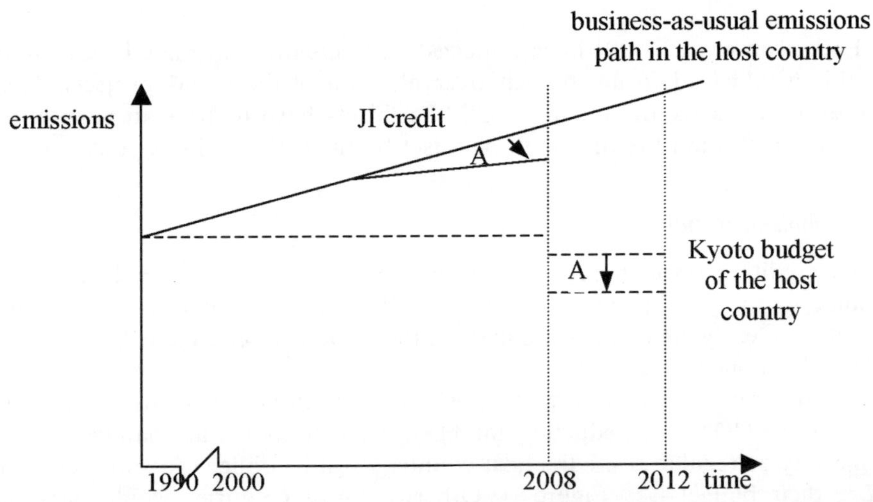

Figure 3. JI early crediting in the host country.

the credited emissions reduction. This was done by the Swiss delegation at the fourth Conference of the Parties in Buenos Aires which circulated a 'non-paper' that argued for early crediting of JI for exactly this reason (Switzerland 1998). This argument holds generally if the baselines for the JI projects are set 'correctly' without cheating. If, however, baseline control for JI is lax (and there are indications that this might be so due to the fact that both partners have an emissions budget) JI in general and early JI in particular could lead to a 'laundering' of 'hot air'. Moreover, if only domestic efforts are considered to

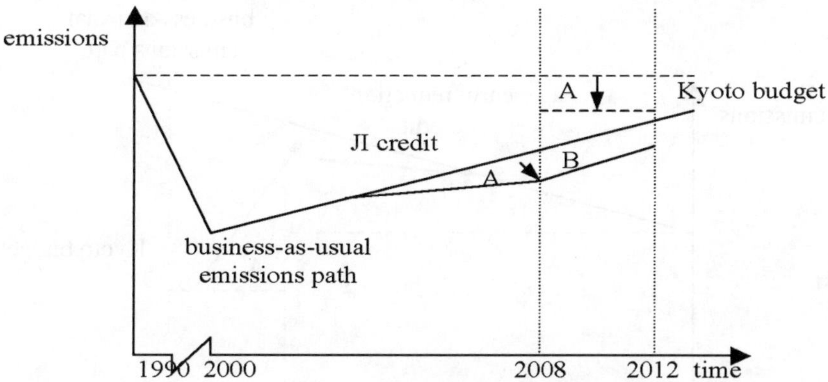

Figure 4. JI early crediting in the host country with 'hot air'.

determine the amount of 'hot air' – which in our view is more appropriate – it could be defined as the amount of allocated permits exceeding the initially forecast emissions for the commitment period. Then, the business-as-usual emissions path would have to be adjusted due to JI activities since JI should be classified as a non-domestic effort (area B in Figure 4) – JI is carried out in the host country but financed by the investor country.

With this definition in mind, a reduction in 'hot air' through early JI would require A > B in Figure 4. This depends on baseline setting, induced change in the future emissions path and especially on the start of the programme. Therefore, we would argue that the amount of 'hot air' is likely to increase (A < B) if early JI is truly additional as the business-as-usual path is shifted downwards. Furthermore, if the programme starts before the year 2003 it runs for longer than the first commitment period, resulting in a longer period of time where early credits can be accumulated.

Only in the case of non-additional JI does the amount of 'hot air' – taking our definition into consideration – seem to fall but this is only due to a 'laundering' of the 'hot air'. Non-additional JI would not be attractive as it would reduce the saleable quantity of 'hot air' by the same amount and presumably entail lower transfers.

Notwithstanding the design of such a programme or whether the amount of 'hot air' will decline or not, as long as there is a subtraction from the host country's target, the total Annex B budget remains unchanged.

Domestically

A domestic system of early crediting would allow actors to create credits by emission reduction or sequestration compared to a baseline (see Figure 5). These credits could be used by the actor to offset future emissions control obligations and would be guaranteed by a proportional set-aside of the

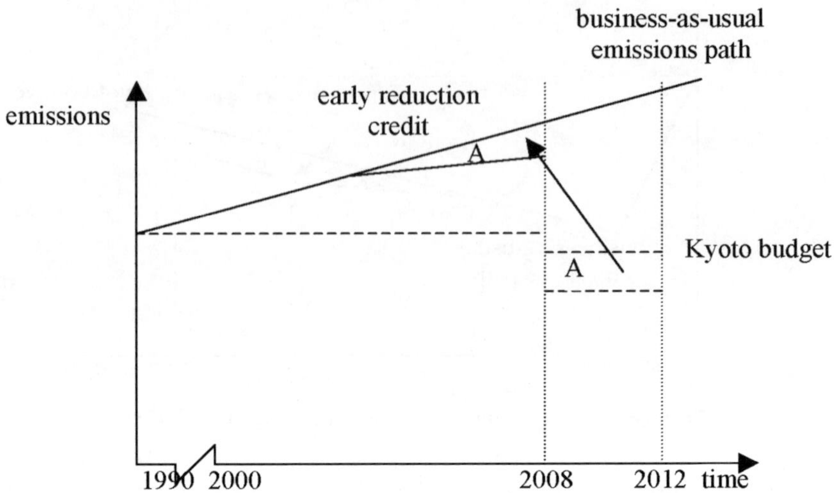

Figure 5. Domestic early crediting.

country's emissions budget for 2008–2012. Otherwise, the country would not be in compliance unless it bought additional permits from other countries to cover the resulting shortfall.

If an early-crediting programme leads to a reduction from business-as-usual by the amount A, the same amount has to be deducted from the Kyoto budget to avoid non-compliance. The reduction in the budget can be quite substantial. If, for example business-as-usual growth is 27% from 1990 to 2008 (which is in the order of magnitude of the forecasts of US emissions), and early action is credited from 2000, the following values would result:

- If early action reduced business-as-usual growth by 0.5 percentage points the Kyoto target would be strengthened by 0.8 percentage points (0.5 multiplied by eight years of early crediting divided by five years of the budget period)
- In the case of a reduction of business-as-usual-growth by 2 percentage points the Kyoto target would be strengthened by 3.2 percentage points.

A general formula in the case of linear business-as-usual growth would be:

$$y = x\frac{z}{5}$$

with:

y: strengthening of the Kyoto target (percentage points)
x: reduction in business-as-usual growth (percentage points per year)
z: number of years in which early crediting applies

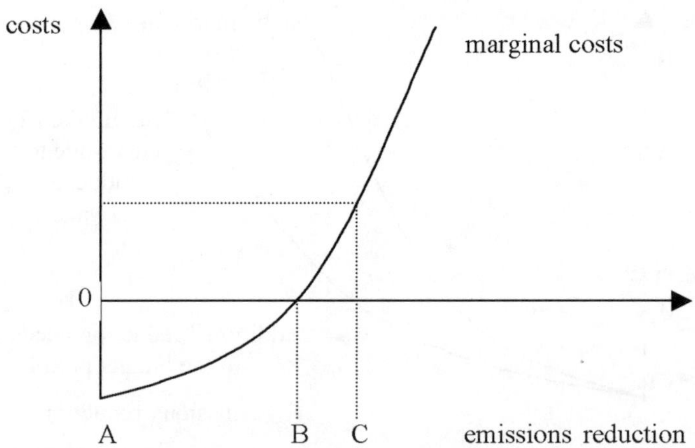

Figure 6. Early crediting and macroeconomic abatement cost before the budget period.

3. Macroeconomic implications of early crediting

Proponents of early crediting argue that it would smoothen emissions reduction requirements as rigid sectors would already have an incentive to use replacement of capital stock to install low-emissions equipment prior to the first commitment period. Without early crediting the actors would follow the business-as-usual scenario because of a lack in reduction incentives with a corresponding growth in emissions. Assuming the Kyoto Protocol will be ratified the country concerned faces binding emissions targets from the year 2008 on. This would impose an abrupt reduction obligation with potentially serious adverse impacts on the economy. Moreover, in case of early crediting, the implicit price of carbon – if actors go beyond no-regret options – would be above zero before the budget period and thus induce innovation.

Are these arguments sufficient to pave the way for early crediting? We discuss them in the framework of macroeconomic costs (see Figure 6).

Point A is chosen under business-as-usual. Point B will be attained if barriers to no-regret measures are removed. Point C will be reached if early crediting is allowed as companies hedge for the budget period. Note that even in the absence of domestic early crediting, companies would invest in CDM projects if the advantages exceeded the costs.

The evaluation of early crediting depends crucially on its influence on innovation (see Figure 7). If there is no innovation, early crediting would lead to a rise of marginal cost in the budget period from point A to B due to the additional reduction necessary. If induced innovation leads to a shift in the cost curve, point C would be reached and thus marginal costs could fall depending on the magnitude of the shift.

Thus, it is critical whether early crediting leads

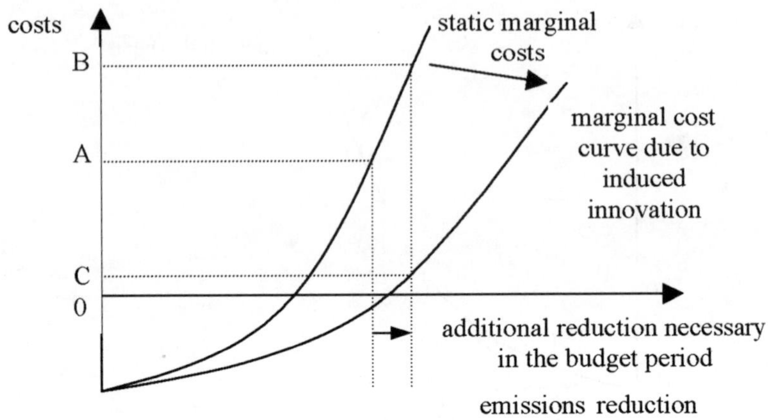

Figure 7. Costs during the budget period.

- to a utilization of the no-regret potential or
- to a shift of the marginal cost curve due to induced innovation

If neither happens, early crediting will lead to a rise in abatement cost!

Domestic early crediting ceteris paribus would lead to a lower demand for CDM reduction prior to the budget period. In the budget period, demand for CDM and JI would rise if the cost curve had not shifted sufficiently to lower marginal abatement costs. In the other case, it would still fall.

4. Microeconomic implications of domestic early crediting

As we have seen in Section 2, early crediting through JI or CDM is quite similar to the regular use of these instruments. Therefore, only domestic early crediting will be discussed in greater detail.

Domestic redistribution

To simplify the analysis we discuss an economy consisting of two companies *X* and *Y*. *X* participates in domestic early crediting, *Y* does not (see Figure 8). The Kyoto budget is translated into caps for the companies during the budget period.

Company *X* gets credits for early mitigation at the amount of A. This amount has to be subtracted from the overall country budget in the first commitment period. Therefore, the initial allocations of permits to companies have to be tightened by a certain amount summing up to A. The distribution of this additional reduction is an issue of equity. One possible way would be a reduction by a same percentage for each actor determined by the share of A in the overall budget. For company *X* this would result in a rise in emissions

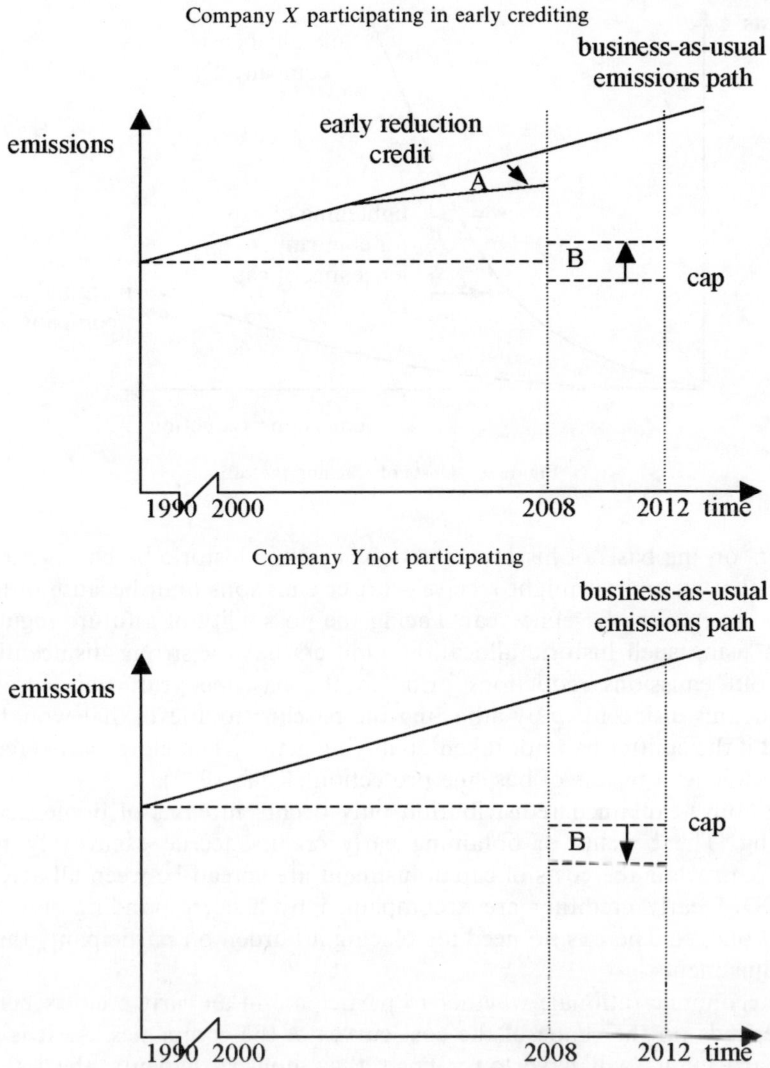

Figure 8. Redistributional consequences of early crediting.

rights by B (A minus cap adjustment). Since only two companies are considered, *Y* faces a corresponding reduction by B (cap adjustment). Thus *Y* will find it more difficult to reach the cap. Early action leads to a redistribution from company *Y* to *X*. Obviously, in a real economy the number of participating companies will be much lower than that of non-participants, thus diluting the effect.

Such a design helps to overcome one major disadvantage of common emissions trading schemes. These schemes usually allocate emissions permits to

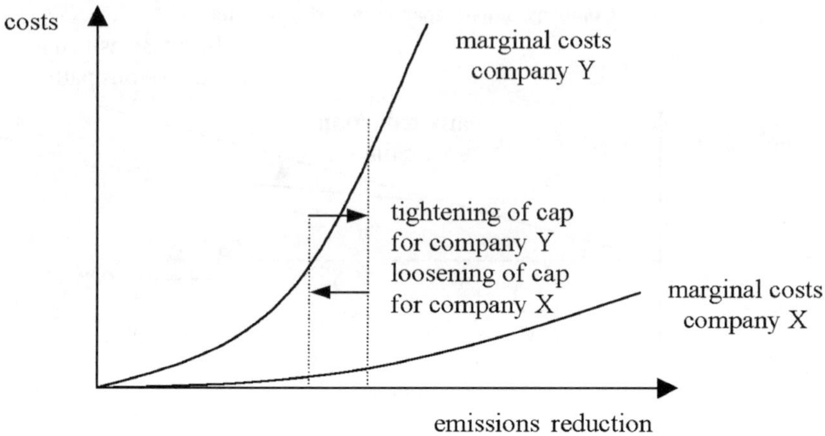

Figure 9. Costs of reaching the cap.

emitters on the basis of historic emissions in some historic baseline year. This means that an emitter might receive a lower emissions limit because of taking action prior to the baseline year. Facing the possibility of a future regulatory system using such historic allocation, emitters have a strong disincentive to carry out emissions reductions prior to the baseline year. Early crediting removes this disincentive by adjusting the baseline to a level that would have existed if the emitter had not taken qualifying action. Therefore, early crediting could serve as a means of baseline protection (Rolfe 1999).

The above-explained redistribution only occurs in cases of domestic early crediting. The benefits of obtaining early credits accrue exclusively to the participant while the costs of cap adjustment are spread between all actors. JI and CDM early crediting are accompanied by a corresponding rise in the overall budget. There is no need for placing a burden on participants through cap adjustments.

The economic rationale whether to participate in an early credit scheme or not depends on the shape of the cost curves of the companies. As it is likely that participants will have lower costs than non-participants, the following situation will occur in a static setting (see Figure 9).

It is obvious that the difference in marginal costs widens through early crediting. If early crediting leads to a shift of the cost curve of company X due to induced innovation the difference even widens further.

Redistribution could also become an issue of interest if an early crediting regime is implemented within a bubble according to Art. 4 of the Kyoto Protocol. Since such a scheme could be seen as being similar to the domestic case (overall budget for the bubble) this could lead (within the EU) to a new round of negotiations of burden sharing. Member states that do not want to participate in early crediting would certainly argue that their share of the

Kyoto budget would not be reduced due to the early crediting in other member states.

Baseline considerations

Many proponents of early crediting state that it might lower the resistance of interest groups against the ratification of the Kyoto Protocol. This can be especially seen in the US where a staunch opponent of climate policy (Senator Mack) is a sponsor of the early credit bill introduced in the Senate in March 1999.[2] His line of argument was that one never knew whether a future Congress might introduce emissions caps so it would be important to allow companies to hedge against this eventuality – that would hurt nobody. This strongly suggests that companies using early crediting should incur no costs, i.e. use only no-regret measures. Thus baselines would have to be chosen in a lax way.

This assumption is confirmed by the design of the early credit bill (US Senate 1999). The proposal allows crediting not only for future domestic emissions reductions but also for past actions identified under §1605(b) of Energy Policy Act of 1992 (EPAct) as well as for projects under the US Initiative on Joint Implementation (USIJI).

EPAct §1605(b) is part of the Climate Wise Program of the Clinton Administration introduced in 1993. Companies can voluntarily report mitigation activities which will then be registered. However, the guidelines are extremely flexible. No verification process exists and in some cases the same reductions are reported by different entities (double counting) (Nordhaus *et al.* 1998, p. 14).

Furthermore, reductions achieved abroad are taken into account, although the programme is designed as a domestic scheme. USIJI is the institutional frame for US activities under the JI pilot phase, the so-called Activities Implemented Jointly (AIJ). As AIJ serves only as a means to collect experiences with the project-based instrument there is a lack of sufficient verification. Reported reductions cannot be used to fulfil the international obligations imposed through the Kyoto Protocol. Therefore, the overall country budget remains unchanged while the individual cap will be increased.

Both options, EPAct §1605(b) and USIJI, put an unnecessary burden on non-participants.

But what are the effects of lax early crediting? It would widen the cost differentials shown in Figure 9. Free riders would capture rents while the other economic actors would face higher costs (Palmisano 1999). This might – paradoxically – lead to higher political pressures in the budget period not to comply.

For these reasons, the bill was criticized quite harshly by the US-based think-tank Resources for the Future (RFF). Kopp *et al.* (1999, p. 1) conclude that it involves the risk of 'distributing too many credits for questionable early reductions'.

Leakage

An early crediting programme will certainly cover only a subset of actors. If a company owns sources participating in the programme as well as sources staying outside, displacing emissions from inside to sources not covered by the programme could be a valuable option. An energy supplier, for example, could easily reduce the emissions of a participating power plant – and be credited – through dispatching electricity to a non-participating plant. If the latter is less energy-efficient or uses a more carbon-intensive fuel the net environmental effect would be negative.

This leakage problem exactly arose in the case of the US Acid Rain Program (ARP). Since it is designed to be implemented in two phases, some energy suppliers used the option of reduced utilization to cut back sulphur emissions of sources already regulated in the first phase.

Whether leakage will turn out to become a big issue depends crucially on coverage of relevant actors as well as on the substitution options of a single company. In the case of the ARP the problem was dealt with through an adjustment in programme design. Sources used for the displacement of emissions were subsequently included in the first phase. However, this raised development and implementation costs of the programme quite substantially since new calculations had to be made not only for the number of allocated permits, but also for the impact on air quality (McLean 1996). Moreover, quite a large number of excess allowances were created (Rico 1995), similar to the phenomenon of 'hot air' in climate negotiations.

It should be noticed that leakage is not a unique issue of domestic early crediting. Generally, it arises if a programme does not cover all potential actors. Therefore, leakage is also an issue of concern within the debate of the design of an international regime of flexible mechanisms (Botteon/Carraro 1997).

5. Conclusions

Early crediting has been hailed by many to be a perfect solution to the fact that Kyoto targets are only binding from 2008. It indeed offers the possibility of reducing compliance costs as capital turnover can be used to install more greenhouse-gas-efficient options. We find a number of risks in the concepts of early action, however, especially in the domestic case, that might lead to heavy distortions and even higher abatement costs. Early crediting clearly has a redistributionary implication. The following safeguards are necessary but not sufficient to guarantee an advantageous system of early crediting:

- Strict baselines or proof of barrier removal
- Incentives for innovation (e.g. discounting of low-tech projects)
- Incentives for rigid sectors to install climate-friendly capital replacement.

However, there exists one advantage of an early crediting scheme which could turn out to become a major benefit: it builds up a framework in which practical experience with the flexible mechanisms can be collected. Since experience with market-based instruments is quite low, such a regime would enable institutional learning with lowering transaction costs.

Notes

1. It should be mentioned that in the case of CDM the wording of the Kyoto Protocol is slightly different.
2. The bill was introduced by Democratic Senator Joseph I. Lieberman together with the two Republican Senators John Chafee and Connie Mack, supported by a further nine Senators as 'Credit for Voluntary Reductions Act' (S.547). It is a slightly modified version of the bill introduced in October 1998 as 105th Congress' 'Credit for Voluntary Early Action Act' (S.2617).

References

Botteon, Michele and Carraro, Carlo (1997). Environmental coalitions with heterogeneous countries: burden-sharing and carbon leakage. Working Paper, Venice.

Center for Clean Air Policy (1998). JI for credit now: establishing early joint implementation programs. Washington, DC.

Dutschke, Michael and Michaelowa, Axel (1998). Creation and sharing of credits through the Clean Development Mechanism under the Kyoto Protocol. HWWA Discussion Paper No. 62, Hamburg.

Kopp, R., Morgenstern, R., Pizer, W. and Toman, M. (1999). A proposal for credible early action in US climate policy. Resources for the Future. Discussion Paper. http://www.weathervane.rff.org/features/feature060.html

McLean, Brian J. (1996). Evolution of marketable permits: the US experience with sulphur dioxide allowance trading. Unpublished manuscript. Washington, DC. Acid Rain Division, US Environmental Protection Agency.

Michaelowa, Axel (1998). Joint Implementation – the baseline issue. Global Environmental Change. 8(1): 81–92.

Nordhaus, R.R., Fotis, S.C. and Van Ness Feldman, P.C. (1998). Early action & global climate change – an analysis of early action crediting proposals. Pew Center on Global Climate Change, Arlington.

Palmisano, John (1999). What are the economic and environmental benefits from early 'Crediting'? Washington, DC.

Parkinson, S., Begg, K., Bailey, P. and Jackson, T. (1999). JI/CDM crediting under the Kyoto Protocol: does 'Interim Period Banking' help or hinder GHG emissions reduction? In: Energy Policy. (Forthcoming).

Rico, Renee (1995). The US allowance trading system for sulfur dioxide: an update on market experience. Environmental and Resource Economics. 5(2):115–29.

Rolfe, Chris (1999). Early crediting and baseline protection. Discussion Paper (unpublished). West Coast Environmental Law, March 1999.

Switzerland (1998). Initial ideas on pre-2008 joint implementation. 'Non-paper' for COP4. Buenos Aires.

US Senate (1999). Credit for Voluntary Reductions Act (S.547). http://thomas.loc.gov/cgi-bin/query

CHAPTER 9

Energy Market Projections and Differentiated Carbon Abatement in the European Union*

Christoph Böhringer, Jesper Jensen and Thomas F. Rutherford

1. Introduction

The Kyoto Protocol documents what many believed to be impossible: negotiators for many of the world's countries agreed on emissions targets for the early part of the 21st century. The protocol commits the Annex I countries to reducing their aggregate CO_2 equivalent emissions by at least 5% below 1990 levels in the period 2008–2012.[1] The agreement reflects wide differences in the willingness to accept emissions constraints: developing countries refused any commitment while the European Union (EU) at the other end agreed to an 8% overall reduction. The burden-sharing negotiations subsequently continued at the EU level when the member states distributed the EU constraint internally. Table 1 shows that the country-specific targets within the EU bubble range from a 28% *reduction* for Luxembourg to a 27% *increase* for Portugal.

Does the EU bubble imply a 'fair' distribution of the burden across EU member states? Economic analysis may inform this debate by translating the emissions constraints into estimates for welfare costs. The cost estimate for a given country will obviously depend on its current economic structure, such as the efficiency of the installed energy-consuming technologies, the fuel mix, and trade relations with other countries. Differences in the starting-point for each country may imply that the same proportional cutback results in high costs in one country and low costs in another.

The expected distribution of costs also depends on differences in the future economic development across countries. That is, the cost estimates will depend on business-as-usual (BAU) projections for GDP, populations, energy efficiency improvements, fuel prices, etc. High economic growth, for example, leads by itself to high energy demands and emissions. This would increase the effective abatement requirement as the Kyoto targets refer to 1990 emissions levels, and higher economic growth will therefore imply higher total abatement costs.

We analyse the agreement on differentiated CO_2 abatement within the EU using the Conventional Wisdom (CW) scenario in European Commission

C. Carraro (ed.), Efficiency and Equity of Climate Change Policy, 199–219.

Table 1. EU member state commitments

Member States	Commitments in accordance with the Kyoto Protocol
Austria	−13%
Belgium	−7.5%
Denmark	−21%
Finland	0%
France	0%
Germany	−21%
Greece	+25%
Ireland	+13%
Italy	−6.5%
Luxembourg	−28%
Netherlands	−6%
Portugal	+27%
Spain	+15%
Sweden	+4%
UK	−12.5%
EU	−8%

Source: Appendix 1 in EU (1998).

(1996) as the BAU projection. We estimate both the total costs and the distribution of costs across EU member states and evaluate the main assumptions in the BAU projections. Specifically, we focus on the assumptions about non-uniform efficiency improvements and fuel shifting embodied in the CW baseline. Finally, we analyse the consequences of uniform abatement targets across the EU for total costs and the distribution of these costs.

The literature has surprisingly little to say about these issues. Most modellers are typically careful in specifying their BAU assumptions but they rarely report results from sensitivity analyses.

A couple of articles have reported results from the GEM-E3 model of analyses of carbon abatement in 11 EU member states. Conrad and Schmidt (1998a) find that a 10% reduction in CO_2 emissions from the EU implies an overall welfare loss of around 0.24% of GDP when each country has to reduce emissions by 10%. Country-specific welfare effects range from a 0.06% gain in Portugal to a 0.55% loss in Denmark. Permit prices for CO_2 range from US$10 per tonne in Greece to US$47 in Denmark (assuming that ECU1 = US$1). If emissions can be traded between countries, they estimate the permit price to be US$23 and the overall welfare loss decreases slightly to 0.21%. In a similar analysis, Conrad and Schmidt (1998b) find a double dividend, i.e., 10% lower carbon emissions *and* positive gross welfare effects from the policy, when they recycle permit revenues via a distortionary labour income tax.[2]

Unfortunately, several important differences in model characteristics make it hard to compare the results with the present analysis. First, and most important, none of the articles based on the GEM-E3 model include BAU projections for GDP, CO_2 emissions, energy prices, etc. Second, the GEM-E3 model includes both a labour–leisure choice and involuntary employment whereas we have a fixed labour supply. Finally, it assumes endogenous balances of trade and exogenous exchange rates. We assume that the balances of trade do not change and let the exchange rates clear the foreign exchange markets.

The present analysis makes three sets of contributions. First, to develop an explicit dynamic model, we extend an existing static, multi-sector model of six EU member states to incorporate representative agents with myopic expectations.[3] Second, we develop a calibration method that allows us to use the CW scenario as our baseline equilibrium. Finally, we use this recursively dynamic model to analyse the EU agreement on differentiated CO_2 abatement.

The following results emerge from our simulations. First, total costs in 2010 for the EU amount to a 0.4% consumption loss and a range of costs for the member states between zero and a loss of 0.7%. CO_2 tax rates vary from US$22 per tonne CO_2 in Spain to US$99 in Italy.[4]

Second, the CW baseline embodies significant fuel shifting in electricity production in all countries and high efficiency gains in the Northern European countries. These assumptions imply large decreases in CO_2 intensity and therefore low effective abatement requirements compared with our alternative baseline of uniform growth in energy efficiency of 1% per annum. With the latter baseline, total costs double and the range of costs changes from a low of 0.1% in Spain to a high of 2.2% in Denmark. The uniform baseline lowers the overall level of energy efficiency gains, and CO_2 tax rates therefore increase in all countries. The North experiences the largest decreases in energy efficiency and therefore also the largest increases in CO_2 tax rates.

Finally, we find that uniform abatement targets within the EU have a small impact on total costs, holding total EU abatement constant. The distribution of costs changes, however. Spain and France would suffer large economic losses as their abatement targets change from +15% and 0%, respectively, to −11%.[5] Uniform targets increase costs for Spain and France, while Denmark and Germany reduce their total abatement costs by around 50%. This scenario, although politically unrealistic, illustrates the size of the implicit transfers involved in the burden-sharing negotiations.

Section 2 discusses the CW baseline for future energy markets. Section 3 describes our model and the steps involved in calibrating it to the CW baseline. Section 4 defines our policy scenarios and Section 5 presents the results. Section 6 offers concluding remarks.

2. Baseline energy market projections

EU member states differ significantly with respect to CO_2 characteristics. Table 2 presents summary statistics revealing the main differences. It focuses on the six countries in our model and shows that the selected countries jointly

Table 2. Benchmark data for 1990

A. Summary statistics

	DE	DK	ES	FR	IT	UK	REU	EU
GDP (% of EU total)	24	2	6	19	15	16	18	100
CO_2 (% of EU total)	32	2	7	12	13	18	17	100
CO_2 (tonnes per capita)	13	11	5	7	7	10	8	9
Electricity (tonnes of CO_2/GWh)	653	892	429	105	565	686	260	462

B. CO_2 emissions sources (%)

	DE	DK	ES	FR	IT	UK	REU	EU
Electricity	36	43	31	12	30	38	23	30
Energy production	3	2	6	5	5	5	5	4
Industry	20	11	21	23	20	14	24	20
Transport	17	26	32	34	25	24	26	24
Final demand	24	17	10	26	20	19	22	21
Total	100	100	100	100	100	100	100	100

Regions

1.	DE	Germany
2.	DK	Denmark
3.	ES	Spain
4.	FR	France
5.	IT	Italy
6.	UK	United Kingdom

Note: The Rest of the European Union (REU) includes Austria, Belgium, Finland, Luxembourg, Greece, Ireland, the Netherlands, Portugal and Sweden.
Source: Calculations based on European Commission (1996; 161–209).

accounted for more than 80% of EU GDP and aggregate EU CO_2 emissions in 1990.[6] The sector contributing most to CO_2 emissions is electricity, whose fuel mix differs significantly across countries. A large share of nuclear-based electricity results in France having a very low CO_2 intensity (tonnes of CO_2 per GWh), whereas large shares of coal-based electricity implies higher intensities in Denmark and Germany. Finally, the rich Northern European countries also have the highest emissions per capita.

What happens to economic growth, energy prices, and CO_2 emissions without climate policy? Cost estimates for the Kyoto commitments crucially depend on answers to this question. CO_2 emissions are directly linked to the combustion of fossil fuels, so it follows that baseline projections of the development of the energy system play a key role in the design of the abatement policies. In particular, projections for energy efficiency improvements and the fuel mix in electricity generation are crucial for the reference level of CO_2 emissions.

Table 3. Conventional Wisdom baseline assumptions

	DE	DK	ES	FR	IT	UK	EU
A. Average GDP growth rate (% p.a.)							
1990–2000	2.8	2.6	2.1	1.9	1.8	1.9	2.1
2000–2010	2.3	2.0	2.7	2.5	2.0	2.1	2.3
B. Average CO_2 growth rate (% p.a.)							
1990–2000	−0.2	1.9	0.6	0.6	0.9	0.3	0.3
2000–2010	0.1	−0.8	1.4	0.1	0.5	0.2	0.5
C. Implied average improvement in CO_2 intensity (% p.a.)							
1990–2000	2.9	0.7	1.5	1.3	0.9	1.5	1.8
2000–2010	2.1	2.7	1.3	2.4	1.5	1.9	1.8
D. Average growth rates for fossil fuel inputs to electricity generation (% p.a.)							
1990–2000							
Coal	−0.8	1.6	−4.9	3.3	6.7	−4.2	−1.0
Oil	8.0	9.3	3.3	−13.1	−0.9	−6.3	−0.1
Gas	4.2	25.4	25.8	6.5	2.2	31.3	8.3
2000–2010							
Coal	−0.5	−3.4	2.1	−11.9	−0.8	−6.0	1.2
Oil	−1.5	−5.1	−1.2	−14.3	−5.4	3.4	−2.2
Gas	6.0	7.3	7.9	10.9	8.5	6.1	6.6

Source: Calculations based on European Commission (1996; 161–209).

The CW scenario in European Commission (1996) suggests how energy markets in Europe might evolve if current policies remain in place.[7] The assumptions of the CW scenario include smooth increases in world energy prices, no changes in current energy taxes, limited penetration of more efficient supply technologies, and no changes in energy market regulation. Renewable, CO_2 free, energy production increases its share in total energy supply but fossil fuels continue to be the main source of energy. Gas and oil consumption increase, whereas the use of nuclear fuels and coal slowly decreases.

Table 3 summarizes the CW baseline with respect to economic growth, the development of CO_2 emissions, and the fossil fuel mix in thermal electricity generation. France and the CO_2-intensive Northern European countries experience the largest decreases in the CO_2 intensity. Overall, the CW baseline

Table 4. Sectors in the model

Production sectors (Eurostat's R59 code in brackets):	
1.	Coal (031,033)
2.	Crude oil (071)
3.	Refined oil (073)
4.	Natural and manufactured gases (075,098)
5.	Electricity and steam (097,099)
6.	Agriculture (010)
7.	Iron and steel (135,136)
8.	Chemical products (170)
9.	Non-ferrous metals (137)
10.	Non-metallic minerals (151,153,155,157)
11.	Machinery (190,210,230,250)
12.	Transport equipment (270,290)
13.	Paper pulp and printing (471,473)
14.	Wood and wood products (450)
15.	Food processing beverages and tobacco (310,330,350,370,390)
16.	Textiles and leather (410,430)
17.	Transport (611,613,617,631,633,650)
18.	Other industries (095,490,510,530,550,570,590,670,690,710,730,750,770,790,810,850,890,930)

embodies a strong decoupling of growth in GDP and CO_2 emissions, as summarized by an average change in the CO_2-intensity of almost 2%. Fuel shifting and significant efficiency improvements account for this result. Section D in Table 3 focuses on the electricity sector where the emissions intensity decreases significantly due to fuel shifting from coal to gas.

3. Analytical framework and baseline calibration

This section presents the main characteristics of a recursively dynamic multi-sector model of the six EU countries. We also discuss the representation of the CW scenario as our baseline and the subsequent sensitivity analysis of the BAU assumptions.

The model is a dynamic extension of a previous static model designed to investigate the economic implications of alternative CO_2 abatement strategies for the European Union.[8] Table 4 gives an overview of the sectoral disaggregation of the current model for the six regions in Tables 2 and 3. The choice of sectors and regions captures key dimensions in the analysis of CO_2 abatement such as differences in CO_2 intensity across sectors, energy goods, and bilateral trade. Appendix A provides an algebraic documentation of the model.

We have developed an explicitly dynamic model to incorporate the time paths for GDP, CO_2 emissions, energy prices, etc., for the CW scenario. We assume constant, region-specific marginal propensities to save consistent with a growth rate of 2% per year in all countries. The path for the economy is represented by a set of connected equilibria where the current period's saving augments capital stocks in the next period. Consumers allocate income between present and future consumption subject to a fixed marginal propensity to save and investment adjusts passively to savings. The intra-period model corresponds to the previous static model. Base year data determine the parameters of the functional forms from a given set of quantities, prices and benchmark elasticities.

A simple calibration of the model along the time path would typically involve calibration to a steady state where all physical quantities (including CO_2 emissions) grow at an exogenous uniform rate while relative prices remain unchanged. The virtue of a steady-state baseline is that it provides a transparent reference path for the evaluation of policy interference: any structural change in the counterfactual can be attributed to the new policy.

In the present analysis we want to incorporate exogenous information from the CW baseline on non-uniform growth rates in GDP, fossil fuel production, fuel mixes in electricity generation, changes in world market energy prices and CO_2 emissions profiles. The calibration procedure involves two steps.[9] First, we scale factor endowments, fuel demands in electricity production and world market energy prices. Second, we incorporate autonomous energy efficiency improvements (AEEI) to match the aggregate emissions profiles. AEEI represents energy efficiency improvements *in addition* to energy demand reductions caused by changes in energy prices. The European Commission (1996) mentions research or changes in public standards as sources of efficiency improvements.

After the calibration the model represents the CW baseline as an equilibrium solution. Figure 1 shows the CW CO_2 emissions profile with an alternative BAU path labelled LOW_AEEI. The latter deviates from CW only with respect to the assumption about AEEI and exogenous fuel shifting in the electricity sector. In the LOW_AEEI baseline we solve the model with all AEEI equal to 1% p.a. and let relative prices determine the fuel mix in the electricity sector.

Table 5 shows that aggregate CO_2 emissions in 2010 exceed the 1990 level by 6% in the CW baseline and 20% in the LOW_AEEI baseline. Two factors explain most of this difference. First, the CW baseline implies more fuel shifting from coal to gas in the electricity sector than the model can explain with changes in relative prices. Second, AEEI exceeds 1% p.a. in many cases in the CW baseline, and in particular Germany and Denmark are expected to experience large energy efficiency improvements.[10]

We do not regard the LOW_AEEI baseline as more realistic than the CW baseline that is based on a compilation of detailed country studies. Rather we think of it as a sensitivity analysis that illustrates the implications of the CW assumptions about relatively high values for AEEI and fuel shifting in electricity generation.

4. Scenario definitions

Our numerical analysis distinguishes three CO_2 abatement scenarios: KYOTO, UNIFORM, and HIGH_BAU. The KYOTO scenario implements the EU

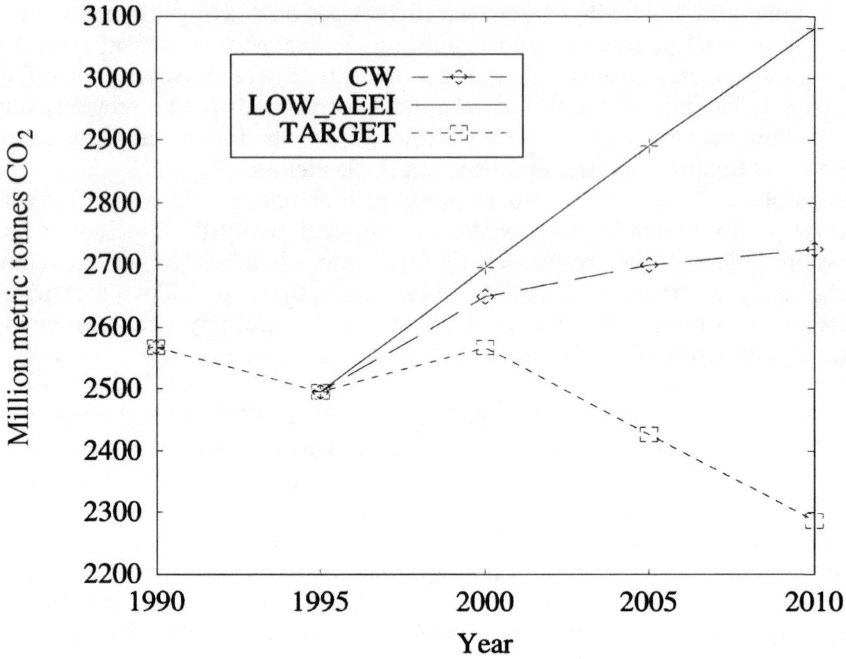

Figure 1. Emissions profiles.

Table 5. Change in emissions in 2010

	Relative to 1990 emissions			Relative to BAU emissions in 2010		
	KYOTO	CW	LOW_AEEI	KYOTO	UNIFORM	HIGH_BAU
DE	−21%	−1%	18%	−21%	−10%	−33%
DK	−21%	12%	45%	−30%	−21%	−45%
ES	15%	22%	28%	−6%	−27%	−10%
FR	0%	7%	19%	−6%	−17%	−16%
IT	−6.5%	15%	15%	−18%	−22%	−18%
UK	−12.5%	5%	22%	−17%	−15%	−28%
EU	−11%	6%	20%	−16%	−16%	−26%

Source: European Commission [1996; 161–209] and own calculations.

agreement on differentiated emissions reductions that distributes the 8% Kyoto commitment by the EU across member states.[11] The KYOTO scenario, which employs the CW baseline as the BAU, represents our base case.

The differentiated cutbacks amount to an average reduction of 11% for the six countries in our model. The UNIFORM scenario requires all six countries to reduce emissions uniformly by 11% and it also uses the CW baseline. This scenario allows us to analyse the implications of the differentiated cutbacks, holding total EU emissions constant.

In the HIGH_BAU scenario we apply the differentiated cutbacks used in the KYOTO scenario but employ different BAU assumptions: All AEEI equal 1% p.a. and relative prices determine the fuel mix in electricity production, i.e., the LOW_AEEI baseline.

The last three columns in Table 5 show the effective abatement requirements in 2010 in the three scenarios. As expected, the choice of baseline has a considerable impact on the required abatement effort.

Several characteristics are common to all three scenarios. First, all scenarios result in the same aggregate CO_2 abatement for the six EU countries in the model. We ignore the issue of carbon leakage and we thus assume that the EU policies do not lead to higher emissions outside the EU. This allows us to compare the results without considering the benefits from CO_2 abatement, i.e., we only consider the gross costs.

Second, we use the same cutback profile for aggregate EU emissions in all scenarios. Specifically, we assume a stabilization of aggregate emissions in the year 2000 at 1990 levels and a linear cutback between 2000 and 2010. In other words, all countries have to meet half their 2010 commitments by 2005.

Third, we do not allow for trade in emissions across countries. Emissions can be traded across sectors within a given country, but all emissions reductions must take place domestically and no emissions can be purchased abroad.

Fourth, we incorporate the emissions constraints as a resource constraint in the model where we interpret the shadow price on the constraint as the CO_2 tax rate. The representative consumer in each region collects all the revenue, and the provision of public goods stays constant.

The model solves for counterfactual equilibriums in five-year steps starting in the year 2000. We report consumption losses and CO_2 tax rates in Figures 2–5. Consumption losses are calculated as the percentage change in final consumption relative to the relevant baseline and CO_2 tax rates (marginal abatement costs) are reported in US dollars at constant 1990 prices per metric tonne of CO_2.

5. Results

The effective emissions targets provide the starting-point for the interpretation of the results (see Table 5) and we begin with our base case, the KYOTO scenario. Figure 2 shows that the high-abatement countries, Denmark, Germany, Italy and the UK, face high carbon taxes whereas the low targets for France and Spain require relatively low tax rates. The stabilization of Germany's emissions in the year 2000 does not require any carbon tax as the unification caused a significant drop in emissions from 1990 to 1995.

The carbon taxes cause efficiency losses, and Figure 3 summarizes the results

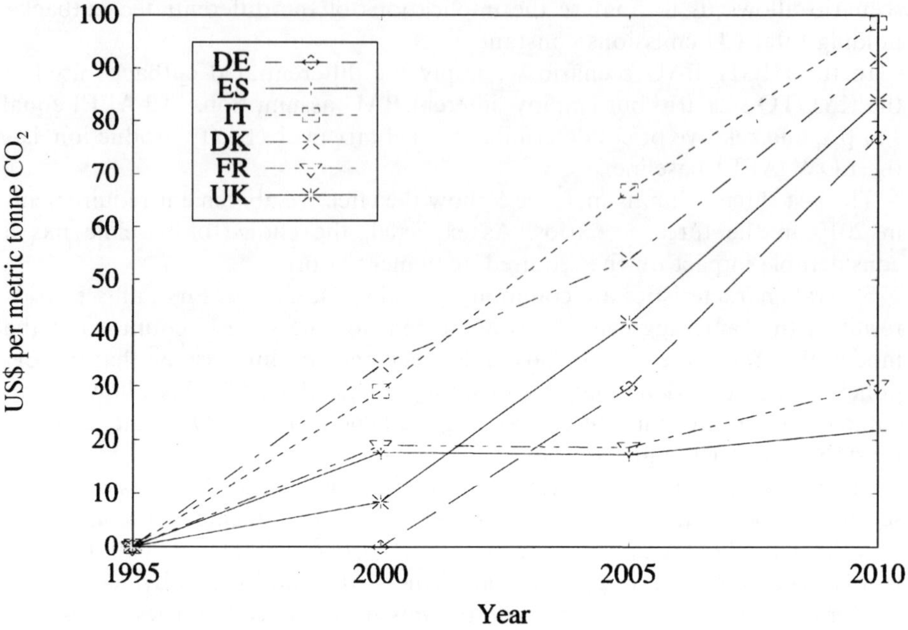

Figure 2. Carbon tax rates in the KYOTO scenario.

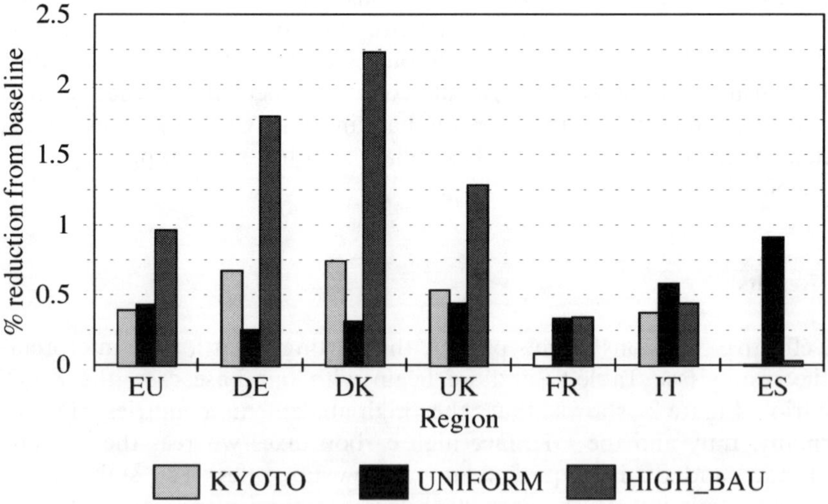

Figure 3. Consumption losses in 2010.

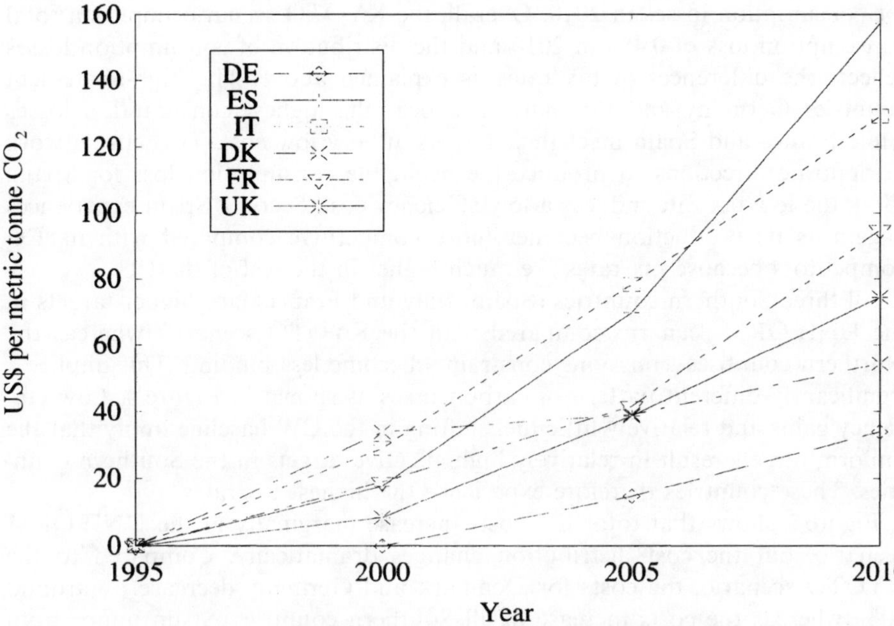

Figure 4. Carbon tax rates in the UNIFORM scenario.

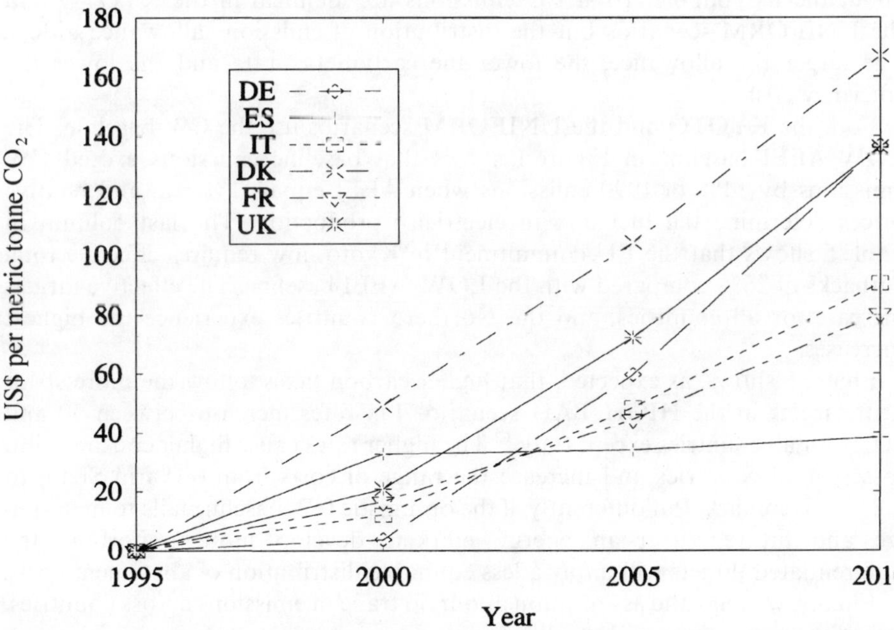

Figure 5. Carbon tax rates in the HIGH_BAU scenario.

for consumption losses in 2010. Overall, the KYOTO scenario implies a total consumption loss of 0.4% in 2010 and the distribution of consumption losses reflects the differences in tax rates as explained above. The high-abatement countries Germany and Denmark also bear the highest consumption losses while France and Spain meet their targets at very low cost. Two effects work in opposite directions to produce the negligible consumption loss for Spain. First, the low tax rate induces a low efficiency cost. Second, Spain experiences a gain as its production becomes more competitive compared with its EU competitors because tax rates are much higher in the rest of the EU.

All three Southern countries (Spain, Italy and France) face higher targets in the UNIFORM scenario compared with the KYOTO scenario, whereas the Northern countries' emissions constraints become less binding. This implies a significantly different pattern of carbon taxes as shown in Figure 4. Low efficiency gains and relatively little fuel shifting in the CW baseline imply that the uniform targets result in relatively high effective targets in the Southern countries. These countries therefore experience the highest tax rates.

Figure 3 shows that total EU costs increase marginally in the UNIFORM scenario, but the cost distribution changes dramatically. Compared to the KYOTO scenario, the costs for Denmark and Germany decrease by around 50% whereas the costs increase in all Southern countries. Spain jumps from virtually zero costs to a 0.9% consumption loss. While this scenario may be politically irrelevant, it shows the implicit transfers involved in the negotiations about the EU bubble. Total EU emissions are identical in the KYOTO and the UNIFORM scenarios but the distribution of emissions allowance differs. The larger the allowance, the lower the carbon tax rate and the lower the efficiency cost.

Both the KYOTO and the UNIFORM scenarios use the CW baseline. The LOW_AEEI baseline in Figure 1 shows that baseline emissions exceed CW emissions by 14% of 1990 emissions when AEEI equals 1% p.a. and relative prices determine the fuel mix in electricity production. The last column in Table 5 shows that the EU commitment in Kyoto now requires effective total cutbacks of 26% compared with the LOW_AEEI baseline. The effective targets increase for all countries, and the Northern countries experience the highest increases.

Figure 5 shows, as expected, that higher carbon taxes follow the more stringent targets in the HIGH_BAU scenario. Tax rates increase between 50 and 100% for all countries except Italy.[12] The higher rates cause higher consumption losses in all countries and increase the range of costs from 0.1% in Spain to 2.2% in Denmark. Put differently, if the optimistic CW baseline fails to materialize and future European energy markets develop more uniformly, the differentiated targets will imply a less equitable distribution of abatement costs.

Finally, we relax the assumption about no trade in emissions across countries. Specifically, we assume that all six countries can trade emissions freely such that the same carbon tax rate applies in all countries. The quantitative effects of this change turn out to be small and we thus conclude that our results are robust with respect to the assumption about emissions trading.

In our model, trade in emissions leads to equalization of the domestic carbon tax rates and this influences the results via two channels: the distortionary effects of carbon taxes change and the values of the emissions endowments change. Countries that previously had carbon tax rates below the new common tax rate will experience higher distortionary costs from higher carbon tax rates but they will also find their emissions endowments more valuable. Conversely, countries with high carbon tax rates will have lower distortionary costs but also lower carbon tax revenues. Emissions trading may also change the international incidence of carbon taxes as the burden of the taxes in a given country now gets shifted differently to foreign and domestic agents.

We use the model to trace all these effects, and the main results can be summarized as follows.[13] First, some countries gain slightly, others experience small losses and the overall EU costs decrease marginally. That is, trading *within* the EU implies small total cost savings. Intuitively, the differences in carbon tax rates before we allow for trading are significant but not large, and this effectively limits the extent of emissions trading. Spain, a net exporter, and Germany, a net importer, gain most from emissions trading. Second, we also conclude that our results regarding energy market projections are robust. We find that the effects of changing projections are similar to the no-trading case, although mostly the effects are quantitatively smaller. Finally, uniform abatement targets continue to show large implicit transfers from Northern European member states to Southern member states. This result holds although the differences in consumption losses decrease between differentiated and uniform abatement targets.

6. Concluding remarks

Burden sharing dominates the climate policy negotiations since the participating countries must view the expected distribution of costs as 'fair' in order for them to be politically acceptable. Negotiators point both to the countries' current economic structure *and* to the projections for the future economic development when they propose 'fair' abatement targets.

We have analysed the EU agreement on differentiated greenhouse gas abatement with a particular focus on the role of baseline projections for future energy markets. We used a dynamic multi-sector, multi-region general equilibrium model for the EU. The following insights emerge from our policy simulations:

(1) The overall costs of the EU bubble following its Kyoto commitment are low when we use the CW scenario provided by the European Commission as the baseline. The range of costs across member states is narrow, and the

necessary CO_2 taxes in the year 2010 are below US$100 per tonne CO_2 in all countries.
(2) The CW baseline implies significant fuel shifting in electricity generation and large energy-efficiency improvements by historical standards. If the projected, strong decoupling of GDP growth and CO_2 fails to materialize, both the overall CO_2 abatement costs and the range of costs across member states may increase significantly.
(3) Uniform abatement hardly increases total costs but it changes the cost distribution between countries considerably. We show that the EU bubble implicitly involves large economic transfers from the Northern European member states to the Southern European member states. This may be consistent with a higher willingness to pay for emissions reductions in the Northern countries.

Finally, we close with a caveat. Our model captures important aspects of bilateral trade and energy consumption in the EU. It is nonetheless a crude approximation of the real world's technologies, preferences, factor endowments, etc., and we therefore caution against too literal an interpretation of the numerical results.

Notes

* Helpful comments have been provided by Ulrich Fahl, Glenn Harrison, Claus Kastberg Nielsen, and participants at the IPCC workshop on *Economic Impacts of Annex 1 Actions on All Countries* in Oslo, August 18–20, 1997. None of the views expressed here should be attributed to any of our employers or research sponsors. Corresponding author: Jesper Jensen, Copenhagen Economics, Kronprinsessegade 46 E, 1306 Copenhagen K, Denmark. E-mail: jj@copenhageneconomics.com. Financial support from the Ministry of Business and Industry is gratefully acknowledged.

1. Article 3 in United Nations (1997). The group of Annex I countries consists of developed economies plus economies in transition, listed in Annex I of United Nations (1992).
2. See Capros et al. (1998) for more results from the GEM-E3 model.
3. The model includes three Northern EU countries (Denmark, Germany and the UK) and three Southern EU countries (France, Italy and Spain).
4. We recycle carbon tax revenues via lump-sum transfers to the representative agents in the model. Also, we assume that emissions cannot be traded across countries.
5. The EU targets for the six countries amount to an average emissions cutback of 11%.
6. The six selected countries also emit roughly 80% of EU emissions in 2010 in the projections reported below.
7. See European Commission (1996; pp. 48–54) for the complete description of this scenario.
8. See Böhringer, Ferris and Rutherford (1998) for an algebraic formulation of the static model, and Böhringer, Harrison and Rutherford (1998) for an application of the static model to carbon abatement and burden sharing within the EU.
9. See Appendix B for details of the calibration procedure.
10. A poll of 22 experts reported in Manne and Richels (1994) results in an average AEEI value of 0.7% p.a.
11. See EU (1998).

12. Italy's CW baseline emissions grow relatively slowly because of the absence of fuel-shifting in the electricity sector and low efficiency improvements.
13. The tables with results are available from the authors.

References

Böhringer, C., Ferris, M., and Rutherford, T.F. (1998). Alternative CO_2 abatement strategies for the European Union. In: Braden and Proost.

Böhringer, C., Harrison, G.W., and Rutherford, T.F. (1998). Sharing the burden of carbon abatement in the European Union. Working Paper 98-5. Boulder: University of Colorado.

Braden, J.B. and Proost, S. (eds.) (1998). *Climate change, transport and environmental policy – empirical applications in a federal system.* London: Edward Elgar.

Capros, P., Georgakopoulos, P., Zografakis, S., van Regemorter, D., and Proost, S. (1998). Coordinated versus non-coordinated European energy/carbon tax solutions analysed with GEM-E3 linking the EU-12 countries.

Conrad, K. and Schmidt, T.F.N. (1998a). 'National economic impacts of an EU environmental policy: an applied general equilibrium analysis.

Conrad, K. and Schmidt, T.F.N. (1998b). Economic effects of an uncoordinated versus a coordinated carbon dioxide policy in the European Union: an applied general equilibrium analysis. *Economic Systems Research.* 10(2):161–82.

European Commission (1996). *Energy in Europe – European energy to 2020 – a scenario approach.* Brussels: European Commission.

European Commission (1992). *European economy: the economics of limiting CO_2 emissions. Special edition No. 1.* Brussels: European Commission.

European Union (1998). *Community strategy on climate change – Council conclusions.* Communiqué from the 2106th Council meeting. 16–17 June 1998. Luxembourg: European Union.

Inter-Governmental Panel on Climate Change (IPCC). *Climate change 1995.* New York: Cambridge University Press, 1997.

Manne, A.S. and Richels, R.G. (1994). The costs of stabilizing global CO_2 emissions: a probalistic analysis based on expert judgements. *The Energy Journal.* 15(1):31–56.

United Nations (1997). *Kyoto Protocol,* FCCC/CP/1997/L.7/Add.1. Geneva: United Nations.

United Nations (1992). *United Nations framework convention on climate change.* Geneva: United Nations.

Weyant, J.P. (1993). Costs of reducing global carbon emissions. *Journal of Economic Perspectives.* 7(4):27–46.

Appendix A

This appendix provides an algebraic summary of the equilibrium conditions for the model. First, we present the main assumptions of the model and introduce notation. We then present the algebraic model.

Nested constant elasticity of substitution (CES) functions characterize the use of inputs in production, and all production exhibits non-increasing returns to scale. Goods are produced with capital, labour, energy and materials, and all sectors produce a single good. We therefore use 'goods' and 'sectors' interchangeably. Firms behave competitively and all markets are perfectly competitive.

A representative agent (RA) in each region is endowed with three primary factors: natural resources (used for fossil fuel production), labor and initial stock of capital. Nested CES functions characterize consumption by the RA (final demand) and the RA has myopic expectations, i.e., he is not forward-looking. The supplies of labour and natural resources are exogenous, and labour and capital can move freely across sectors within each region but cannot move between regions. Natural resources are sector specific.

All goods, except coal, crude oil and gas, are differentiated by region of origin. Constant elasticity of transformation functions characterize the differentiation of production between production for the domestic markets and the export markets. Regarding imports, nested CES functions characterize the choice between imported and domestic varieties of the same good (Armington).

The rest of the world is represented with horizontal export demand and import supply schedules, i.e., the EU regions behave as a small open economy. The balance of payments for each region with respect to the rest of the world is exogenous.

A constant marginal propensity to save characterizes the RA's allocation of income between savings, i.e., investments and current consumption. Investments begin to provide capital services one period after installation, and the capital stock depreciates at a constant rate. Growth augments the labour endowment at a constant rate.

Finally, lump-sum transfers finance the exogenous government demands, and the government in each region transfers all carbon tax revenues to the RA in the region.

The myopic expectations imply that the economies effectively reduce to a set of connected equilibria, where the current period's investments augment the capital stock in the next period. That is, all equilibrium conditions are strictly intra-period and we can thus omit time indices without loss of generality.

We characterize the model with two classes of equilibrium conditions: zero profit conditions and market clearance conditions. The former determines activity levels and the latter determines price levels. Tables A.1 and A.2 show the activity and the price levels in the model and Table A.3 explains the set notation.

The algebraic exposition of the model begins with the profit functions. We then derive the market clearance conditions by exploiting Shephard's lemma:

Table A.1. Activity variables

Y_{ir}	Production in sector i and region r
E_{ir}	Aggregate energy input in sector i and region r
M_{ir}	Aggregate imports in good i and region r
A_{dir}	Armington aggregate for demand category d of good i in region r
I_r	Aggregate investment in region r
G_r	Aggregate public output in region r
C_r	Aggregate household consumption in region r
D_r	Aggregate household energy consumption in region r

Table A.2. Price variables

p_{ir}	Output price of good i produced in region r
p_{ir}^E	Price of aggregate energy in sector i and region r
p_{ir}^M	Import price aggregate for good i imported to region r
p_{dir}^A	Price of Armington aggregate for demand category d of good i in region r
p_r^I	Price of investment demand in region r
p_r^G	Price of government demand in region r
p_r^C	Price of aggregate household consumption in region r
p_r^{EC}	Price of aggregate household energy consumption in region r
w_r	Wage rate in region r
v_r	Price of capital services in region r
q_{ir}	Rent to natural resources in region r ($i \in FF$)
p^w	Real exchange rate with the rest of the world
t_r^c	Carbon tax in region r

Table A.3. Sets

i	Sectors and goods
j	Aliased with i
r	EU regions
s	Aliased with r
EG	All energy goods: Coal, crude oil, refined oil, gas and electricity
FF	Primary fossil fuels: Coal, crude oil and gas
LQ	Liquid fuels: Crude oil, refined oil and gas
d	Demand categories: Y = intermediate, C = household, and I = investment

differentiation of the profit functions with respect to input and output prices yields compensated demand and supply functions. The exposition and our empirical implementation uses calibrated CES functions as the key element. Tables A.4–A.6 explain the notation for the calibrated model parameters.

Zero profit conditions:

1. Production of goods except fossil fuels:

$$\Pi_{ir}^Y = \left(\sum_s \theta_{irs}^X p_{is}^{1-\eta} + \theta_w^X p^{w^{1-\eta}} \right)^{1/(1-\eta)} \sum_{j \notin E} \theta_{jir} p_{Yjr}^A - \theta_{ir}^{KLE}$$
$$\times [\theta_{ir}^E p_{ir}^{E^{1-\sigma_{KLE}}} + (1-\theta_{ir}^E)(w_r^{\alpha_{ir}^L} v_r^{1-\alpha_{ir}^L})^{1-\sigma_{KLE}}]^{1/(1-\sigma_{KLE})} = 0, \quad i \in FF$$

2. Production of fossil fuels:

$$\Pi_{ir}^Y = p^w - \left[\theta_{ir}^Q q_{ir}^{1-\sigma_Q} + (1-\theta_{ir}^Q) \right.$$
$$\left. \times \left(\delta_{ir}^L w_r + \delta_{ir}^K v_r + \sum_j \theta_{jir}^{FF} p_{Yjr}^A \right)^{1-\sigma_Q} \right]^{1/(1-\sigma_Q)} = 0, \quad i \in FF$$

3. Sector-specific energy aggregate:

$$\Pi_{ir}^{E} = p_{ir}^{E} - \left\{ \theta_{ir}^{CO} p_{\{Y,COA,r\}}^{A^{1-\sigma_{CO}}} + (1-\theta_{ir}^{CO}) \left[\theta_{ir}^{EL} p_{\{Y,ELE,r\}}^{A^{1-\sigma_{EL}}} + (1-\theta_{ir}^{EL}) \left(\prod_{j \in LQ} p_{Yjr}^{A^{\beta_{jir}}} \right)^{1-\sigma_{EL}} \right]^{(1-\sigma_{CO})/(1-\sigma_{EL})} \right\}^{1/(1-\sigma_{CO})} = 0, \quad i \notin FF$$

4. Armington aggregate except fossil fuels:

$$\Pi_{dir}^{A} = p_{dir}^{A} - \{[\theta_{dir}^{A} p_{ir}^{1-\sigma_A} + (1-\theta_{dir}^{A}) p_{ir}^{M^{1-\sigma_A}}]^{1/(1-\sigma_A)} + t_r^c a_{dir}^c\} = 0, \quad i \notin FF$$

5. Armington aggregate for fossil fuels

$$\Pi_{dir}^{A} = p_{dir}^{A} - (p^w + t_r^c a_{dir}^c) = 0, \quad i \in FF$$

6. Aggregate imports across import regions except fossil fuels:

$$\Pi_{ir}^{M} = p_{ir}^{M} - \left(\theta_{ir}^{W} p^{w^{1-\sigma_M}} + \sum_s \theta_{isr}^{M} p_{is}^{1-\sigma_M} \right)^{1/(1-\sigma_M)} = 0, \quad i \notin FF$$

7. Investment:

$$\Pi_r^I = p_r^I - \sum_i \theta_{ir}^I p_{Iir}^A = 0$$

8. Public good production:

$$\Pi_r^G = p_r^G - \sum_i \theta_{ir}^G p_{Iir}^A = 0$$

9. Household consumption demand:

$$\Pi_r^C = p_r^C - \left[\lambda_r^E p_r^{EC^{1-\sigma_{EC}}} + (1-\lambda_r^E) \left(\prod_{i \in EG} p_{\{C,i,r\}}^{A^{\gamma_{ir}}} \right)^{1-\sigma_{EC}} \right]^{1/(1-\sigma_{EC})} = 0$$

10. Household energy demand:

$$\Pi_r^{EC} = p_r^{EC} \left\{ \lambda_r^{EL} p_{\{C,ELE,r\}}^{1-\sigma_{EL}} + (1-\lambda_r^{EL}) \times \left[\left(\sum_{i \in EG|_{\{ELE\}}} \lambda_{ir}^{E} p_{\{C,i,r\}}^{A^{1-\sigma_{NE}}} \right)^{1/(1-\sigma_{NE})} \right]^{1-\sigma_{EL}} \right\}^{1/(1-\sigma_{EL})} = 0$$

Market Clearance Conditions

11. Labour:

$$\overline{L_r} = \sum_i Y_{ir} \frac{\partial \Pi_{ir}^{Y}}{\partial w_r}$$

12. Capital:

$$\overline{K_r} = \sum_i Y_{ir} \frac{\partial \Pi_{ir}^{Y}}{\partial v_r}$$

13. Natural resources:

$$\overline{Q_{ir}} = Y_{ir} \frac{\partial \Pi_{ir}^{Y}}{\partial q_{ir}} \quad i \in FF$$

14. Output except fossil fuels:

$$Y_{ir} \frac{\partial \Pi_{ir}^{Y}}{\partial p_{ir}} = \sum_{d,j} A_{dir} \frac{\partial \Pi_{djr}^{A}}{\partial p_{ir}} + \sum_{s} M_{is} \frac{\partial \Pi_{is}^{M}}{\partial p_{ir}}, \quad i \notin FF$$

15. Sector-specific energy aggregate:

$$E_{ir} = Y_{ir} \frac{\partial \Pi_{ir}^{Y}}{\partial p_{ir}^{E}}$$

16. Import aggregate except fossil fuels:

$$M_{ir} = \sum_{d} A_{dir} \frac{\partial \Pi_{dir}^{A}}{\partial p_{ir}^{M}}, \quad i \notin FF$$

17. Armington aggregate:

$$A_{dir} = \sum_{j} Y_{jr} \frac{\partial \Pi_{jr}^{Y}}{\partial p_{dir}^{A}} + C_{r} \frac{\partial \Pi_{r}^{C}}{\partial p_{dir}^{A}} + I_{r} \frac{\partial \Pi_{r}^{I}}{\partial p_{dir}^{A}} + G_{r} \frac{\partial \Pi_{r}^{G}}{\partial p_{dir}^{A}}$$

18. Household consumption:

$$p_{r}^{C} C_{r} = (1 - \mu_{r}) \left(w_{r} \overline{L_{r}} + v_{r} \overline{K_{r}} + \sum_{j \in FF} q_{jr} \overline{Q_{jr}} + t_{r}^{c} \overline{Z_{r}} - p_{r}^{G} \overline{G_{r}} - p^{w} \overline{B_{r}} \right)$$

19. Savings:

$$p_{r}^{I} I_{r} - \mu_{r} \left(w_{r} \overline{L_{r}} + v_{r} \overline{K_{r}} + \sum_{j \in FF} q_{jr} \overline{Q_{jr}} + t_{r}^{c} \overline{Z_{r}} - p_{r}^{G} \overline{G_{r}} - p^{w} \overline{B_{r}} \right)$$

20. Aggregate household energy consumption:

$$D_{r} = C_{r} \frac{\partial \Pi_{r}^{C}}{\partial p_{r}^{EC}}$$

21. Balance of payments:

$$\sum_{i} Y_{ir} \frac{\partial \Pi_{ir}^{Y}}{\partial p^{w}} + \sum_{r} \overline{B_{r}} = \sum_{i \in FF} \frac{\partial \Pi_{ir}^{M}}{\partial p^{w}} + \sum_{i \in FF,d} \frac{\partial \Pi_{dir}^{A}}{\partial p^{w}}$$

22. CO_2 emissions:

$$\overline{Z_{r}} = \sum_{d,i} A_{dir} \frac{\partial \Pi_{dir}^{A}}{\partial p_{r}^{c}}$$

23. Government output:

$$\overline{G_{r}} = G_{r}$$

Table A.4. Cost shares

θ_{irs}^{X}	Share of exports from sector i in region r to region s ($i \in FF$)
θ_{ir}^{W}	Share of exports from sector i in region r to the rest of the world ($i \in FF$)
α_{ir}^{L}	Share of labour in value added in sector i and region r ($i \in FF$)
θ_{jir}	Share of intermediate good j in sector i and region r ($i \in FF$)
θ_{ir}^{KLE}	Share of KLE aggregate in sector i and region r ($i \in FF$)
θ_{ir}^{E}	Share of energy in the KLE aggregate of sector i and region r ($i \in FF$)
θ_{ir}^{Q}	Share of natural resources in sector i and region r ($i \in FF$)
θ_{ir}^{FF}	Share of intermediate good j in sector i of region r ($i \in FF$)
δ_{ir}^{L}	Share of labour in sector i and region r ($i \in FF$)
δ_{ir}^{K}	Share of capital in sector i and region r ($i \in FF$)
θ_{ir}^{CO}	Share of coal in energy demand by sector i in region r ($i \in FF$)
θ_{ir}^{EL}	Share of electricity in non-coal energy demand by sector i in region r ($i \in FF$)
β_{jir}	Share of fossil fuel j in energy demand by sector j in region r ($i \in FF, j \in LQ$)
θ_{isr}^{M}	Share of imports of good i from region s to region r ($i \in FF$)
θ_{dir}^{A}	Share of domestic variety i in Armington aggregate for demand category d for good i in region r
θ_{ir}^{I}	Share of good i in investment in region r
θ_{ir}^{G}	Share of good i in government demand in region r
λ_{r}^{EC}	Share of energy in aggregate household consumption in region r
γ_{ir}	Share of non-energy good i in non-energy household consumption demand in region r
λ_{r}^{EL}	Share of electricity in aggregate household energy consumption in region r
λ_{ir}^{E}	Share of non-electric energy good i in the non-electric household energy consumption in region r
μ_{r}	Marginal propensity to save in region r

Table A.5. Endowments and emissions coefficients

α_{djr}^{c}	Carbon emissions coefficient for fossil fuel j in demand category d of region r ($j \in FF$)
$\overline{L_r}$	Aggregate labour endowment in region r
$\overline{K_r}$	Aggregate capital endowment in region r
$\overline{Q_{ir}}$	Endowment of natural resource i in region r ($i \in FF$)
$\overline{G_r}$	Aggregate government demand in region r
$\overline{B_r}$	Balance of payment surplus in region r
$\overline{Z_r}$	Endowment of carbon emissions rights in region r

Table A.6. Elasticities

η	Transformation between production for the domestic and export markets	4
σ_{KLE}	Substitution between energy and value added in non-fossil fuel production	0.3
σ_{Q}	Substitution between natural resources and other inputs in fossil fuel production	
ρ	Supply in fossil fuel production	1
σ_{CO}	Substitution between coal and the non-coal energy in non-fossil fuel production	0.5
σ_{EL}	Substitution between electricity and the non-coal fossil fuels in non-fossil fuel production	0.3
σ_{M}	Substitution between imports from different regions	8
σ_{A}	Substitution between the import aggregate and the domestic good	4
σ_{EC}	Substitution between energy and non-energy in household consumption	0.3
σ_{EL}	Substitution between electricity and the non-electric energy in household energy consumption	0.3
σ_{NE}	Substitution between non-electric energy in household energy consumption	0.5

Appendix B

This appendix describes the calibration of our model to the CW baseline. In two steps, we incorporate exogenous information on non-uniform growth rates in GDP, fossil fuel production, fuel mixes in electricity generation, changes in world market energy prices and CO_2 emissions profiles.

First, we fix the time profile of fossil-fuel supplies from the EU to the exogenous baseline projections by making supplies inelastic and scaling sector-specific resources with the exogenous growth rates in fossil-fuel production. This allows us to partially control the emissions profile from the supply side (except the effects of imported fuels). On the demand side, we incorporate AEEI on the energy demands by consumers and production activities. We also incorporate exogenous fuel shifting in the electricity sector, and exogenous, region-specific GDP growth rates determine the size of labour endowments. Finally, we adjust import demand and export supply functions with respect to the rest of the world to account for exogenous changes in world market energy prices. We solve the model with these changes in parameter values and obtain country-specific estimates for CO_2 emissions.

These estimates deviate from the CO_2 emissions profile in the CW baseline by changes in the fossil fuel supplies from the rest of the world. We therefore include a second step in the calibration procedure where we scale the non-electric AEEI factors to reduce energy demands and, hence, CO_2 emissions. The scaling takes place country by country to match the model as closely as possible to the CW emissions profiles. We solve the model again to verify that it represents the CW baseline as an equilibrium solution. Finally, we use the equilibrium prices to calibrate the fossil fuel production functions to a price elasticity of supply equal to one.

CHAPTER 10

Green Electricity Certificates – A Supplement to the Flexible Mechanisms of the Kyoto Protocol

Lene Nielsen and Tim Jeppesen

1. Introduction

There are at least two good reasons why common policies should be implemented to ensure the introduction and development of renewable energy sources in the European Union.

First, in 1997, the EU made a commitment to reduce greenhouse gas emissions by 8% in the period 2008–2012 compared to the 1990 emissions. This commitment was made under the auspices of United Nation's Framework Convention on Climate Change in Kyoto. The European Commission envisages that one possible means to fulfil its commitment is to increase the share of renewable energy sources in the generation of electricity. The Commission has given its suggestion as to how this increase may actually be achieved in its White Paper on renewable energy (COM(97) 599). In this White Paper, the Commission suggests setting an indicative objective of 12% for the contribution by renewable sources of energy to the European Union's gross inland energy consumption by 2010. Currently, electricity generated from renewable sources is more expensive than electricity from competing fuels. Therefore, co-ordinated initiatives to promote renewable energy production will be necessary.

The second main reason for implementing harmonized European policies on renewable energy generation is to avoid potential distortions to trade within the Internal European Electricity Market. Trade distortions may arise due to different national schemes to promote renewable energy sources already introduced by several Member States. The absence of common European rules does not create any distortions to electricity trade in a non-liberalized electricity market because trade is limited and strictly regulated. In such a market, consumers have no alternative foreign suppliers. With the internal electricity market, starting off in 1999, some co-ordination of the regulatory schemes is required.[1] In such a market, subsidies to domestic renewable-energy producers may distort competition. Moreover, in the internal electricity market, consumers can freely choose between domestic and foreign renewable-energy producers.

C. Carraro (ed.), Efficiency and Equity of Climate Change Policy, 221–244.

Therefore, different national renewable-energy policies will affect domestic and foreign renewable-energy producers differently and hence affect renewable electricity trade. Co-ordination of European renewable energy policies must restrict trade distortions in renewable electricity, distortions of competition and general trade distortions.

The purpose of this paper is to examine the possibilities for implementing harmonized measures in the EU. These measures in particular must promote the use and development of renewable electricity and in addition effectively co-ordinate national initiatives. Thus, the questions we will try to answer are: Which instruments are available and appropriate to use in order to introduce and develop renewable energy sources in the context of the internal electricity market? How do these instruments perform compared to the flexible mechanisms identified in the Kyoto Protocol? Moreover, the harmonization of national measures touches upon aspects of sovereignty, distribution of powers, and subsidiarity. Therefore, the influence of the principle of subsidiarity on the development of harmonized measures will also be discussed.

In this paper we introduce Green Certificates as a new regulatory tool. We argue that a European system of Green Certificates is a cost-efficient method of inducing renewable energy production and thus of reducing CO_2 emissions. A system of Green Certificates combines the advantages of command-and-control (CAC) regulation with market-based incentive regulation. Like CAC regulation, Green Certificates are based on a politically determined target, while the market ensures cost-efficient behaviour by the agents. The system of Green Certificates implicitly implies the harmonization of different national subsidy schemes supporting renewable energy generation. Therefore, the system may contradict the principle of subsidiarity and thus be opposed by some Member States. Green Certificates have a number of similarities with the flexible mechanisms of the Kyoto Protocol. Green Certificates are, however, solely directed towards the electricity sector while the flexible mechanisms in principle cover all energy-producing and energy-consuming sectors. Therefore, Green Certificates may interfere more with national electricity planning policies than the flexible mechanisms do.

The structure of the paper is as follows: In Section 2 we introduce Green Certificates as a regulatory instrument to assure the implementation of renewable energy sources in a liberalized electricity market. In 1998, a somewhat similar system was voluntarily implemented by the Dutch electricity distributors. In Section 2 we therefore also examine experiences acquired with the Dutch system. In Section 3 we identify and analyse some of the obstacles that remain to be settled in order to implement a system of Green Certificates in the EU. This analysis focuses in particular on the functioning of the markets for electricity and Green Certificates. In Section 4 we compare Green Certificates with the flexible mechanisms of the Kyoto Protocol and discuss the issue of subsidiarity. Conclusions are given in Section 5.

2. Implementing renewable energy sources in a liberalized electricity market

The costs of renewable energy generation exceed the costs of conventional energy generation. This is the situation today, and this will probably also be the case in the near future. The Internal Electricity Market Directive allows producers to collect additional costs of renewable energy generation via the general electricity prices.[2] Although this puts renewable and conventional electricity on an equal footing, producers are unlikely to make the necessary investments without additional incentives to do so.

Member States have taken different approaches to increase the share of renewable energy sources (Haas *et al.*, 1999). Whereas some countries have implemented support schemes to initiate investments, others have done very little to increase the share of renewables. The different levels of effort may result in an inexpedient situation with some Member States facing high average electricity prices due to a high share of renewable energy production. Other Member States, not pursuing the increase in renewable energy generation, may face relatively low average electricity prices and therefore be better equipped to compete on the liberalized electricity market. Thus, the internal market for electricity may bring disincentives to invest in renewables. Therefore, in order to realize the goal formulated in the White Paper on renewable energy, it is important that electricity producers and consumers in the European market face similar incentives.

The EU has several instruments at its disposal to make each Member State contribute to the goals of the White Paper. The EU has, in a White Paper (COM(97) 599), suggested that each Member State supply renewable energy sources corresponding to 12% of its electricity consumption. If this goal was to be achieved by production quotas it would, however, result in very different costs of implementation across Member States and thus probably lead to political opposition. In addition, production quotas may be viewed as inconsistent with the principle of subsidiarity because the EU directly influences domestic electricity production.

A more acceptable solution for the Member States would be to initiate increased renewable energy generation by granting subsidies, tax exemptions and/or price guarantees. These types of regulation have been, and still are, used by several Member States such as Sweden, Denmark, Germany, the Netherlands, Austria and the UK.

Subsidies and price guarantees may be inconsistent with Art. 92 of the EU Treaty regarding state aid. Until now, the Commission has taken a favourable position as regards aid to renewable energy production.[3] This situation may, however, change with the liberalization of the electricity market as the discriminatory impact of state aid will become decisive under the new circumstances of competition. Community law does not forbid Member States to maintain subsidy schemes provided these do not discriminate between foreign and domestic electricity producers. This implies, for instance, that German investors, setting up wind turbines in Greece, must receive the same subsidies as the domestic investors. In order to avoid the situation of subsidizing foreign electricity producers, Member States may decide to abolish subsidy schemes and implement different regulatory schemes in order to fulfil the commitments of the Kyoto agreement as well as the Community strategy on renewable energy sources.

A system of Green Certificates is a substitute to subsidies, price guarantees and exemptions. The Green Certificate is a policy instrument available to the Member States as well as to the Commission. This system may ensure that the use of renewable energy sources accounts for 12% in 2010 and harmonizes different national policies. Thus, provided the system is implemented efficiently, it contributes to the fulfilment of the EU commitments made in Kyoto, and improves the functioning of the internal electricity market by removing different national subsidy schemes.

2.1. A system of Green Certificates

Regulation by Green Certificates requires consumers to purchase a fixed share of electricity based on renewable energy sources. As such, all electricity consumers are obliged to purchase for example 12% of their total electricity consumption from renewable sources.

Green Certificates differs from typical quota arrangements in that they prescribe a minimum consumption quota to be implemented on the demand side. Usually, quotas are implemented as maximum quotas on either production input, production output or by-products like emissions or effluents. The fact that regulation is moved away from the production side towards the demand side may prove to be a particular strength of the certificate system because the politically sensitive supply sector is not subject to direct regulation. Green Certificates do not directly affect the profit function of the utilities, but affects the utility function of the households instead.

A Green Certificate is a document certifying that a given producer has generated a specified amount of renewable electricity. If a producer generates for example 100 MWh of renewable electricity per year, four certificates could be issued per year, each representing the value of 25 MWh. The certificate is issued either by an authority controlling the producer or by the producer under the control of an appropriate authority.

Each consumer must prove that he complies with his minimum consumption quota. This is proved by the ownership of Green Certificates, representing the specified amount of consumption. Consumers may purchase certificates directly from the producers. This may, however, make overall transaction costs very high due to the large number of small-scale consumers. In order to reduce transaction costs, distributors may instead purchase certificates on behalf of consumers who consume less than for example 100 GWh per year.[4] This solution would reduce the number of actors on the market. It should be noted though, that it would also increase the potential for market dominance and the formation of price cartels.

Electrons produced by renewable energy sources cannot be separated from electrons produced by conventional sources. Thus, separate markets for renewable and conventional electricity will be unduly complicated and expensive. With Green Certificates, two other markets emerge: a new financial market for Green Certificates with certificates representing a certain amount of renewable electricity; and a physical market for electricity, with green and conventional electricity traded at similar market prices. The additional cost of producing renewable electricity will be covered by the price paid for Green Certificates.

Power producers will be attracted to the market for Green Certificates due to the government/EU induced demand for renewable electricity. In a perfect market, producers will choose to supply renewable electricity and satisfy demand as long as it is profitable to do so. Consequently, consumption quotas initiate a new type of demand making renewable energy sources more valuable and inducing market prices on renewable energy sources to increase. This, again, initiates capacity expansions into renewable energy sources and increases the development of renewable technologies. Ideally, expansions will continue until demand equals supply. However, renewable energy sources are often unpredictable compared to conventional energy because of, for example, fluctuating climatic conditions. The equilibrium amount of renewable energy capacity may thus be hard to identify as it differs from time to time. Consequently, certificate prices are likely to fluctuate widely, making investments more risky. We will return to the issue of risk management in Section 3.

Like in any other well-functioning market, the price in the certificate market is determined by supply and demand. However, in this market, demand is induced by the obligation to purchase the specified amount of certificates, and not *per se* by a willingness to pay the additional costs of renewable energy. Therefore, the demand curve will be vertical up to a maximum price determined by the settlement rate that consumers must pay if they are not in compliance with their quota. The politically determined settlement rate represents the maximum price consumers are willing to pay for Green Certificates and thus the upper boundary of the demand function, cf. Figure 1.

Supply is determined by a number of factors such as costs, technology mix and expectations. The supply curve is relatively inelastic as costs of production are low once the technology is established (Skytte, 1999). Consequently, the supply curve is steep but upward sloping and thus intersects the demand curve as illustrated in Figure 1. If the supply of Green Certificates is relatively inelastic (S_1 in Figure 1), the market price (P1) is determined at the horizontal part of the demand curve and equals the settlement price. If on the other hand the settlement price is high (or if supply is relatively elastic (S_2)), new investments in renewable electricity are initiated resulting in increased supply. Consequently, a market price (P2) will be established below the settlement rate.

As mentioned, Green Certificates could be traded either bilaterally between

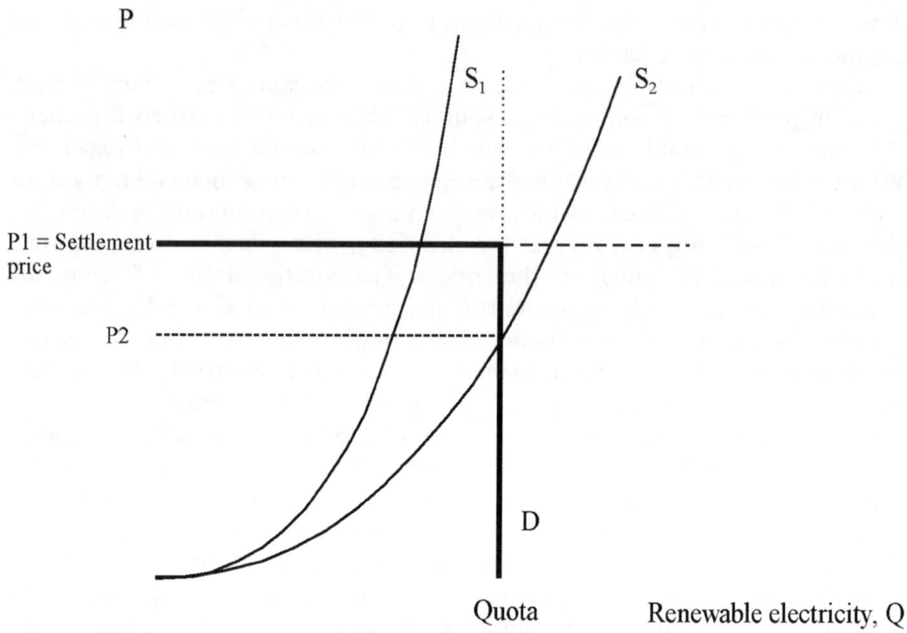

Figure 1. Supply and demand at a market for Green Certificates (Source: Skytte (1999)).

distributors and producers, or centrally at a pool. In order to gain transparency, a centralized certificate pool would have to be established. A centrally organized pool will lower search costs as well as informational rents. Certificate trade could be organized either through the existing stock exchanges or through the electricity pool that will mature with the liberalization of the electricity market.

Figure 2 illustrates the market structure for 'green' as well as 'conventional' electricity and the market for Green Certificates. It is assumed that certificate trades are carried out via the power pool and that distributors manage the purchase obligation on behalf of the consumers. National or EU authorities carry out the functions of controlling that distributors are in compliance with minimum quotas and issue the Green Certificates.

Control may be very simple in this system as the amount of watt-hours delivered to the net from different sources is easily monitored. A Green Certificate documents that green electricity has been delivered to the net, and that the holder of the certificate is in compliance with the obligation to consume the specified amount of green electricity. Consumers/distributors must possess the required number of certificates on the day when inspection is carried out. Inspection should be carried out regularly, for instance once a year, by the appropriate Member State authority. Each certificate should be provided with a serial number containing information regarding the name of the producer and issuer, the technology used in production, and the date of issue and of

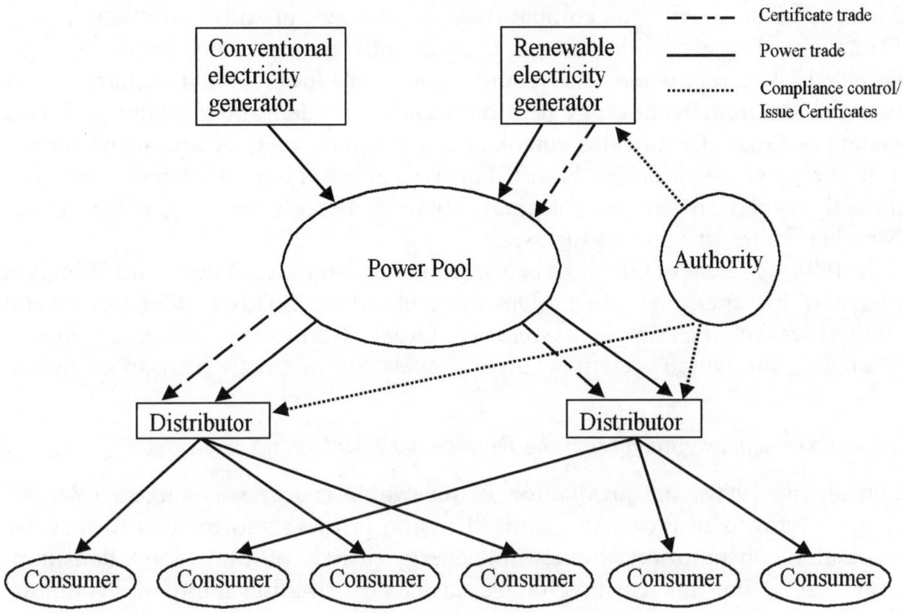

Figure 2. Market structure of a centrally organized market for Green Certificates.

expiry. This information should be registered centrally in order to avoid any cheating or double counting. At the date of expiry, the certificate is registered as used, and thus loses its value. In the following period, the consumer must purchase new certificates.

Consumers or distributors not holding sufficient certificates at the time of inspection may be sanctioned in several ways. One way would be to let those not in compliance make a compensation payment into a renewable fund. This fund may be used for grants supporting R&D in renewable technologies or for investment support for renewable plants. Alternatively, sanctions could be carried out by making distributors failing to meet their quota buy Green Certificates at a high settlement rate from distributors in compliance.

The European Commission has several instruments at its disposal to achieve the goals of the Kyoto Protocol and its own White Paper on renewable energy. Compared to simple production quotas or unco-ordinated national tax and subsidy schemes, a European system of Green Certificates would have several advantages. Green Certificates combine the advantages of Command-and-Control regulation with the advantages of market-based incentive regulation as the overall goal is defined by the authorities and responses are determined by the actors in the market. This induces cost-minimizing behaviour. The system of Green Certificates implies that renewable energy is produced by the most cost-efficient technologies. In Europe this has the obvious implication that, for instance, electricity from wind turbines is primarily generated in

Member States with a comparative advantage in this production (e.g. Denmark). These Member States consequently may produce more than the specified EU target while other countries may produce less and perhaps by the use of a different technology (e.g. biomass in Sweden and Finland). Thus, a system of Green Certificates equalizes the marginal costs of producing renewable energy across Member States. This is analogous to market-based environmental regulation where marginal abatement costs are equalized across Member States or across sources.

In 1998, a system of Green Labels was introduced to regulate the introduction of renewable energy production in the Netherlands. This system has several similarities to the simple system of Green Certificates described above. Therefore, the Dutch system of Green Labels will be briefly introduced below.

2.2. Experiences gained from the Dutch market of Green Labels

During the 1990s, the production of renewable energy was subject to wide-ranging reforms in the Netherlands. The aim of these reforms was to increase the Dutch consumption of renewable energy to 10% of total energy consumption by 2020. For this reason, Dutch electricity distribution companies voluntarily committed themselves to investment in renewables. Distributors believed that a system of Green Labels could be an intriguing method to induce those investments and therefore agreed to implement a system of Green Labels in the period 1998–2000 in order to test the system. The system is co-ordinated by the Association of Energy Distribution Companies in the Netherlands (EnergieNed).[5]

Dutch distributors have agreed to ensure that 3.2% of the total electricity consumption is based on renewable energy sources before the year 2000. In the first phase of the programme, 1998–2000, the Green Labels market is based on voluntary purchase of 'green power' at a higher price. From 2001, demand will be driven by binding minimum consumption quotas.

Renewable-energy producers receive a Green Label for each 10 000 kWh of renewable electricity they produce. Green Labels are valid for one year only. This implies that excess labels issued in one year cannot be used to fulfil the quota in subsequent years. All electricity is sold at the same market price which implies that electricity from renewable sources is not priced higher than electricity from conventional sources. The higher production costs of electricity from renewable sources are covered by selling Green Labels to the distributors. Thus, two markets have evolved: one in which electricity is sold, and one in which Green Labels are sold.

The market for Green Labels is expected to be divided into a market for Green Labels already issued (the spot market) and a market for Green Labels to be issued (the futures market). The futures market enables producers and distributors to close contracts on renewable energy that will be produced in the future. Trade is bilateral, i.e., directly between distributors and producers.

In order to improve trading conditions, EnergiNed has opened a Web site on the Internet which serves as a marketplace where supply and demand for Green Labels can meet online. The primary purpose of the Web site is to facilitate the search for available labels and to identify potential buyers. In addition, the Web site will contain information regarding the distribution of available certificates on the different renewable technologies, and information about the difference between the quota and the number of labels sold.

If a distributor fails to fulfil the requirement by the end of the year he will be sanctioned by compulsory purchase of Green Labels, at a previously fixed settlement rate. These labels must be purchased from a company in compliance with its minimum quota. The fixed settlement rate is based on the principle that it should be more profitable to purchase the Labels on the free market or to generate renewable electricity than to risk compulsory settlement. The Dutch Green Labels are supplemented by a number of financial compensation and bonus arrangements further inducing renewable energy generation.[6] Therefore, compliance is monitored by the tax department.

The Dutch market for Green Labels was launched on 1 January 1998. Due to the recent start there is little information about how the system has performed. EnergieNed has, however, gathered information about 55% of the trading transactions performed by August 1998. In this sample, 34 000 Labels were traded, corresponding to 20% of the goal for 1998. Some 70% of the labels traded were based on renewable energy from wind turbines, 14% from water and gas,[7] and 2% from biomass. All Green Labels were traded as long-term contracts. 67% of these were long-term contracts for 10 years or more. The average price for a label was Dfl0.04 per kWh, or approximately US$0.02 per kWh (Drillisch, 1998).

If the system of Green Labels proves to be a cost-effective method of inducing investments in renewable energy generation, distributors plan to suggest the passing of a mandatory system in the Netherlands (the second phase). In a mandatory system, the Green Labels will be replaced by Green Certificates, and annual quotas for renewable electricity will be announced for a period of the following five years.

The Dutch system of Green Labels is quite similar to the system of Green Certificates presented in Section 2.1, but there are a number of differences as well. First, the bilateral trade with Green Labels is substituted by Certificate trading at the national electricity pool. This is an advantage especially for small producers because it is cumbersome to approach and negotiate with a number of different distributors. A national pool should only be approached once. Second, in a system of Green Certificates several of the schemes supporting and inducing renewable energy will be phased out or changed. When prices are determined on the spot markets, a guaranteed price cannot be maintained. A Dutch price subsidy will, however, be established from 2002 and phased out in a five-year period. Furthermore, it is envisaged that tax deductions and speedy write-off provisions are to be replaced by schemes subsidizing the installation of new plants. Dutch energy taxes are planned to double. This will not, however, affect the absolute price of renewable electricity because energy taxes will continue to be reimbursed for renewable electricity.[8]

3. Obstacles to a market in Green Certificates

Currently, a system of Green Certificates is being considered by the European Commission (COM(98) 167 final). Considerations are, however, in a very early phase and it is still uncertain how such a system may be designed. So far, the Commission has outlined the basic principles of the system, leaving several questions to be answered before a European system of Green Certificates emerges (COM(98) 167 final, pp. 5, 6). It is, for instance, not settled how certificates should be traded and how the market should be organized. Neither is it decided for how long certificates should be valid, nor how the credibility of the system is secured and the risks reduced in order to initiate investments. Nor is the level of detail settled that should be applied in a policy covering all EU Member States. Hence, a system of Green Certificates may be very specific in the regulation of national energy policies. This may not be in accordance with the EU principle on subsidiarity.

The Dutch system of Green Labels is likely to clarify some, but not all, of these questions. The system of Green Labels has just been put into operation on a relatively small and voluntary scale. It is therefore not clear whether the Dutch experiences are applicable to a European system of Green Certificates.

The success of a market for Green Certificates is dependent on the creation of a transparent market based on a high degree of trust that non-compliers will be punished and that punishment will be severe enough to avoid disobedience. Thus, the first precondition for a well-functioning market is a credible system to ensure compliance. The second condition is that the environment for investments is safe enough to initiate investments.

3.1. Ensuring compliance

Potential investors naturally hesitate to enter the market if they find the system of Green Certificates unreliable. If producers assume that no demand will be initiated by the quotas, no market will be created. Consumers may decide not to fulfil their quotas for two reasons: either because sanctions are too weak or because the probability of being caught cheating is too low. Effective and credible monitoring procedures, control, and sanctions are therefore decisive for an efficient system of Green Certificates.

In a system of Green Certificates, control and monitoring must be directed towards the 'green' power producers and distribution companies. The control directed at the producer must ensure that electricity delivered is actually based on renewable energy sources. The control directed at distribution companies must ensure that they comply with their consumption quotas.

Monitoring producers is straightforward in a market of Green Certificates. Every watt-hour delivered from the different sources to the net is registered. Therefore, it is easy to control whether the amount of renewable electricity supplied equals the number of certificates issued. Each certificate may be provided with a serial number making identification of origin possible. By registering certificates and trades centrally, double counting is avoided and control is made easier. Registering and control may be carried out by national authorities and through the establishment of certificate pools.

It is up to Member States to sanction consumers not fulfilling their quotas. This may be done as in the Dutch system, by making non-compliers buy certificates from over-compliers at a higher settlement rate. This system may, however, be difficult to implement in a European context if domestic non-compliers had to buy certificates from selected foreign producers. How should certificates, which are to be sold at a higher rate, be picked out if the number of excess certificates throughout Europe is high? And what if there are no excess certificates? How should sanctions then be carried out? Due to these problems, it may be better to implement another type of sanctions as envisaged in Denmark.

In Denmark, a majority in Parliament agreed on the principles of a new electricity reform in March 1999 (Energistyrelsen, 1999). This reform includes provisions regarding the establishment of a market for Green Certificates. In the Danish model it is envisaged that sanctions should consist of penalty payments earmarked for use in the development of renewable energy (Energistyrelsen, 1998a). However, using penalty payments raises the question of whether the EU or the Member States should collect the penalty payments and redistribute the resulting revenue.

In the liberalized electricity market envisaged in Denmark, transmission and distribution will remain state-regulated natural monopolies in order to assure third-party access to the net and in order to safeguard that 'public service obligations' are observed. The regulated transmission and distribution companies have a long tradition for monitoring electricity producers while Member State authorities have a long tradition for controlling production as well as transmission. Sanctions may therefore be more effectively implemented at Member State level than at the EU level. The argument for decentralized control is further strengthened if bilateral trade in certificates is allowed because bilateral trade is not automatically registered at the power exchange. In the case of bilateral trade, certificates trading is carried out directly between large-scale consumers and producers, and this increases the need for detailed information about the local agents in the market.

3.2. *Managing the risks*

The second precondition for an effective system of Green Certificates is that the environment in which investments are carried out is stable and provides potential investors with clear expectations about price developments. If prices fluctuate widely, the risk of investing is too high, preventing at least some investors from entering the market. Then the question is how to minimize the risks of a market in Green Certificates.

One reason why renewable energy investments are uncertain is the changing conditions of production. Most renewable energy sources are difficult to predict due to climatic variations. If the rain is not falling, the sun is not shining, and the wind is not blowing, very little renewable energy is produced. Despite these changing conditions, fluctuations in the price of renewable electricity have been remarkably small in a number of Member States. This is, however, mainly due to the fact that the price of renewable electricity has been supported by subsidies and price guarantees ensuring the producer a minimum price. When a system of Green Certificates is implemented, minimum prices will be replaced by Green Certificates. The price of the certificates is, of course, highly uncertain. Therefore, it has been claimed that Green Certificates will make investments in renewable energy more uncertain compared to the existing price guarantees. There are, however, also arguments for the opposite to be true and ways to minimize the risk.

In a spot market, the price rises when supply is short. Therefore, in years when climatic conditions are unfavourable, the supply of Green Certificates will be low relative to demand, putting an upward pressure on certificate prices. In a system of subsidized prices, prices are guaranteed no matter the level of kWh produced, but the volume of kWh is still unpredictable. Therefore, a system of Green Certificates may create a more stable investment environment than subsidy schemes as higher prices on certificates counterbalance reduced earnings due to lower production. Therefore, it is possible that over time, earnings are more stable with a system based on Green Certificates than with a system based on price guarantees.

The argument presented above assumes that certificate trade can only take place in a climatically and technologically homogeneous market. A market like that of the EU is, however, not homogeneous. The EU covers a wide area and if the wind is weak in Southern Europe it may be strong in Northern Europe. If we assume an EU system of Green Certificates being implemented, then the heterogeneous market conditions have the consequence that the certificate price in low-wind areas (e.g. Southern Europe) will not rise because certificates can be purchased from high-wind areas (e.g. Northern Europe). Consequently, higher certificate prices may not counterbalance decreased supply of certificates due to climatic variations for the local producers specializing in one technology. One way to counterbalance the fluctuating prices of certificates would be to spread the risk by diversifying production into different technology types and different locations. In this case, higher prices of wind power in low-wind years may be counterbalanced by higher prices of hydro power if it is a wet year. This is, however, not a possible solution for small-scale producers, and insurance may be needed in order to deal with this type of risk.

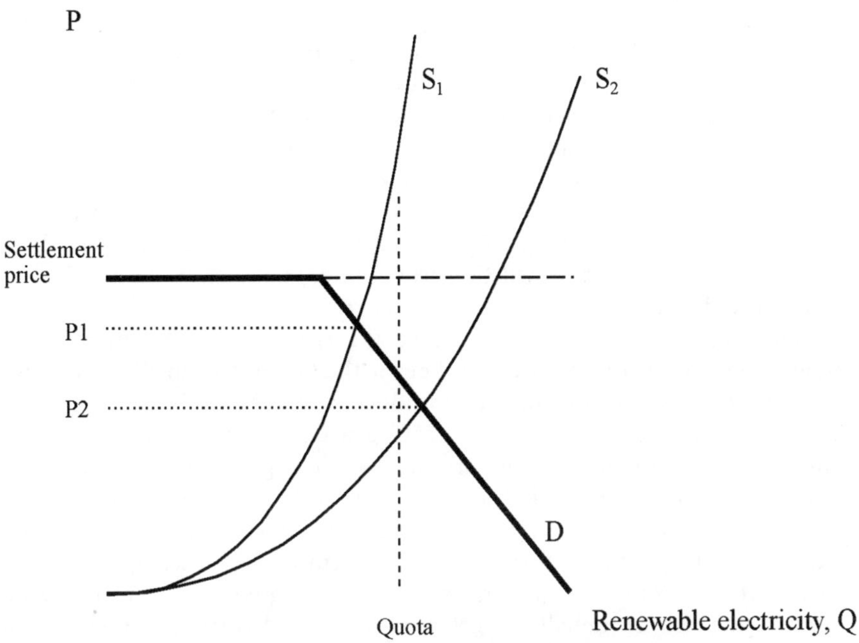

Figure 3. Supply and demand at a market for Green Certificates when banking is allowed.

In the Netherlands, Green Labels initially were valid for one year only. The unfavourable consequence of short-lived credits is that trade is not possible in the beginning of the year when no credits are yet produced. However, consumers may expect price elasticity on the supply side to rise infinitely at the end of the year because the credit is worthless in the following year. This may cause the certificate price to drop to zero if the market is characterized as a 'buyer's market'. If this is recognized by the consumers there will be no trade during the year and all trade will take place at the end of the year. Thus, the market for certificates with finite validity is likely to be inefficient. If consumption quotas are mandatory and the market is characterized as a 'seller's market', certificate prices may also be very close to the fixed settlement price. In a seller's market, producers may hold back certificates in order to put an upward pressure on certificate prices. If purchasers regard penalty prices as a real threat and if they fear under-supply, then they will be willing to pay a price very close to the fixed settlement price. These problems are avoided if certificates are valid for more than one year. This means that holders should be allowed to store unused certificates and use them in the following years. This will make demand more elastic as shown in Figure 3. Compared to Figure 2, the demand curve in Figure 3 is more elastic because of the banking opportunity. This narrows the span between P1 and P2 and hence reduces fluctuations in the certificates price. In a windy year (S_2), for instance, consumers can buy certificates relatively cheaply (P2) and use them in a subsequent less windy year (S_1) when the price is relatively high (P1). Moreover, in windy years producers will store certificates and sell them in less windy years when the price is higher.

A further risk aspect, which may hinder an efficient certificate market, is that investors are unfamiliar with such a market. Because the market is new to investors, risks and uncertainties are perceived to be very high, and this may make potential investors reluctant to invest in or to enter the market. If this is the case, the supply curve will be vertical and, in the short term, placed to the left of the demand curve.

An important way in which to reduce the risk of investing in renewable technologies is to set up a futures market for the Green Certificates along with the spot market. By selling futures, producers of renewable energy commit themselves to providing the buyer with Green Certificates at a fixed price in the future. This is of benefit to the producer as he is guaranteed the sale of his production and his investment becomes safer. A futures market also benefits the purchaser of credits as purchasers may be uncertain whether enough certificates will be supplied in the future for them to fulfil their quotas and to avoid penalty. Overall, a futures market would bring continuity to the system and reduce the risk on behalf of both the supply and the demand sides. Such a market is well know in electricity markets, where futures, and forward and option contracts have had a significant effect in reducing price fluctuations.

3.3. Incentives and disincentives in a market for Green Certificates

In Section 2, it was claimed that minimum consumption quotas may ease implementation as the production sector is not directly regulated by these quotas. However, what may be a problem is that quotas are tied to consumption and that 'green' and 'conventional' electricity are complementary goods.

Assuming that minimum consumption quotas of renewable electricity are fixed at 50% of total electricity consumption, and assuming stable demand as shown in Figure 4 this results in the consumption of Q_t kWh.

Assuming now that electricity based on renewable sources becomes cheaper due to the introduction of a new efficient technology, the new technology changes the total marginal costs from MC_{total} to MC^n_{total}, increasing demand for electricity to Q^n_t.

Due to complementarity of demand, demand for conventional electricity increases along with demand for renewable electricity. Consequently, the introduction of new, efficient renewable technology types may increase demand for conventional electricity in a system of Green Certificates because quotas are implemented on the demand side and electricity is sold in the same market irrespective of the fact that renewable and conventional electricity may be regarded as two different commodities. Hence, a system of Green Certificates may have contradictory properties: if the system is successful, investments and

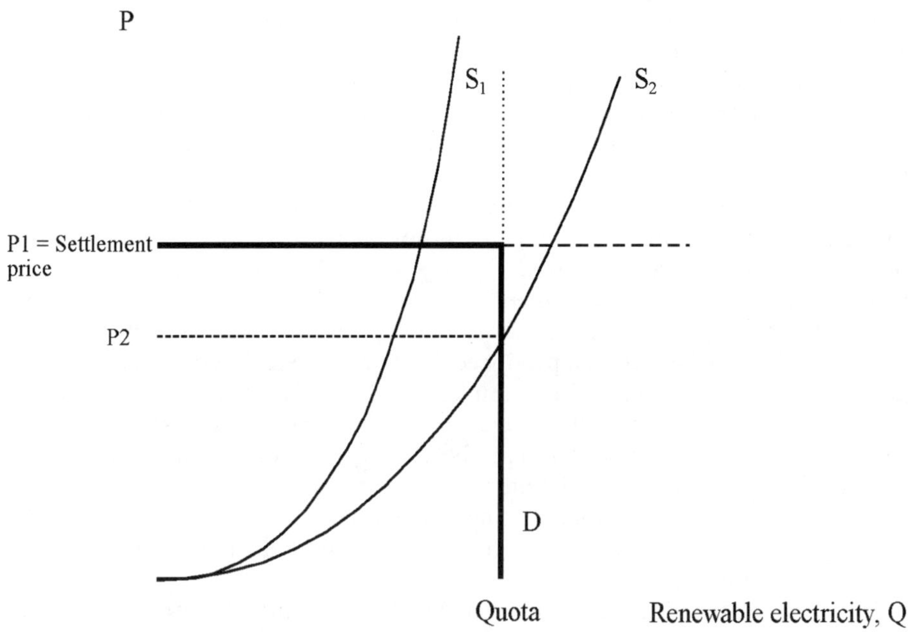

Figure 4. Complementarity in the electricity market when renewable and conventional electricity is regarded as one commodity.

innovation of renewable energy sources will rise, and the costs of renewable energy generation will fall, lowering the general electricity price; subsequently, demand for conventionally produced electricity will rise, resulting in adverse environmental effects.

The adverse environmental effects are, however, not unavoidable. A natural response to falling prices for renewable electricity would be to increase the consumption quota of renewable electricity. Naturally, this reduces the adverse environmental effect of Green Certificates.

3.4. *Relations to the principle of subsidiarity*

The principle of subsidiarity means that the EU should take action only if the action can be attained more sufficiently at the EU level than at the Member State level. According to the fifth action plan, subsidiarity should be implemented in a way that ensures the mixing of actors and instruments at the appropriate levels.[9] The Commission has implemented subsidiarity as a new law-making procedure, with the EU determining the goal and leaving it to the Member States to determine the means to achieve the goal.[10] In a European system of Green Certificates, the EU is determining the goal (e.g. 12% of electricity consumption must be renewable), but it does not necessarily leave it to the Member States to decide how to achieve it.

Traditionally, each Member State has decided the share and types of renewable technologies to be implemented. However, as quotas in a Green Certificates system are implemented on the demand side, it is left to the suppliers to decide which types of renewable technologies to implement. Evidently, the most competitive technologies, being the ones with the lowest certificate prices, will be implemented first.

Seen from an innovative perspective, this may not be optimal as it may delay the innovation of technologies which are not competitive in the short run, but potentially competitive in the long run (e.g. wave and solar energy). Therefore, in order to secure the development of uncompetitive renewable technologies some Member States may regard it as appropriate to determine specific consumption levels of electricity produced by specific technologies. The Danish government may determine, for example, that 25% of renewable energy should be produced by biomass technologies in order to induce domestic producers to innovate those technologies. The fixing of quotas on certain technologies may also be implemented by defining a different value on certificates generated by the technology in question. Certificates generated via solar energy may, for example, be worth 50% more than certificates generated via wind turbines due to political priorities.

However, in the Internal Electricity Market with an EU system of Green Certificates this option is no longer appropriate. In this situation, the national government cannot force electricity producers to use specific technologies because it would induce unfair competition among the producers.

As mentioned, due to natural monopoly characteristics of transmission and distribution companies, these remain subject to state regulation in the Internal Electricity Market. Therefore, the Member States have the possibility of forcing distributors to buy electricity produced by specific technologies in order to induce innovation. Such a policy would, however, not necessarily result in more innovation by domestic producers, because certificates may also be purchased from other Member States. Distributors are free to choose between producers, and would choose the producer with the lowest price. In this situation, foreign producers may oust domestic producers. Thus, if the Danish distributors are to buy renewable electricity produced by biomass technologies, Danish biomass producers (high-cost producers) may be forced out of the market by, for instance, Finnish producers (low-cost producers). Such detailed regulation of technology choice would of course not be cost-efficient but it might nevertheless be desired by some Member States. These Member States may obstruct a European system of Green Certificates due to subsidiarity reasons because their potential to determine the mix of renewable technologies is reduced.

4. Green Certificates and the flexible mechanisms of the Kyoto Protocol

In 1997 the third Convention of the Parties to United Nation's Framework Convention on Climate Change (UNFCCC) was held in Kyoto, Japan. At this summit, perceptible progress was made towards a global-warming agreement containing legal rights and obligations for the Parties signing and ratifying the Protocol. The Kyoto Protocol contains three flexible mechanisms for cost-efficient greenhouse gas (GHG) abatement. Below, these mechanisms will be briefly introduced in order to discuss and compare their differences and similarities as well as their compatibility with a European market of Green Certificates.

4.1. The flexible mechanisms of the Kyoto Protocol

Though the flexible mechanisms are all incorporated in the Kyoto Protocol, it is not specified how they are to be designed and implemented. For all three instruments the Protocol refers to the Convention of the Parties (CoP) or to the CoP serving as the Meeting of the Parties (CoP/MoP) to elaborate technical, institutional and procedural details.[11] Since the flexible mechanisms have not been thoroughly defined, the comparisons made with Green Certificates will be based on several assumptions regarding the specific implementation of the mechanisms.

The flexible mechanisms introduced with the Kyoto Protocol are Joint Implementation (JI), Emissions Trading (ET) and the Clean Development Mechanism (CDM).

Joint implementation

Joint Implementation allows an Annex B[12] country or an enterprise in that country to carry out a part of its commitment to reduce GHG concentrations in another Annex B country. The parties may transfer or acquire Emission Reduction Units (ERUs) resulting from projects and activities implemented in another Annex B country. A JI transaction involves an agreement between an 'investor' and a 'host' in which the host provides for ghg abatement services and transfers a quantity of this GHG abatement to the investor. ERUs gained by JI investments may be reported to the relevant institution of the FCCC and thereby included in the national GHG account. In return for credits, investors provide host Parties with financial resources, energy technologies, human capital or any transfer the Parties may agree on.

Emissions trading

Emissions Trading allows an Annex B country with an excess of emissions units to sell its credits to another Annex B country unable to meet its commitments. Credits are generated when a country reduces emissions below the commitment level. The main difference between JI and ET is that JI is based on bilateral trading while ET is based on a fully integrated multilateral trading system. Under a JI system, investors will finance specific projects and, depending on the organizational structure, negotiations and transactions will be carried out directly between project partners. Project partners accordingly will agree on a price per tonne of GHG reduction and the amount of GHG abatement to transfer. Under an ET system, participants will purchase fungible disaggregated credits at a pool, whereas the market price will be determined by demand and supply.

A second significant difference between JI and ET is that the JI policy seeks to pursue other goals than GHG abatement, although abatement is the primary objective. Other objectives in this respect could be closing the efficiency gap or increasing the living standard of the host countries. These considerations are usually based upon distributional aspects and once the large economic differences of the Annex B countries have been smoothed out, JI may lose its relevance. However, the fact that JI takes multiple aspects into consideration might make the mechanism more politically acceptable than ET.

Clean Development Mechanism

Clean Development Mechanism allows governments or private entities in industrialized countries to implement emissions reduction projects in developing countries in order to meet their commitments. The donor countries receive credits in the form of Certified Emissions Reductions (CERs). The main difference to JI would be that host countries are those which are developing and are not committed to fulfilling a reduction target under the Protocol. In contrast to JI and ET, the Kyoto Protocol furthermore identifies three institutional aspects to be implemented under the CDM. These aspects include the establishment of operational entities certifying emissions reductions resulting from each project. In addition, the CDM is to be supervised by an executive board. Finally the CDM is to assist in arranging funding of other CDM projects.

4.2. Green Certificates and the Kyoto mechanisms

Green Certificates and the Kyoto mechanisms possess several common features and complement each other in a number of respects.

Green Certificates equalize the marginal costs of producing renewable electricity while the Kyoto mechanisms equalize the marginal abatement costs. For that reason, all four instruments therefore present cost-efficient means to reduce GHG abatement. The Kyoto mechanisms and the Certificates, however, differ with respect to environmental certainty. As such, the Kyoto mechanisms are applied under a specific target for GHG emissions and the environmental impact is known in advance. Green certificates, on the other hand, define a minimum target for renewable electricity consumption. Therefore, the environmental impact of certificates is dependent on which type of conventionally produced electricity the renewable sources substitute.

In the Kyoto Protocol it is stated that the flexible mechanisms must be supplemental to domestic abatement. During CoP4, the EU strongly defended a strict definition of the criteria of supplementarity. As such, the EU suggested that at least 50% of each Party's commitment must be fulfilled by domestic GHG abatement. A system of Green Certificates contributes to fulfilling the criteria of supplemental action as the share of renewable energy sources is increased throughout the territory of the EU.

The Kyoto mechanisms may be implemented either by a common UN policy or by each Party defining its own criteria for the basis on which the instruments will be implemented. The EU Member States have signed the Kyoto Protocol both as individual Member States and as one Party under the auspices of the EU. Therefore the EU may decide to implement the Kyoto mechanisms within a common EU framework. In terms of EU and the Kyoto mechanisms, ET is the most relevant mechanism to implement as a common EU policy including Member States only. CDM and probably also JI include countries outside the EU, and policies regarding these instruments must be based on negotiations between all relevant Parties. ET may be carried out between Member States in an EU market, and assigned amounts have already been defined by the internal EU burden-sharing agreement.[13]

Both Green Certificates and ET are most appropriately organized through a centralized market – a pool. In order to provide transparency and to reduce transaction costs, a futures market should be organized along with the spot market. Another similarity regards the fact that price elasticities may be infinite if countries are not allowed to transfer units produced or bought in one year to the subsequent years. Therefore, the trade of emissions units and Green Certificates should be organized similarly through a pool with banking opportunities and a futures market. However, it is clear that Green Certificates and emissions units are not directly tradable with each other, as the units contained in a Green Certificate consist of renewable watt-hours while the units in ET are CO_2 emissions units.

It is, however, very likely that the two markets have a mutual impact. For instance, if a system of Green Certificates imposes on each consumer the purchase of an increased share of renewable electricity, higher demand increases the certificate price, attracting more investments to the production of renewable electricity. When the consumption of renewable electricity increases, the consumption of conventionally produced fossil-fuel electricity presumably decreases correspondingly. Consequently, CO_2 emissions decrease and parts of the assigned amounts are released from the national GHG accounts. These emisssions units are tradeable on the ET market and may be traded either in the EU or on the world market. Producers of renewable electricity may thus gain a double bonus via the ET system because Green Certificates trigger trade on the ET market.

Of course an efficiently implemented ET system would be sufficient in order to achieve the desired level of emissions. In this respect it need not be supplemented by a system of Green Certificates. A system of Green Certificates is, however, designed to regulate the electricity sector specifically. Therefore, a system of Green Certificates may supplement an ET system if it is an explicit political objective to increase the share of renewables in the production of electricity. An ET system alone may not regulate a specific sector in detail.

The fact that a system of Green Certificates can be used as a detailed energy sector regulation may cause some Member States to oppose to the system. In some Member States a system of Green Certificates may be viewed as a system which assign powers to the EU level which Member States should have. In an efficiently implemented system of Green Certificates it is no longer possible for Member States to maintain the different support schemes that promote renewable energy. Therefore, a system of Green Certificates implicitly harmonizes national renewable energy policies. An ET system does not require such harmonization. On the contrary, in an ET system the EU only determines the environmental target and leaves it for the Member States to choose the means to achieve it. Consequently, Member States can maintain different support schemes promoting renewable energy.

Market size is another factor which seems to show the benefit of an ET system over Green Certificates. It is well known that the success of a market depends on market size. The ET market is obviously larger than the market for Green Certificates, as the certificate market only constitutes one part (electricity generation) of the total ET market. The ET market involves large CO_2 emitting sectors outside the electricity sector such as transport, industry, forestry and agriculture.

A potential barrier of an ET system compared to Green Certificates relates to the initial distribution of rights and obligations. In a system of ET, rights must be distributed initially with regard to emissions. The EU recently agreed to a burden-sharing agreement. This agreement allows the southern Member States (and Ireland) to increase GHG emissions in the period 2008–2012 compared to 1990 emissions while most northern Member States (except Sweden) are to carry out substantial emissions reductions. Although Member States have agreed to the burden-sharing agreement, it is rather questionable if they will actually respect this initial allocation if a European ET system is implemented. Will Member countries with energy-efficient production sectors agree to the fact that they must bear the GHG abatement costs of the countries that have done very little to increase energy efficiency? This question may imply widespread political controversies inside the EU. A system of Green Certificates partly solves this issue as regulation is implemented on the demand side, and no Member State is given an obligation to reduce GHG emissions by a certain amount. Obligations rather are placed with the individual EU citizen, and changes in production will take place due to changes in demand. As such, production changes will take place where these are implemented at the least costs, and marginal costs of compliance are equalized throughout the EU.

5. Concluding remarks

With the liberalization of the European electricity market, new instruments are searched for in order to co-ordinate climate change initiatives with competition.

In this paper we have argued that a European system of Green Certificates is a cost-efficient method to induce renewable energy production and to reduce CO_2 emissions. A system of Green Certificates combines the advantages of command-and-control regulation with market-based incentive regulation as the goal is determined centrally by for instance the EU, and the market ensures cost-efficient behaviour by producers and consumers.

In a system of Green Certificates, producers receive certificates corresponding to the amount of renewable electricity produced. Consumers are obliged to purchase certificates representing a share of for instance 12% of their consumption. Thus, producers sell certificates to consumers, and the certificate price covers the cost difference between conventionally produced electricity, and electricity produced by renewable technologies.

A system of Green Certificates implicitly harmonizes different national subsidy schemes supporting renewable energy. Because price guarantees contradict a liberalized market, these must be abolished. However, Member States may maintain subsidy schemes provided these are non-discriminatory, in which case governments may have to subsidize foreign companies. Green Certificates imply that consumers instead of the exchequer pay the additional costs of renewable production through the certificate price and that governments therefore avoid the situation of subsidizing foreign producers. However, Green Certificates also imply that governments lose the power of controlling the market through issuing subsidies. Therefore, the system may contradict the principle of subsidiarity and thus be opposed by some Member States.

Although Green Certificates may be a promising alternative to price guarantees, subsidies and production quotas, some obstacles do exist which impede the introduction of a European Certificate market. One of these obstacles includes the risk that consumers may find the threat of sanctions too weak and therefore choose not fulfil their quotas. This implies that producers have no incentive to increase production of renewable electricity and that no market for Green Certificates is developed. In order to safeguard the credibility of the certificate market, efficient procedures for monitoring, control and sanctions must be established at Member State level.

A second factor which implies that a European market for Green Certificates may not develop is the risk characterizing the generation of electricity based on renewable sources. The price of Green Certificates is difficult to predict due to the fact that many production technologies are dependent upon climatic circumstances. In a market with production based on homogeneous technology types, low production of renewable electricity in years with poor climatic conditions is counterbalanced by higher certificate prices. However, in a market where electricity generation is based on several competing renewable technologies, producers may experience great losses due to bad climatic conditions and due to low certificate prices as other technology types may perform well in these years. This type of risk may be dealt with through geographic and technological diversification of investments or through appropriate insurance arrangements.

A third potential obstacle concerns the lifetime of the certificates. Experiences from the Dutch market illustrate that finite validity of certificates is inefficient and may cause the price elasticity to rise to infinity. In order to provide for an efficient market without drastic price fluctuations at the end of the year, banking should be allowed in order to save up certificates for years of low supply. As such, banking implies that renewable energy may be 'stored' as financial assets.

In addition to banking, a futures market for Green Certificates may reduce the risk in the market. By fixing the price of future certificates, producers become familiar with the yield of their investment, provided production is carried out as predicted. A futures market brings continuity into the system and reduces the potential risk on behalf of both consumers and suppliers.

A forth obstacle is related to the fact that renewable and conventional electricity are supplied as complementary goods at the same market price. This relationship implies that a reduced price on renewable electricity increases demand for fossil-fuel-based electricity, and thus have an adverse environmental effect. This adverse effect could be counterbalanced by increasing the consumption quota on renewable electricity. Hence in order to gain the intended environmental effect from Green Certificates, marginal costs of generation must be observed and the quota adjusted in accordance with these.

The principle of subsidiarity presents a final potential obstacle to a European market for Green Certificates. Member States traditionally have had the power to select a specific technology mix after their own choice. A good reason to preserve this system is that the innovation of certain technology types is given priority and thereby speeded up. In a European Certificate market such detailed regulation is still possible but no longer appropriate. Member States may specify the quotas of the consumers as regards technology shares. However, this does not necessarily influence domestic R&D as certificates from specific technology types may be cheaper abroad and no incentives will be given to domestic producers. Due to the fact that governments may no longer define the domestic technology mix in production, Green Certificates may be objected under the argument of subsidiarity.

A system of Green Certificates has a number of similarities with the flexible mechanisms of the Kyoto Protocol. First of all, the instruments are cost-efficient within their given framework. Green Certificates equalize the marginal costs of generating renewable electricity, while the flexible mechanisms equalize marginal costs of greenhouse gas abatement. Second, Green Certificates are compatible with the criteria of supplementarity as defined in the Framework Convention on Climate Change. Green Certificates, when restricted within the EU, cause GHG emissions to decrease in Europe and therefore supplement the Kyoto mechanisms. In fact, the implementation of the Kyoto mechanisms increases the value of renewable energy sources, because they produce renewable electricity and at the same time generate GHG emissions reductions.

Regulation by Green Certificates is directed towards the electricity sector alone. The Kyoto mechanisms have a broader focus and cover all GHG-emitting sectors. Due to the narrow focus, some states may view Green Certificates as too much interference with national policies, and due to that prefer Emissions Trading as a regulatory instrument. Thus, Green Certificates supplement the flexible mechanisms of the Kyoto Protocol, but do not qualify as a substitute.

Notes

1. Directive 96/92/EC *Official Journal* L 027, 30/01/1997.
2. Due to public and environmental service obligations, as stated in Art. 3(2) of the directive.
3. See Community guidelines on state aid for environmental purposes (*Official Journal* C72, 10.3.1994, page 3).
4. In a fully liberalized market, electricity traders could also manage the obligation.
5. EnergieNed: 'The Green Label'. Publication from EnergieNed. For more information see www.groenlabelned.nl.
6. Several schemes supporting the Green Labels are in place: Renewable electricity is ensured a guaranteed price above conventional electricity, certain 'green funds' are exempted from paying taxes, companies can write off 'green investments' faster than other investments, tax deductions may be claimed for large renewable-energy installations, energy taxes are reimbursed.
7. Methane from dung and waste.
8. Nevertheless, renewable electricity production will be affected through a change in the relative prices.
9. Environmental Action Programme 1993–2000, 'Towards Sustainability', *Official Journal* C138, 17 May 1993.
10. See for instance the proposals for new water policy directives: Proposal for a new bathing water directive, COM(94) 36 final, proposal for a directive on the ecological quality of water, COM(93) 680 final, Proposal for a new drinking water directive, COM(94) 612 final, and proposal for a directive establishing a framework for Community action in the field of water policy, COM(97) 49 final.
11. UN FCCC, *the Kyoto Protocol*, 1997
12. Annex B countries include the OECD member countries, a number of former Soviet republics and the Central and Eastern European countries.
13. With the Kyoto Protocol 'bubbles' were introduced as a fourth flexible instrument. Bubbles are not based on trade and so far the EU is the only Party to the Protocol officially forming a bubble. A bubble policy implies that countries within the bubble may redistribute GHG commitments internally.

References

COM(97) 599 final. Energy for the future: renewable sources of energy. White Paper for a Community Strategy and Action Plan.

COM(98) 167 final. Report to the Council and the European Parliament on harmonization requirements. Directive 96/92/EC Concerning Common Rules for the Internal Market in Electricity.

Drillisch, J. (1998). Quotenregelung für erneuerbare Energien und Zertifikatshandel auf dem niederländischen Elektrizitätsmarkt. *Zeitschrift für Energiewirtschaft*. 4:247–63.

Energistyrelsen (1998a): Notat om det grønne elmarked, herunder overgangsordning. (Note on the green market, including a transitional period.) Copenhagen, The Danish Energy Agency.

Energistyrelsen (1998b). Notat om de hollandske erfaringer med et grønt marked. (Note on the Dutch experiences with a green market.) Copenhagen, The Danish Energy Agency.

Energistyrelsen (1999). Aftale om ny elreform. (Agreement on a new electricity reform.) Copenhagen, The Danish Energy Agency.

Haas, R., N. Wohlgemuth, and C. Huber (1999). Promotion strategies for renewable energy systems in liberalised electricity markets. Paper presented at the Forty-Seventh International Atlantic Economic Conference. Vienna, 16–23 March.

Skytte, Klaus (1999). Modelling markets for green certificates and interaction with other electricity markets. Paper presented at the workshop: Co-ordination in deregulated electricity markets. 22–24 March, Ivalo, Finland.

PART 3

International and Intergenerational Equity Issues

CHAPTER 11

A Dynamic Analysis of the Efficiency and Equity of Tradeable Greenhouse Gas Emissions Permits*

Adam Rose and Brandt Stevens

1. Introduction

In December 1997, 38 industrialized and transitional countries agreed to the Kyoto Protocol, which committed them to targets and timetables to reduce six greenhouse gases (GHGs). In November 1998, in Buenos Aires, at the Fourth Conference of the Parties to the Framework Convention on Climate Change (COP-4), efforts were made to improve upon Kyoto by expanding the number of co-operating countries and enhancing the range of policy instruments to implement its objectives. However, COP-4 was a failure, as only two developing countries made commitments, and little progress was made in increasing the flexibility of policy implementation.

Why is the failure of the Buenos Aires meeting so important? First, a truly global problem such as climate change cannot be adequately addressed by a relatively small subset of the world's nations. Although industrialized countries currently generate the major proportion of GHGs, projections of future growth in developing countries indicate they as a group will be the major emitters after a few decades. Second, developing countries have the relatively lower-cost options for mitigating GHGs. In the context of flexible policy instruments, a system of globally tradeable GHG emissions permits has the potential to decrease global mitigation costs considerably. At the same time, policy makers need to be sensitive to alternative concepts of fairness, both for their own sake and for pragmatic reasons of forging a global agreement. Many industrialized countries view as unfair their obligation to shoulder the entire burden under the Kyoto Protocol. On the other hand, developing countries point to the fact that they are responsible for only a small portion of the stock of GHGs in the atmosphere, and that the opportunity costs of diverting resources from current needs and future development are immense. A mechanism for reconciling these two opposing views is a must for progress in climate change policy.

The purpose of this paper is to show how a system of globally tradeable GHG emissions permits can overcome the current policy stalemate. Cost-effectiveness

C. Carraro (ed.), Efficiency and Equity of Climate Change Policy, 247–271.

can be enhanced by bringing in developing countries and by making the permits tradeable over time. The equity dilemma can be overcome by initial permit allocations that impose little or no cost on developing countries and simultaneously lower the costs to industrialized nations. Further equity issues can be addressed by subsequently adjusting the allocation of permits across all countries.

The entire situation, however, is fraught with uncertainty, especially with regard to future availability of energy resources and new technologies. This has a major bearing on both users and producers of energy resources in terms of mitigation costs, and has major implications for the feasibility of a truly global treaty.

A dynamic non-linear programming model is used to evaluate the efficiency and equity implications under alternative conditions of permit allocations, permit trading over space and time, timing of commitments, and technological change. Special attention will be placed on the implications for Western Europe, Eastern Europe, the former Soviet Union, and major developing countries. However, the analysis is limited to only one of the GHGs – carbon dioxide.

The model is a dynamic extension of that formulated by Rose *et al.* (1998). It is structured so that the implications of increasing flexibility in permit trading is apparent both theoretically and empirically. For example, for the 14 major world countries and country groupings and over a 25-year time horizon, dynamic optimization under intra-country permit banking involves 364 constraints. The number of constraints is reduced progressively as trading extends temporally and spatially. Analogously, global abatement costs decrease in a progressive manner as well. Another feature of the model is that the benefit functions are tied to CO_2 concentration levels through the Peck and Teisberg (1992) reduced-form climate equations. Furthermore, sensitivity analyses are performed for a range of projections regarding technological change and resource availability.

The relative prominence of each refinement of the permit trading mechanism has important policy implications. Climate change negotiations are a source of tension between countries, and advance knowledge of which factors are most effective at reducing costs would help make these options the major focus of attention. In addition, various improvements in policy instrument design for the sake of improving efficiency are likely to have differing implications for equity. For example, permit banking opportunities will favour some countries, while borrowing will favour others. The timing of abatement commitments for developing countries will influence inter-country equity as well. Often these implications will accentuate the tensions, and advance knowledge could help reduce them by the specification of additional transfers or a manipulation of permit allocations.

2. Advantages of marketable permits

The agreement reached at the Third Conference of the Parties to the Framework Convention on Climate Change (COP-3) in Kyoto commits most of the industrialized nations to emission caps for each of six major GHGs by the year

2010, as shown in the second column of Table 1. In addition, the agreement allows individual nations to achieve even lower emissions and to bank their excess reductions for later use. Countries can also develop projects for reduction of GHG emissions in other countries through bilateral agreements in the form of Joint Implementation (JI) and the Clean Development Mechanism (CDM) (FCCC, 1997).[1]

Tradeable GHG permits, a form of multilateral co-operation, had been proposed at prior Framework Convention Meetings (see, e.g., Barrett *et al.*, 1992; Tietenberg and Victor, 1994), though only a few limited features of this instrument were included in the Kyoto Protocol. Economists familiar with permit trading will recognize the banking provision of the agreement as a limited form in which individual nations can 'sell' permits to themselves to be used in the future. JI and CDM could also be viewed as another limited form of permit trading, where all trades are along bilateral lines rather than in a market with many buyers and sellers.

Potentially sizeable gains in both efficiency and equity could arise from establishing more open and complete permit markets. First steps toward an unrestricted market would include permit transfers between at least the Annex B countries, credits for sponsoring mitigation in developing countries, and then including developing countries as signatories. Further steps would be to allow interregional trading and trading of permits across time (including borrowing as well as banking).[2]

The case for international trading of GHG permits rests on a well-known theorem by Nobel laureate Ronald Coase (1960). In this context, suppose that a specific quantity of transferable greenhouse gas permits, determined according to a GHG Protocol, is distributed without charge to its signatories. The stock of permits effectively sets an overall cap on emissions. Next assume that countries are allowed to exchange permits in an open market. This effectively gives each country a stake in the environment, which is valued by the market. A country whose marginal abatement cost is higher than the market permit price will buy permits from those countries whose marginal abatement costs are below it. When trading concludes, each country will hold the economically efficient amount of permits (i.e., marginal abatement costs across nations will be equal).[3]

The equity implications of the Coase Theorem are ambiguous, however. Overall global wealth is maximized and is identical regardless of how the permits are initially distributed, but equity is highly dependent on the permit distribution (as well as on mitigation costs and benefits). This separability of efficiency and equity is another major advantage of tradeable permits. In many other approaches to policy making, pursuit of equity will lead to disincentive effects, thereby undercutting efficiency. At the same time, separability has been used as an excuse to ignore equity, since it has no bearing on the pursuit of what economists consider the primary goal in most contexts.

Table 1. Basic socioeconomic and emissions data

Country (Area)	Mitigation percentages				GDP[c] (2010)		Population[d] (2010)		CO_2 emissions[e] (2010)	
	1990[a]	2010[a]	2020[a]	2035[a]	Level (US$ billion)	Growth rate	Level (million)	Growth rate	Level (million t C)	Growth Rate
Africa	0.0	0.0	0.0	0.0	393.7	4.0	741.2	3.0	296.2	4.6
Aus/NZ	−7.0	1.2	8.8	19.0	382.0	1.2	23.3	1.3	82.0	0.8
Canada	6.0	10.6	14.9	21.1	642.4	1.2	28.7	0.8	117.3	0.5
China	0.0	0.0	0.0	0.0	653.5	6.0	1290.0	1.3	817.5	2.6
E-Europe	6.0	12.3	18.3	26.4	337.9	1.9	126.1	0.2	290.9	0.7
FSU	0.0	6.8	13.0	21.7	575.7	1.9	151.7	0.2	1084.5	0.7
India	0.0	0.0	0.0	0.0	414.6	5.0	1005.5	1.7	267.8	3.8
Japan	6.0	13.2	19.8	28.9	3315.7	1.2	127.3	0.3	310.9	0.8
L-America	0.0	0.0	0.0	0.0	1345.2	3.3	509.8	1.8	374.2	3.2
M-East	0.0	0.0	0.0	0.0	685.5	5.0	287.9	2.9	310.9	3.8
Asian Tigers	0.0	0.0	0.0	0.0	797.4	5.0	74.7	0.4	161.5	3.4
S-Asia	0.0	0.0	0.0	0.0	591.6	4.1	835.9	1.6	241.4	5.1
US	7.0	11.5	15.8	21.9	6075.3	1.2	270.7	0.8	1354.9	0.5
W-Europe	8.0	14.2	20.0	27.9	7825.7	1.2	468.2	0.8	977.2	0.7
Total	3.5[b]	6.9	9.2	10.9	24036.2	1.9	5941.0	1.6	6687.2	1.7

[a] Base case where developing countries have not joined Kyoto Protocol.
[b] Reduction percentage for Annex B countries alone is 5.0%.
[c] IPCC (1996); Nordhaus and Yang (1996).
[d] World Bank (1996).

There are cases, such as international treaties, however, where global efficiency may not be a rallying-point and where equity instead may be paramount, as in global climate change, where efficiency pursuits alone may lead to major disparities in costs and benefits between countries (see, e.g., Rose, 1992; Manne and Richels, 1995). Individual country welfare is not addressed directly by global efficiency, but is subservient to what is often considered an abstract goal. Countries are more likely to join in an accord that addresses their own welfare and its relation to that of other countries directly. Therefore, we can stand on its head the traditional rationale for ignoring equity. If efficiency is not affected by how permits are distributed, then let us fine-tune the distribution in any way necessary to pursue equity objectives. Interestingly, this pursuit will actually promote efficiency as well. We have taken for granted the issue of the comprehensiveness of the number of parties to an agreement. If more parties can be enticed into it by appealing to their equity concerns, environmental quality will be further enhanced. If these parties have relatively lower abatement costs than the original signatories, overall economic efficiency will be improved.

Note also that the Coase Theorem generalizes to the dynamic context. Permit trading over time promotes even higher levels of efficiency by capitalizing on abatement cost differentials in different periods. In this case the permit price is determined not only by benefit and cost considerations in the pure efficiency case but also by the fact that the permits represent a fixed stock. Thus, there is an intertemporal opportunity cost (user cost) that influences their value. As in the basic model of the extraction of a non-renewable resource following Hotelling, the price of a permit will appreciate in conformance with the market rate of interest. Trading of GHG emissions permits over time will be consistent with this outcome. Moreover, the dynamically efficient (post-trading) outcome is unique and is insensitive to alternative permit distributions.

The stock of permits at any time depends on the global abatement requirement, which is derived from emissions targets. As long as costs do not exceed benefits of increased tightening of this requirement overall, a sustainability definition of intergenerational equity can be applied globally. Alternatively, this sustainability criterion might be achieved by varying initial permit allocations or making additional transfers.

3. The rationale for permit trading

The exchange of transferable greenhouse gas permits is a function of differences in marginal abatement costs, which can stem from several sources. The most obvious source of cost differentials is different marginal cost of abatement (MC_A) functions among countries. However, even if two countries had the same MC_A functions, a cost differential would arise if they had different emissions reduction requirements (placing them at different points on the function).

Thus trading would arise on this basis alone among Annex B countries because of differences in Kyoto emissions cap commitments.

Of course, there are differences in abatement cost functions among Annex B nations, though these functions differ more significantly between the group of developing nations and the industrialized group (see, e.g., IPCC, 1996; Halsnaes, 1996). Permit exchanges among Annex B countries would reduce the group's total GHG abatement costs incurred by the Kyoto targets. It therefore follows that, if developing nations have MC_A functions that are generally lower than those of Annex B nations, finding a mechanism for admitting them to the permits market, even by giving them permits equal in number to their emissions every year (the 'No Harm' rule), could decrease global costs.

There are also likely to be differences in abatement costs over time. With the relatively long period over which mitigation would occur, technological change could reduce these costs, perhaps substantially. On the other hand, rising abatement percentages over time (i.e., abatement levels relative to uncontrolled greenhouse emissions) will increase MC_A temporally. In addition, the discount rate will affect the perception of future costs. In the absence of technological change and rising abatement percentages, any positive discount rate will reduce future MC_A.

These dynamic considerations suggest the possibility that shifting abatement from times when it is relatively expensive to times when it is relatively cheap could reduce the present discounted value of global abatement costs. Interperiod trading of greenhouse gas permits could have two forms of outcome. The first would allow for early abatement above the required targets; the permits created by these excess emissions reductions could be 'banked' and used in some future period. This type of exchange could be restricted so that the country creating the banked permits could be the only nation to use them later, or they could potentially be sold for future use by other nations. From an economic perspective, banking only makes sense if discounted MC_A is lower in the present than in the future. The second option, permit 'borrowing,' would result from the converse, i.e., discounted MC_A being greater in the present than in the future, perhaps due to technological change. Under this outcome, it is cheaper to defer abatement to the future, which can occur when permits are transferred to the present.[4]

Another dynamic consideration is the timing of commitments themselves. Individual country welfare will, ceteris paribus, be higher, the longer emissions targets are deferred. This ties timing to fairness as well. However, total global welfare would not be affected by timing of commitments, if all countries are involved in permit trading during each year of the planning horizon and as long as the cumulative stock of permits is fixed.

4. The simulation model

Our simulations were performed with a dynamic non-linear programming model consisting of the abatement cost and benefit functions for the 14 countries or country groups listed in Table 1. The empirical base of the model is an enhancement of that used in Rose and Stevens (1993) in three major ways: (1) an extension to the entire globe and a finer delineation of countries, (2) an updated set of marginal abatement cost functions, and (3) an updated set of marginal benefit functions. The conceptual basis is an advance over that presented in Rose *et al.* (1998), in that a dynamic formulation is established by relaxing constraints on permit trading balances to reflect different types of intertemporal trading. Also, the benefit functions are tied to CO_2 concentration levels through the Peck and Teisberg (1994) reduced-form climate equations.

To facilitate the exposition, we will present the models in a cost-effectiveness context. This takes a global target of emissions reductions as given (as in the Kyoto targets) and minimizes the cost of achieving it across countries. Benefits are calculated separately through a set of individual country functions exhibiting diminishing returns to damage reduction. Thus, the objective function of our model can be specified as:

$$\text{minimize } TC = \sum_{i=1}^{n} \left[\frac{(1-R_i)\,ln(1-R_i)+R_i}{\alpha_i} \right] E_i \qquad (1)$$

where

$TC \equiv$ total global abatement cost (endogenous)
$R_i \equiv$ percentage abatement for country i (endogenous) $0 \leq R_i \leq 1$
$E_i \equiv$ gross (unabated) CO_2 emissions (in tonnes) for country i (exogenous)
$\alpha_i \equiv$ slope parameter of abatement cost function for country i (exogenous)

Our illustrative total abatement cost function yields a marginal cost function of the form:

$$-\ln(1-R_i)/\alpha_i$$

For positive α_i, the function exhibits positive and increasing marginal cost as percentage abatement increases, consistent with theoretical expectations and empirical findings on diminishing returns for pollution control.

For criteria oriented directly to permit allocations, as opposed to welfare outcome (see Rose *et al.*, 1998), we minimize the sum of global abatement costs, TC, subject to the following constraints:

$$(1-R_i)E_i - P_i \leq \bar{P}_i \qquad i=1,\ldots,n \qquad (2)$$

$$\sum_{i=1}^{n} P_i = 0 \qquad (3)$$

where

$\bar{P}_i =$ permit allocation (in tonnes) for country i (exogenous)
$P_i \equiv$ permit purchases or sales (in tonnes) for country i (exogenous)

and other variables are as previously defined.

Equation (2) enables us to examine the consequences of alternative permit allocations on mitigation levels and costs. The constraint requires that each country's initial stock of permits, as designated by a permit distribution criterion (equity-based or otherwise) be greater than or equal to its gross annual CO_2 emissions minus emissions abated minus emissions permits bought or sold. Suppose, for example, that a treaty is based on the Sovereignty criterion requiring abatement of each country's emissions equal to 20%; this defines a permit distribution of 80% of each country's gross emissions, E_i. Relative marginal abatement costs would determine each country's decision to add additional permits to its stock and abate less than 20% or to sell permits from its stock and abate more than 20% of initial emissions.

In equation (2), we specify individual permit allocations, $\bar{P}_i$, and the model solves for the individual country equilibrium (post-trading) abatement percentage, R_i, and permit trades, P_i, so as to minimize the sum of global abatement costs. The model equalizes the marginal abatement costs of all nations at a level equal to the equilibrium permit price, which it calculates as well. Equation (3) requires permit purchases to equal permit sales (negative values of P_i).

Major features of the solution to this problem are invariant to the permit distribution. Consistent with the Coase Theorem, different distributions of a fixed, global permit supply should not affect the global permit price, the ith country's optimal percentage abatement or each country's post-trading stock of permits. Alternative initial distributions of permits do affect the number of permits bought and sold by each country and individual country net benefits. All of these considerations will be illustrated by the analysis below.

Incorporating banking and borrowing is readily accomplished by substituting the following constraint sets in sequence. The flexibility of a truly generalized permit trading system is indicated by the fact that the number of constraints decreases from 364 to 1 in the process.

Banking only (no borrowing/no interregional)	net deposits of permits in each time period must be zero or positive for each country	(3a)

$$\left.\begin{array}{l}\sum_{t=1} P_{i,t} \geq 0 \\ \sum_{t=1}^{2} P_{i,t} \geq 0 \\ \vdots \\ \sum_{t=1}^{26} P_{i,t} \geq 0\end{array}\right\} \text{ for all } i \qquad [26 \text{ (years)} * 14 \text{ (country/area) constraints}]$$

Interperiod Trading only (banking and borrowing/ no interregional)	sum of purchases and sales of permits must be zero at end of time horizon for each country	(3b)

$$\sum_{t=1}^{26} P_{i,t} = 0 \quad \text{for all } i \qquad \text{(14 constraints, one for each country/area)}$$

Interperiod and Interregional Trading	sum of purchases and sales of permits must be zero at end of time horizon	(3c)
$\sum_{i=1}^{14} \sum_{t=1}^{26} P_{i,t} = 0$	(one constraint only)	

5. Empirical analysis

In order to evaluate the relative gains from improvements in policy design, we performed several step-wise simulations. They begin with the most basic institutional arrangement – CO_2 emissions quotas for Annex B countries as set forth in the Kyoto Protocol – and progress to the ultimate form – a globally comprehensive and fully dynamic CO_2 emissions trading system.

1. No-trading (Kyoto-based emissions quotas for Annex B countries)
2. Trading by Annex B countries only (Kyoto-based tradable quotas)
   a. trading among countries or groups (interregional trading)
   b. banking
   c. banking and borrowing (complete interperiod trading)
   d. interregional and interperiod trading
   e. interregional and interperiod trading and intergenerational equity
3. Trading among Annex B countries and developing countries (equity-based quotas for developing countries)
   a. interregional trading (No Harm to developing countries)
   b. banking
   c. banking and borrowing
   d. interregional (No Harm) and interperiod trading
   e. timing of developing-country emissions reduction commitments
   f. interregional and interperiod trading and intergenerational equity

The timing simulations merit further explanation, as we examine welfare impacts of alternative commitment dates for two groups of developing countries based on their per capita income (PCI): Group A (PCI > US\$1000) Middle East, Asian Tigers and Latin America; and Group B (PCI < US\$1000) China, South Asia, India and Africa. We ran two timing simulations: (1) All developing countries commit to quotas based on 2020 emissions, and (2) Group A developing-country quotas and the start of their commitments is in 2010, and Group B developing-country quotas and the start of their commitments is in 2020. Only selected results are presented in this paper, with an emphasis on implications for Western Europe, Eastern Europe, and the former Soviet Union (FSU).

6. Simulation results

6.1. *Individual cases*

The results of selected simulations are presented in Tables 2–6.[5] The results in Table 2 represent the most basic outcome-fixed Kyoto target emissions levels

Table 2. Benefits and costs of CO_2 mitigation without permit trading: Kyoto target levels for Annex B countries and no mitigation for developing countries (in 2010 present discounted value billions of dollars)

Country	Mitigation benefit			Mitigation cost			Net benefit		
(Area)	2010	2020	2035	2010	2020	2035	2010	2020	2035
Africa	2.77	3.96	4.85	0.00	0.00	0.00	2.77	3.96	4.85
Aus/NZ	0.11	0.16	0.19	0.00	0.07	0.22	0.11	0.09	−0.03
Canada	0.19	0.27	0.33	0.21	0.30	0.37	−0.02	−0.03	−0.04
China	6.44	9.21	11.28	0.00	0.00	0.00	6.44	9.21	11.28
E-Europe	0.20	0.28	0.35	0.23	0.37	0.50	−0.03	−0.11	−0.15
FSU	0.12	0.17	0.21	0.39	1.07	1.89	−0.27	−0.90	−1.68
India	2.81	4.02	4.92	0.00	0.00	0.00	2.81	4.02	4.92
Japan	0.96	1.37	1.68	0.90	1.53	2.10	0.06	−0.16	−0.42
L-America	3.34	4.77	5.84	0.00	0.00	0.00	3.34	4.77	5.84
M-East	3.67	5.24	6.42	0.00	0.00	0.00	3.67	5.24	6.42
Asian Tigers	1.67	2.38	2.91	0.00	0.00	0.00	1.67	2.38	2.91
S-Asia	4.31	6.17	7.55	0.00	0.00	0.00	4.31	6.17	7.55
US	1.63	2.32	2.86	2.62	3.57	4.18	−0.99	−1.24	−1.33
W-Europe	2.26	3.23	3.96	3.30	4.84	6.01	−1.03	−1.61	−2.05
Total	30.47	43.54	53.34	7.65	11.75	15.28	22.82	31.79	38.06

Table 3. Benefits and costs of CO_2 mitigation with permit banking and borrowing for each Annex B country: Kyoto target levels for Annex B countries and no mitigation for developing countries
(in 2010 present discounted value billions of dollars)

Country (Area)	2010							2020				2035			
	Pre-trading			Post-trading				Post-trading				Post-trading			
	Benefit	Cost	NB	Benefit	Cost	Permit US$	NB	Benefit	Cost	Permit US$	NB	Benefit	Cost	Permit US$	NB
Africa	2.77	0.00	2.77	2.88	0.00	0.00	2.88	3.89	0.00	0.00	3.89	4.94	0.00	0.00	4.94
Aus/NZ	0.11	0.00	0.11	0.12	0.06	−0.06	0.12	0.16	0.09	−0.01	0.08	0.20	0.14	0.05	0.01
Canada	0.19	0.21	−0.02	0.19	0.19	0.01	−0.01	0.26	0.28	0.02	−0.03	0.33	0.44	−0.06	−0.04
China	6.44	0.00	6.44	6.70	0.00	0.00	6.70	9.04	0.00	0.00	9.04	11.50	0.00	0.00	11.50
E-Europe	0.20	0.23	−0.03	0.21	0.28	−0.02	−0.05	0.28	0.37	0.00	−0.10	0.35	0.50	0.00	−0.14
FSU	0.12	0.39	−0.27	0.12	0.82	−0.24	−0.46	0.17	1.13	−0.03	−0.93	0.21	1.58	0.17	−1.54
India	2.81	0.00	2.81	2.93	0.00	0.00	2.93	3.95	0.00	0.00	3.95	5.02	0.00	0.00	5.02
Japan	0.96	0.90	0.06	1.00	0.96	−0.05	0.09	1.35	1.42	0.09	−0.17	1.71	2.38	−0.25	−0.42
L-America	3.34	0.00	3.34	3.47	0.00	0.00	3.47	4.68	0.00	0.00	4.68	5.96	0.00	0.00	5.96
M-East	3.67	0.00	3.67	3.81	0.00	0.00	3.81	5.14	0.00	0.00	5.14	6.55	0.00	0.00	6.55
Asian Tigers	1.67	0.00	1.67	1.73	0.00	0.00	1.73	2.34	0.00	0.00	2.34	2.97	0.00	0.00	2.97
S-Asia	4.31	0.00	4.31	4.49	0.00	0.00	4.49	6.05	0.00	0.00	6.05	7.71	0.00	0.00	7.71
US	1.63	2.62	−0.99	1.70	2.29	0.23	−0.83	2.29	3.26	0.23	−1.20	2.91	5.11	−0.80	−1.40
W-Europe	2.26	3.30	−1.03	2.35	3.02	0.23	−0.89	3.18	4.41	0.35	−1.58	4.04	7.27	−1.16	−2.07
Total	30.47	7.65	22.82	31.69	7.62	0.10	23.97	42.76	10.95	0.65	31.16	54.42	17.41	−2.05	39.05

Note: Permit price (real terms and discounted) in all years: Aus/N.Z.: US$13.48/t C; Canada: US$24.62/t C; E-Europe: US$6.03/t C; FSU: US$7.36/t C; Japan: US$28.23/t C; US: US$23.56/t C; W-Europe: US$28.12/t C.
Permit trading selected years: 2010: −26 million t C; 2020: 18 million t C; 2035: −46 million t C.
Permit trading balance for Annex B countries: 2010–2035 = 0 t C.

Table 4. Benefits and costs of CO_2 mitigation with interregional permit trading among Annex B countries: Kyoto target levels for Annex B countries and no mitigation for developing countries
(in 2010 present discounted value billions of dollars)

Country (Area)	2010						2020			2035		
	Pre-trading			Post-trading			Post-trading			Post-trading		
	Benefit	Cost	NB	Cost	Permit US$	NB	Cost	Permit US$	NB	Cost	Permit US$	NB
Africa	2.77	0.00	2.77	0.00	0.00	2.77	0.00	0.00	3.96	0.00	0.00	4.85
Aus/NZ	0.11	0.00	0.11	0.08	−0.13	0.16	0.14	−0.08	0.10	0.20	0.02	−0.03
Canada	0.19	0.21	−0.02	0.11	0.08	0.00	0.19	0.11	−0.03	0.26	0.09	−0.03
China	6.44	0.00	6.44	0.00	0.00	6.44	0.00	0.00	9.21	0.00	0.00	11.28
E-Europe	0.20	0.23	−0.03	0.78	−0.73	0.15	1.26	−1.17	0.19	1.62	−1.51	0.23
FSU	0.12	0.39	−0.27	2.00	−2.29	0.41	3.33	−2.95	−0.21	4.49	−3.24	−1.04
India	2.81	0.00	2.81	0.00	0.00	2.81	0.00	0.00	4.02	0.00	0.00	4.92
Japan	0.96	0.90	0.06	0.29	0.43	0.24	0.51	0.74	0.12	0.73	1.00	−0.05
L-America	3.34	0.00	3.34	0.00	0.00	3.34	0.00	0.00	4.77	0.00	0.00	5.84
M-East	3.67	0.00	3.67	0.00	0.00	3.67	0.00	0.00	5.24	0.00	0.00	6.42
Asian Tigers	1.67	0.00	1.67	0.00	0.00	1.67	0.00	0.00	2.38	0.00	0.00	2.91
S-Asia	4.31	0.00	4.31	0.00	0.00	4.31	0.00	0.00	6.17	0.00	0.00	7.55
US	1.63	2.62	−0.99	1.44	1.01	−0.82	2.42	1.04	−1.13	3.29	0.84	−1.28
W-Europe	2.26	3.30	−1.03	0.92	1.64	−0.29	1.58	2.34	−0.69	2.23	2.81	−1.08
Total	30.47	7.65	22.82	5.63	0.00	24.84	9.42	0.00	34.12	12.82	0.00	40.52

Note: Permit price (real terms and discounted) in selected years: 2010: US$25.16/t C; 2020: US$26.55/t C; 2035: US$22.81/t C.
Permit trading balance in each year equals 0 t C.

Table 5. Benefits and costs of CO_2 mitigation with interregional trading among Annex B countries and developing countries: Kyoto target levels for Annex B countries and no harm for developing countries (in 2010 present discounted value billions of dollars)

Country (Area)	2010							2020				2035			
	Pre-trading			Post-trading				Post-trading				Post-trading			
	Benefit	Cost	NB	Benefit	Cost	Permit US$	NB	Benefit	Cost	Permit US$	NB	Benefit	Cost	Permit US$	NB
Africa	2.77	0.00	2.77	2.77	0.11	−0.22	2.88	3.96	0.18	−0.37	4.15	4.85	0.24	−0.48	5.09
Aus/NZ	0.11	0.00	0.11	0.11	0.02	−0.02	0.12	0.14	0.02	0.04	0.08	0.19	0.01	0.08	0.10
Canada	0.19	0.21	−0.02	0.19	0.02	0.09	0.07	0.23	0.03	0.12	0.08	0.33	0.02	0.12	0.19
China	6.44	0.00	6.44	6.44	0.30	−0.60	6.74	9.21	0.42	−0.85	9.64	11.28	0.40	−0.81	11.69
E-Europe	0.20	0.23	−0.03	0.20	0.16	0.07	−0.03	0.24	0.18	0.16	−0.10	0.35	0.13	0.24	−0.03
FSU	0.12	0.39	−0.27	0.12	0.39	−0.01	−0.27	0.15	0.46	0.48	−0.80	0.21	0.33	0.90	−1.02
India	2.81	0.00	2.81	2.81	0.10	−0.20	2.91	4.02	0.15	−0.31	4.18	4.92	0.18	−0.36	5.11
Japan	0.96	0.90	0.06	0.96	0.06	0.33	0.57	1.19	0.07	0.49	0.63	1.68	0.05	0.52	1.11
L-America	3.34	0.00	3.34	3.34	0.29	−0.60	3.65	4.77	0.43	−0.90	5.24	5.84	0.45	−0.93	6.33
M-East	3.67	0.00	3.67	3.67	0.31	−0.65	4.00	5.24	0.49	−1.02	5.77	6.42	0.55	−1.15	7.02
Asian Tigers	1.67	0.00	1.67	1.67	0.04	−0.09	1.71	2.38	0.07	−0.14	2.45	2.91	0.07	−0.15	2.99
S-Asia	4.31	0.00	4.31	4.31	0.19	−0.39	4.51	6.17	0.34	−0.70	6.53	7.55	0.46	−0.95	8.05
US	1.63	2.62	−0.99	1.63	0.27	1.13	0.23	2.02	0.32	1.47	0.24	2.86	0.22	1.44	1.19
W-Europe	2.26	3.30	−1.03	2.26	0.17	1.15	0.94	2.80	0.21	1.54	1.06	3.96	0.15	1.53	2.28
Total	30.47	7.65	22.82	30.47	2.43	0.00	28.04	42.51	3.36	0.00	39.15	53.34	3.25	0.00	50.09

Note: Permit price (real terms and discounted) in selected years: 2010: US$10.79/t C; 2020: US$9.35/t C; 2035: US$5.63/t C.
Permit trading balance in each year equals 0 t C.

Table 6. Benefits and costs of CO_2 mitigation with interperiod and interregional permit trading among Annex B countries and developing countries: Kyoto target levels for Annex B countries and all developing countries commit to quotas based on year 2020 emissions. Sovereignty criterion allocation vs. egalitarian allocation for developing countries
(in 2010 present discounted value billions of dollars)

Country	Sovereignty 2035				Egalitarian 2035			
	Post-trading				Post-trading			
(Area)	Benefit	Cost	Permit US$	NB	Benefit	Cost	Permit US$	NB
Africa	7.02	0.82	2.60	3.61	7.02	0.82	0.47	5.74
Aus/NZ	0.28	0.05	0.12	0.11	0.28	0.05	0.12	0.11
Canada	0.47	0.06	0.19	0.22	0.47	0.06	0.19	0.22
China	16.34	1.39	4.43	10.52	16.34	1.39	5.45	9.49
E-Europe	0.50	0.45	0.05	0.00	0.50	0.45	0.05	0.00
FSU	0.30	1.16	0.69	−1.55	0.30	1.16	0.69	−1.55
India	7.14	0.61	1.94	4.59	7.14	0.61	−1.68	8.20
Japan	2.43	0.18	0.90	1.36	2.43	0.18	0.90	1.36
L-America	8.47	1.45	0.61	6.41	8.47	1.45	1.95	5.07
M-East	9.30	1.69	−0.23	7.83	9.30	1.69	2.02	5.59
Asian Tigers	4.22	0.26	1.24	2.73	4.22	0.26	3.35	0.62
S-Asia	10.95	1.48	0.62	8.85	10.95	1.48	−0.36	9.83
US	4.14	0.80	2.18	1.16	4.14	0.80	2.18	1.16
W-Europe	5.74	0.54	2.63	2.57	5.74	0.54	2.63	2.57
Total	77.30	10.92	17.95	48.42	77.30	10.92	17.95	48.42

Note: Permit price (real terms and discounted) in all years: US$11.73/t C.
Permit trading selected years: 2010: −471 million t C; 2020: −640 million t C; 2035: 1531 million t C.
Permit trading balance: 2010–2035 = 0 t C.

(quotas) without any flexibility that would arise from permit trading over time and space.[6] The present discounted value of the global net benefits of achieving the Kyoto targets are increasingly positive over time from this basic institutional arrangement. Mitigation costs increase, not because MC_A functions shift, but because emissions caps under economic growth conditions require movement to higher points on the functions. Our results are based on the premise that the marginal benefits of GHG abatement increase more rapidly than the marginal costs, owing to the exponentially shaped damage functions associated with radiative forcing.[7] Net benefits are negative for each of the Kyoto signatory groups, though this figure never exceeds US$2 billion for any of them until 2035 in the case of Western Europe.[8] The coexistence of these results with positive global net benefits stems from the fact that the majority of these benefits are experienced by developing countries (see especially China and South Asia), who at the same time do not incur the costs of GHG mitigation.

Banking reaps economic advantages as well, though the results are somewhat difficult to assess from just the three selected years.[9] They are better illustrated in the present value display for the entire time horizon presented below in Table 7. The reason is that banking (or borrowing) alters the pattern of gross and net benefits over time, thus making year-by-year comparisons with Table 2 difficult. The results indicate that all Annex I countries find it advantageous to shift permits to the future in 2010. This raises net benefits in that year, but lowers them in comparison to other years for the previous cases examined.[10]

Banking differs from a more flexible interperiod trading institution because it requires a zero or positive permit account balance at all times, whereas full interperiod trading allows individual nations to have a negative permit account balance (borrowing). The results of simulating complete interperiod trading among Annex B countries in Table 3 indicates that taking permits from the future and using them in the present is advantageous, primarily for Western Europe, Canada, Japan and the US. Abatement costs typically decline over the 25-year time horizon for each of these nations, implying that they can reduce current costs by using permits that can be generated more cheaply at a later date. However, the global economic gains from this trading institution, compared to banking only, are modest at best, in part because the reduction in costs affects only a few nations, and each nation's discounted gross benefits are reduced in early time periods and increased in later time periods.

Allowing permit trading among only Annex B countries improves the situation from both the global and the individual country perspective (see Table 4). Gross benefits for each country stay the same, but mitigation costs are lowered for each Annex B country through trading. In 2010, all countries experience positive net benefits except Western Europe and the US, although these countries' net benefit deficit is lower than in the inflexible quota case. Net benefits become negative for other Annex B countries over the time horizon, but they are all still better off than without intercountry permit trading. The US, Western Europe and Japan are the permit buyers in 2010, and the other three Annex B

Table 7. Summary of net benefits in permit trading simulations, 2010 to 2035[a]

Country (Area)	Kyoto (Fixed Quotas Annex B)	Kyoto + Interperiod (Annex B Only)	Kyoto + Interregional (Annex B Only)	Kyoto + Interregional All (No Harm DCs)	Kyoto + Interperiod + Interregional All (No Harm DCs)	Kyoto + I'period + All DCs 2020 (Sovereignty)	Kyoto + I'period + All DCs 2020 (Egalitarian)	Kyoto + I'period + (Group A 2010 Group B 2020)
Africa	104.8	104.7	104.8	109.2	111.8	137.1	166.1	146.9
Aus/NZ	1.5	1.7	1.8	2.7	2.4	4.1	4.1	4.5
Canada	−0.9	−0.9	−0.6	3.6	3.1	5.5	5.5	7.4
China	243.8	243.6	243.8	252.9	258.2	330.0	309.0	364.3
E-Europe	−2.6	−2.7	4.8	−1.0	−2.2	−0.5	−0.5	−5.6
FSU	−26.7	−26.7	−11.0	−19.7	−23.8	−28.1	−28.1	−41.7
India	106.5	106.4	106.5	110.0	112.3	144.7	201.6	157.6
Japan	−5.3	−5.3	2.7	23.2	21.2	34.9	34.9	45.9
L-America	126.3	126.3	126.3	136.4	140.9	182.7	160.5	163.3
M-East	138.8	138.7	138.8	150.7	156.0	207.6	171.7	182.8
Asian Tigers	63.0	63.0	63.0	64.5	65.7	85.2	53.7	87.9
S-Asia	163.3	163.2	163.3	171.9	176.5	238.1	262.9	240.3
US	−32.1	−31.6	−28.7	20.2	13.8	31.0	31.0	49.4
W-Europe	−43.0	−42.6	−18.1	44.9	38.7	67.6	67.6	97.4
Global Net Benefits	837.4	837.8	897.4	1069.4	1074.5	1439.7	1439.7	1500.3
Incremental NB Increase	—	0.4	59.6	172.0	5.2	365.3	0.0	60.6
Cumulative NB Increase	—	0.4	60.0	231.9	237.0	602.3	602.3	662.9

[a] In current discounted 1990 US dollar terms summed over the entire time horizon 2010–2035 (4% discount rate).

groups are sellers. Approximately US$3.1 billion of permits are exchanged. This trading pattern is the same until the 2030s when Australia/New Zealand become permanent buyers. The global gains from trade are US$2 billion in the first year (compare the total US$24.84 billion in Table 4 with US$22.82 billion in Table 3). The gain dips in 2020 and then reaches more than US$2.5 billion (in constant dollars) in 2035.

Space limitations also do not allow us to display the full simulation results for the case of both interregional and interperiod trading among Annex B nations. These results are quite interesting in comparison to Table 4, the interregional trading only case. For example, the present discounted value of permits declines to US$18.36 per tonne of carbon. The decrease in permit prices reduces the value of permits sold by the FSU and increases the number (and hence the value) of permits purchased by large buyers, both interregion and interperiod, such as Western Europe and the US. The opportunity to borrow permits from the future brings down the price of permits, thereby improving the net benefits of buyers and diminishes it for sellers, even though, for example, the FSU sells about 100,000 more tonnes of carbon (in comparison to the results in Table 4) in both 2010 and 2035. Globally, net benefits are higher by about US$.5 to US$1 billion dollars annually.

The results of the inclusion of developing countries into permit trading are extremely interesting as presented in Table 5. Global gains and net benefits become quite large, reaching more than US$50 billion in 2035, a US$12 billion increase over the base case in Table 2. Interestingly, however, some Annex I countries stand to be worse off from the admission of developing countries into the permit market in comparison to the counterpart Annex B only case (Table 3), since they lose their comparative advantage in permit selling. For example, Eastern Europe and the FSU now for the most part become permit buyers rather than permit sellers. They still are better off than under the inflexible quota system; however, their gains from lower mitigation costs are not as high as their now forgone gains from permit selling. Still, this should not overshadow the large global gains. Essentially, net benefits are increased dramatically by bringing developing countries (DCs) into the permit trading process. At the same time, DCs can gain experience in permit trading at 'No Harm' to themselves, i.e., they can choose not to sell any permits if they are apprehensive. However, they will eventually realize that some permits can be sold safely with a net gain to themselves.

The simulation results for interregional and interperiod trading among the Annex I and developing nations (not shown) indicate net benefits for nations such as the US and the FSU are lower for all three years, and higher for nations and regions such as China and Latin America, in comparison to previous results. For example, China's net benefits increase by US$2.52 billion and US$1.4 billion in 2010 and 2035, while Latin America's net benefits are great by US$1.93 and US$2.79 billion, respectively. Globally, net benefits are greater in 2010 and 2020, though they are lower in 2035.

Table 6 shows simulation results in 2035 from two different distributions of permits to developing countries, based on 2020 emissions cap for these countries as a whole, while Annex I countries continue to meet the Kyoto targets. A Sovereignty equity criterion distributes permits to developing nations such that each nation's percentage reduction in greenhouse gases is equal to the percentage reduction of the entire group of developing nations. The Egalitarian equity criterion distributes the stock of permits available to the developing nations according to relative population share after 2020. Table 6 shows that countries or regions with relatively high population growth rates, such as Africa (3% population growth) are better off relative to countries such as China (1.3% growth) or the Asian Tigers (0.4% growth). For example, China's net benefits are lower by US$1.03 billion in 2035 under the Egalitarian allocation than under the Sovereignty allocation. Since equilibrium abatement costs are unaffected by different permit distributions (recall the Coase Theorem), all of the decline is due to China's smaller initial permit holdings and the subsequent increase in the purchase of permits.

The timing of greenhouse gas abatement commitments for non-Annex I countries is a potentially contentious issue. We simulate a case in which the Middle East, the Asian Tigers and Latin America, all regions with per capita incomes exceeding US$1000 per year, are brought into a greenhouse permits market in 2010 with permit allocations equivalent to emissions in 2010. The remaining developing nations are brought into the permits market in 2020, with initial permit allocations equal to emissions in 2020. Both interregional and interperiod permit trading are allowed. One immediate result (not shown) from this global agreement is that it increases the global supply of permits, relative to those cases where all developing nations are brought into the permits market in 2020 because the early commitment by Group A countries adds their emissions from 2010 to 2020 to the global permit stock. The increased permit supply drives permit prices down, and globally net benefits rise. In comparison to Table 6, which differs only by the timing of greenhouse commitments, net benefits for Latin America and the Middle East are lower in 2010, 2020 and 2035, with the disparity the largest in 2035. For example, Latin America's net benefits are reduced by US$1.72 billion and the Middle East is worse off by US$2.28 billion. The Asian Tigers would experience an increase in net benefits, for example, in 2035, net benefits would rise by US$0.49 billion.

Our final policy simulation – incorporation of an intergenerational equity criterion – did not yield a feasible solution. This is due to a combination of factors affecting Eastern Europe and the former Soviet Union, which are such that it is impossible for these regions to achieve positive net benefits for every year of the planning horizon. The factors include the stringency of emissions reduction requirements, low level of benefits, and high level of costs relative to those of other countries/regions. It appears that if a strict definition of intergenerational equity is imposed, it would be necessary to provide some relief from the Kyoto emissions caps for these countries or to provide them with cash transfers.

6.2. Summary

The marketable permits approach to global warming policy can have many variants depending on the number of countries included, whether permits can be traded over time, the timing of mitigation commitments themselves, and the manner in which the permits are distributed among countries. The latter two features address equity head on, although the first two features, usually motivated by efficiency gains, have equity implications as well. This is important from a normative standpoint but also in regard to the political viability of any greenhouse gas treaty proposal.

Table 7 and Figure 1, which summarize our main simulations, illustrate these points. Comparisons are facilitated by the fact that net benefits are summarized over the entire policy time horizon rather than the previous presentations of just three selected (and sometimes unrepresentative) years.

First, modifying the permits institution to allow trading across time for Annex B countries is better than a system of fixed quotas, but only slightly. One of the reasons is that several countries find it personally advantageous to defer emissions reductions to a later date, thereby reducing near-term benefits to everyone else.

When trading is extended across Annex I countries, global net benefits rise because trading reduces the cost of meeting the Kyoto targets. Much larger gains from interregional permit trading are available if developing nations are granted transferable permits in an amount equivalent to their emissions in 2010 (and subsequent years). The developing nations are made no worse off, even if they choose not to sell any of their permits, and they gain the opportunity to profit from permit sales from ready trading partners among the Annex I nations. These incremental gains are US$231 billion over the model time horizon, again due to the reduction in abatement costs among the industrialized nations.

The two simulations for different permit distributions according to Sovereignty and Egalitarian equity rules indicate another large global gain due primarily to requiring developing countries to commit to GHG emissions targets before the end of the policy horizon, and at the same time delaying their entrance into the permits market (the simulation does not include the No Harm option from 2010 to 2020). These simulations differ from the Kyoto interregional trading case in that we have also incorporated interperiod permit trading. However, this factor alone accounts for little more than a US$5 billion incremental increase in net benefits; hence, the inclusion of developing country mitigation commitments is responsible for most of the incremental gain of US$370 billion. Obtaining these commitments, even though delaying the entrance of developing nations into the permits market, effectively reduces the

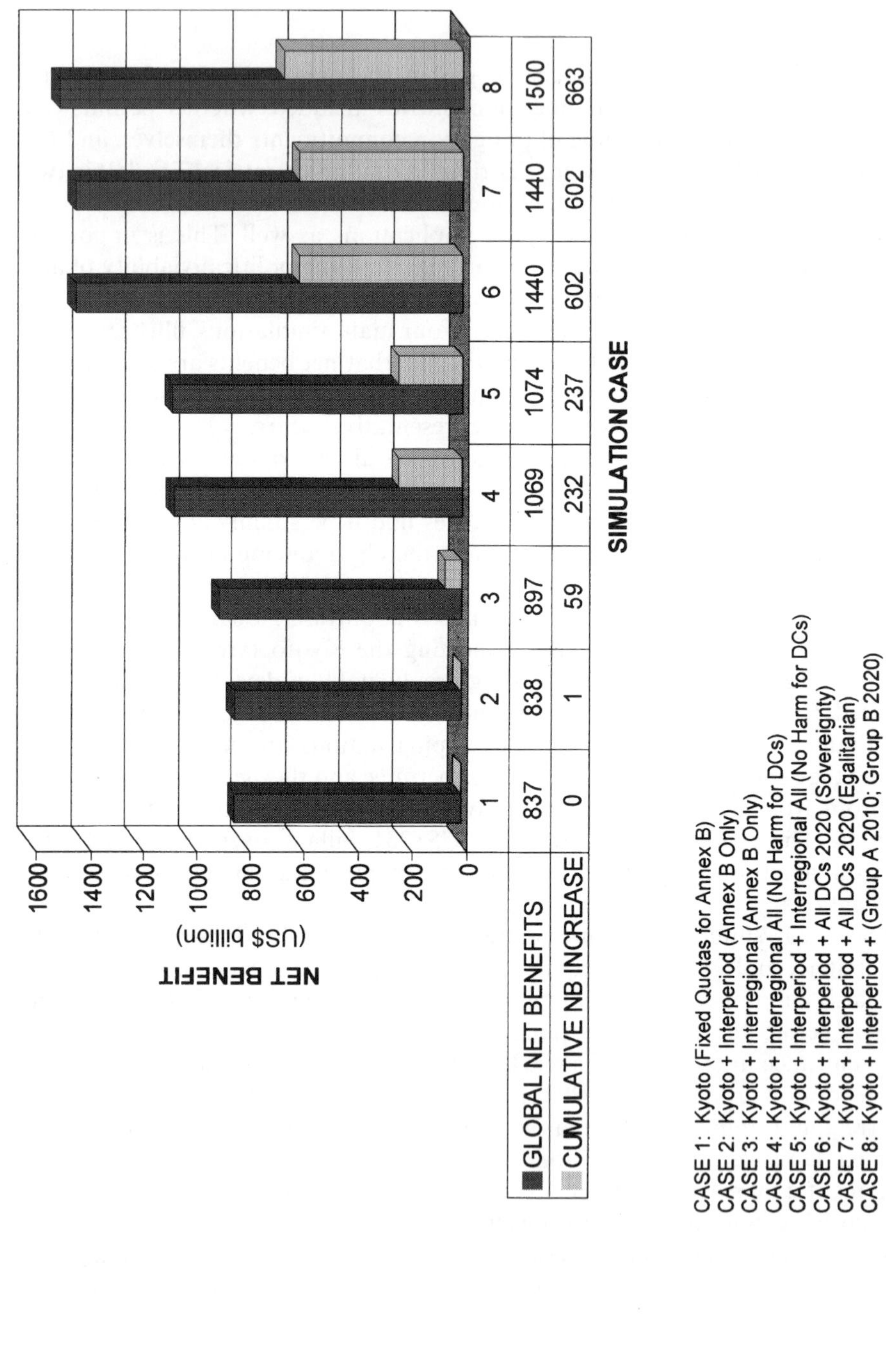

Figure 1. Summary of permit trading simulations, 2010–2035.

total supply of permits over the model time horizon by 10 years' worth of emissions from these countries. The ensuing abatement increases benefits more than it increases costs in comparison to the previous cases.

There is a slight gain of US$61 billion in securing commitments from Latin America, the Middle East and the Asian Tigers in 2010 instead of 2020. We take this to imply that, while it is important to consider the timing of including developing nations in a greenhouse protocol, fine-tuning the dates among non-Annex I nations may only result in a small increase in global net benefits. The reason is that the Group A developing countries (those with per capita incomes exceeding US$1,000) have relatively high abatement cost functions.

The results for selected countries are presented in Figure 2. China receives the highest net benefits across all eight cases, owing to its relatively low abatement costs, its relatively high gross benefits from avoiding global warming, and its ability to sell permits. The net benefits for India are also high, averaging about 75% of those of China. Among Annex B countries, Western Europe stands to gain the most from increasing the flexibility of the permits instrument. Net benefits for Eastern Europe do not change drastically over the simulations because this region typically finds its marginal cost of mitigation near the world permit price, and hence has very little to gain by entering the market. FSU generally fares worse as flexibility increases, since developing country permit offers undercut its ability to be the major permit seller.

Finally, we note that the foregoing analysis is based on a constancy of mitigation costs for each country/region over time. The reader might suspect that this is a major explanation for why the intertemporal trading simulations do not yield major gains. Alternatively, we simulated cases in which mitigation costs decreased by 1% per year over time (an optimistic estimate). Interestingly, this did not improve the relative gains at all for Case 5 and in fact decreased them by $4 billion for Case 2. The reason is that technological change provides an even greater incentive to defer mitigation, and, while a given country may be better off as a result, this action lowers the net benefits recieved by all others. Thus, allowing banking and borrowing (the latter being the preferred alternative for most countries) is still not likely to reap nearly as great a reward as the other improvements in policy instrument design and may even result in some perverse effects.

7. Conclusion

The marketable permits approach to global warming policy can have many variants depending on the number of countries included, whether permits can be traded over time, the timing of mitigation commitments themselves, and the manner in which the permits are distributed among countries. The latter two features address equity head on, although the first two features, usually motivated by efficiency gains, have equity implications as well.

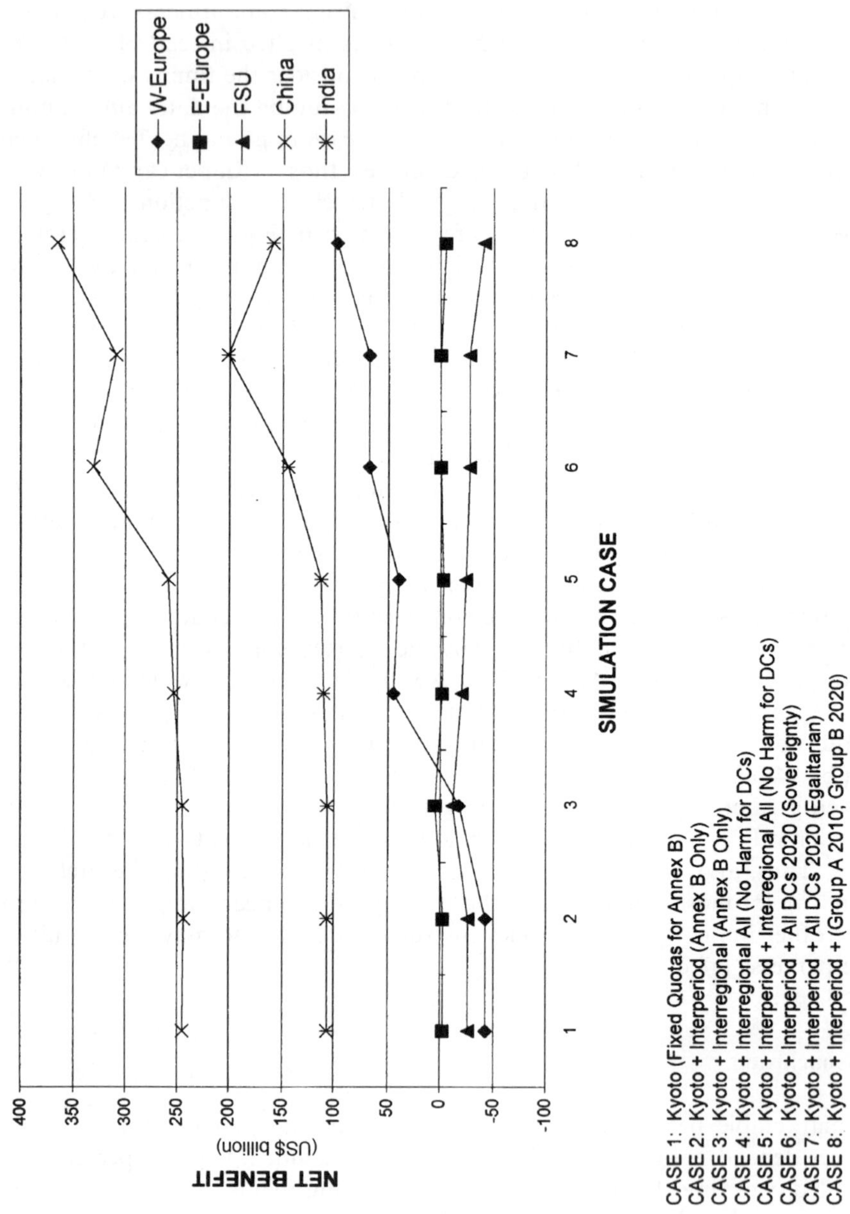

Figure 2. Summary of permit trading simulations, 2010–2035 (selected regions).

In summary, the greatest efficiency gains in the design of a GHG tradeable permits policy stem from utilizing the low-cost mitigation options of developing countries (even if no additional mitigation is forthcoming from this group itself) and then from requiring emissions reductions from developing countries at some future date. Interestingly, interperiod permits trading (banking or borrowing) yields relatively small incremental gains. The first of the more worthwhile extensions of the Kyoto accord are consistent with many nations' conception of equity at present (i.e., the 'No Harm' rule), while the second is likely to become embroiled in some equity controversies, which will be decided to a great extent on the basis of the timing of these additional commitments.

We have also uncovered two ironies in the dynamic context. The first relates to the fact that a given country acting in pursuit of its own objectives might shift its mitigation temporally and thereby unintentionally decrease net benefits to other countries. Also, net benefits could be increased more by having the poorest developing countries commit to emissions caps first, since they have relatively lower mitigation costs and higher mitigation benefits, and then later including the top tier of developing countries, such as Latin America and the Middle East. However, equity considerations are likely to override such a timing decision.

Notes

* Adam Rose is Professor and Head of the Department of Energy, Environmental, and Mineral Economics, The Pennsylvania State University, University Park, PA. Brandt Stevens is an Economist with the Demand Analysis Office, California Energy Commission, Sacramento, CA. The authors wish to thank Shu Yi Liao for his assistance in compiling some of the data and in running the simulations. The views expressed here are solely those of the authors and do not necessarily represent the positions of the institutions with which they are affiliated.

1. JI now refers to bilateral co-operation between Annex B countries. Its extension to developing countries is referred to as the CDM (see also Rose *et al.*, 1999).
2. Interperiod trading, whether it involves banking or borrowing, adds one complication not present in the interregional trading case. Interperiod trading changes the time path of greenhouse gas emissions, and hence changes when a warming will occur. Again, for any positive discount rate, this would affect the sum of the present discounted value of global benefits over time, hence net benefits will change.
3. This conclusion assumes that the transaction costs of permit trades are negligible and that each country's income is largely unaffected by the distribution of permits. Both assumptions are reasonable in the context of transferable GHG permits (see Rose *et al.*, 1998; Chao and Peck, 1998).
4. This stems from the fact that GHG abatement costs change over time; implicitly abatement cost differentials are required in the static context to stimulate trading as well.
5. There is an obvious problem associated with setting emissions quotas on the basis of a future year, since this provides no incentive to reduce emissions between now and that date, and, in fact, every incentive to steer toward higher emissions. It might be asked why we did not use an earlier reference date such as 2000. The reason relates to one of the major developing-country concerns about contributing to the effort to reduce GHGs – that Annex B countries were able

to achieve a high level of industrialization before they undertook the burden of GHG control, and it would be unfair to ask developing countries to do so before they reached a similar level.

6. The numerical results are in 1990 constant dollars, discounted to the Year 2010 by an assumed 4% real discount rate. Stating the results in 2010 dollars would require a forecast of inflation between the present and 2010, which would likely be of dubious accuracy. The year 1990 was chosen because it is the base year for several key variables (e.g., emissions caps). Adjustment to 1998 dollars would result in figures approximately 25% higher than those presented. Of course, none of these considerations would change the *relative* differences of outcomes between countries or between different permit trading institutions.
7. Net benefits are calculated as total benefits minus total abatement costs minus the value of permit transfers. Since a permit sale is a source of revenue, it is given a negative value so its subtraction yields a net addition to benefits. A permit purchase is treated as a cost or a net reduction in benefits.
8. We realize that this premise is debatable, and again emphasize the importance of the relative results between our simulations rather than the absolute numbers of any simulation itself.
9. For permits trading institutions in which permit banking or borrowing is allowed, we follow the convention established for permit sales and purchases. A banked permit has a negative value; that is, it is treated as a sale to oneself or to a permit broker, and thus it is an addition to net benefits. The country or group of countries buys the permit back at some future date, when it is treated as a cost.
10. Note also that interperiod trading results in a constant discounted permit price over the planning period. Again, this represents the intertemporal opportunity cost of utilizing a permit at any given time. The standard arbitrage equation of intertemporal optimization requires that this price be constant, thus ensuring that any potential gains from shifting permits across years have been exhausted.

References

Agarwal, A. and S. Narain (1991). *Global warming in an unequal world: a case of environmental colonialism.* New Delhi: Centre for Science and Environment.

Barrett, S. (1992). Acceptable allocations of tradeable carbon emission entitlements in a global warming treaty. In: S. Barrett *et al. Combating global warming.* Geneva, Switzerland: UNCTAD.

Barrett, S., M. Grubb, K. Roland, A. Rose, R. Sandor and T. Tietenberg (1992). *Combating global warming: a global system of tradeable carbon emission entitlements.* Geneva, Switzerland: UNCTAD.

Chao, H. and S. Peck (1998). Pareto optimal environmental control and income distribution with global climate change. Discussion Paper, Electric Power Research Institute, Palo Alto, CA.

Cline, W. (1992). *Economics of global warming.* Washington, DC: Institute for International Economics.

Coase, R. (1960). The problem of social cost. *Journal of Law and Economics.* 3:1–44.

Edmonds, J., M. Wise and D. Barns (1995). Carbon coalitions: the cost and effectiveness of energy agreements to alter trajectories of atmospheric carbon dioxide emissions. *Energy Policy.* 23:309–35.

Eyckmans, J, S. Proost and E. Schokkaert (1993). Efficiency and distribution in greenhouse negotiations. *Kyklos.* 46:363–97.

Fankhauser, S. (1994). The social costs of greenhouse gas emissions: an expected value approach. *Energy Journal.* 15:155–84.

Framework Convention on Climate Change, Conference of the Parties (1997). Kyoto Protocol to the United Nations Framework Convention on Climate Change. Kyoto, 10 December.

Halsnaes. K. (1996). The economics of climate change mitigation in developing countries. *Energy Policy.* 24:917–27.

Intergovernmental Panel on Climate Change (IPCC) (1996). *Climate change: economic assessment.* Cambridge: Cambridge University Press.

Larsen, B. and A. Shah (1994). Global tradable carbon permits, participation incentives and transfers. *Oxford Economic Papers.* 46:841–56.

Manne, A. and R. Richels (1991). Global CO_2 emission reductions: the impacts of rising energy costs. *Energy Journal*, 12:87–107.

Nordhaus, W. and Z. Yang (1996). A regional dynamic general-equilibrium model of alternative climate-change strategies. *American Economic Review.* 86:741–65.

Peck, S. C. and T. J. Teisberg (1992). CETA: a model for carbon emissions trajectory assessment. *Energy Journal.* 13:55–77.

Rose, A. (1992). Equity considerations of tradeable carbon entitlements. In: S. Barrett *et al. Combating global warming.* Geneva: UNCTAD.

Rose, A. and B. Stevens (1993). The efficiency and equity of marketable permits for CO_2 emissions. *Resource and Energy Economics.* 15:117–46.

Rose, A. and B. Stevens (1998). A dynamic analysis of fairness in global warming policy: Kyoto, Buenos Aires, and Beyond. *Journal of Applied Economics.* 1:329–62.

Rose, A., E. Bulte and H. Folmer (1999). Long-run implications of joint implementation for developing countries. *Environmental and Resource Ecoomics.* (Forthcoming).

Rose, A., B. Stevens, J. Edmonds and M. Wise (1998). International equity and differentiation in global warming policy: an application to tradeable emission permits. *Environmental and Resource Economics.* 12(1):25–51.

Stevens, B. and A. Rose (1998). A dynamic analysis of the marketable permits approach to global warming policy: a decomposition of sources of advantage. Department of Energy, Environmental, and Mineral Economics, The Pennsylvania State University, University Park, PA.

Stevens, B., A. Rose and S. Y. Liao (1998). An analysis of the efficiency of marketable permits when benefits are endogenous. Department of Energy, Environmental, and Mineral Economics, The Pennsylvania State University, University Park, PA.

Tietenberg, T. and D. Victor (1994). Possible administrative structures and procedures for implementing a tradable entitlement approach to controlling global warming. In: T. Tietenberg *et al. Combating global warming: possible rules, regulations, and administrative arrangements for a global market in CO_2 emission entitlements.* New York: United Nations.

Tol, R. (1995). The damage costs of climate change: towards more comprehensive calculations. *Environmental and Resource Economics.* 5:353–74.

Van der Mensbrugghe, D. (1995). *Regional carbon tax required for equiproportional CO_2 emission reduction from baseline.* Paris: OECD.

CHAPTER 12

Equitable Cost–Benefit Analysis of Climate Change

Richard S.J. Tol

1. Introduction

Greenhouse gas emissions and vulnerability to climate change show a strong negative correlation. This is the moral issue at the heart of the climate problem. However, little attention has been paid to this in the literature. See Banuri *et al.* (1996) for an overview. The literature on equity issues in climate change is largely confined to the distribution of emissions reduction targets. This literature takes the need to reduce greenhouse gas emissions for granted, and takes a parametric approach to the required reductions. On the other hand, the literature that tries to derive how much emissions abatement is desirable largely ducks the equity issue. This paper is an attempt to combine the two problems.

Broadly speaking, there are two approaches to advice on desirable emissions abatement. One approach follows the Framework Convention on Climate Change and tries to define a safe, maximum atmospheric concentration of greenhouse gases. The 'safe emissions corridor' approach, advocated by RIVM (e.g., Alcamo and Kreileman, 1996), bases its target concentrations on half-baked ecological considerations (e.g., Azar and Rodhe, 1997; Swart and Hootsmans, 1991; Swart *et al.*, 1989). It thus avoids any discussion about equity. The 'tolerable windows' approach, championed by PIK (e.g., Toth *et al.*, 1997), bases its target concentrations on geological considerations of the German Advisory Council on Global Change (1995). This takes the naturalist fallacy to the extreme: The future may not fall outside past experiences, because the past was the best of times. Both approaches ignore the fact that trying to avoid inequities of climate change may invoke more serious inequities of emissions abatement.

Other authors simply take a concentration target as given, and derive a cost-effective path towards that concentration (e.g., Ha Duong *et al.*, 1997; Manne and Richels, 1995, 1996, 1998; Peck and Teisberg, 1996; Tol, 1999b,c; Wigley *et al.*, 1996). Cost-effectiveness has an appeal to justice in that it minimizes total costs so that, in principle, everyone can be made better off. Rose and

C. Carraro (ed.), Efficiency and Equity of Climate Change Policy, 273–290.

Stevens (1993; see also Rose *et al.*, 1998) propose ten different interpretations of an equitable, international sharing of the burden of meeting a particular concentration target. Tol (1998a) complements cost-effectiveness with inter-temporal equity. However, these papers all ignore the equity implications of selecting a concentration target.

The other approach to deriving emissions and concentration targets includes such trade-offs, at least in principle. However, attempts to derive greenhouse gas emissions reductions so as to maximize human welfare are without exception based on a narrow neo-classical interpretation of justice (e.g., Maddison, 1995; Manne *et al.*, 1995; Nordhaus, 1991, 1992, 1993, 1994; Nordhaus and Yang, 1996; Peck and Teisberg, 1991, 1994, 1995; Tol, 1997, 1999a). 'Maximum welfare' is interpreted to mean 'Pareto optimal'. That is, the status quo (no climate policy) is the base situation and climate policy needs to make everybody better off, at least potentially (cf. Farrow, 1998). This is another form of the naturalist fallacy: The world without policy intervention is a pretty good world. The inequities of a 'do nothing' policy have no place in this framework. In fact, the analysis operates under the 'victim pays principle': countries that suffer most from climate change are expected to convince large emitters to abate (Tol, 1997).

However, the cost–benefit approach is closer to including equity than is the safe concentration approach. Therefore, I try in this paper to extend welfare maximization to considering justice. Roemer (1996) and Sen (1982, 1987) champion this at a theoretic level. I take a more pragmatic approach.

Three alternatives are presented, and their results demonstrated with *FUND*, an integrated assessment model (cf. Weyant *et al.*, 1996, for an overview of such models). The first alternative derives from the basic message of Emanuel Kant (do not to others what you do not want them to do to you) with a Rawlsian flavour (the 'other' is the least well-off region). The second alternative is based on the thought that, for all regions for all times, the sum of costs of emissions reduction and the costs of climate change should be equal. Thus, the inequities of the no-climate-change scenario are maintained (whereas, in a no-policy scenario, inequities would deteriorate). Such relative no-envy solutions often prove a pragmatic way out in everyday policy making. In the third alternative, a global welfare function is maximized that explicitly includes a distaste for inequity. This alternative has strong roots in neo-classical economics, but cannot distinguish between inequities of climate change, inequities of emissions reduction, and inequities of other causes.

The next section presents the model. Sections 3–5 present the results for the three alternatives in the above order. Section 6 concludes.

2. The model

The model used is Version 1.6 of the *Climate Framework for Uncertainty, Negotiation and Distribution* (*FUND*) (cf. Tol, 1999b,c).[1] Essentially, *FUND* consists of a set of exogenous scenarios and endogenous perturbations. The model is specified for nine major world-regions: OECD-America (excl. Mexico); OECD-Europe; OECD-Pacific (excl. South Korea); Central and Eastern Europe and the former Soviet Union; Middle East; Latin America; South and Southeast Asia; Centrally Planned Asia; and Africa. The model runs from 1950 to 2200, in time steps of a year. The prime reason for starting in 1950 is to initialize the climate change impact module. In *FUND*, climate impacts are assumed to depend on the impact of the year before, to reflect the process of adjustment to climate change. Thus, climate impacts (both physical and monetized) are misrepresented in the first decades. This would bias optimal control if the first decades of the simulation coincided with the first decades of emissions abatement. Similarly, the period 2100–2200 is there to provide the forward-looking agents in the 21st century with a proper time horizon. The calculated emissions reductions in 2100–2200 have little meaning in and of themselves.

The *IMAGE* database (Batjes and Goldewijk, 1994) is the basis for the calibration of the model to the period 1950–1990. Scenarios for the period 2010–2100 are based on the EMF14 Standardised Scenario, which lies between IS92a and IS92f (cf. Leggett *et al.*, 1992). Note that the original EMF14 Standardised Scenario had to be adjusted to fit *FUND*'s nine regions and yearly time-step. The period 1990–2010 is a linear interpolation between observations and the EMF14 Standardised Scenario. The period 2100–2200 is an extrapolation of the EMF14 Standardised Scenario.

The scenarios concern the rate of population growth, urbanization, economic growth, autonomous energy efficiency improvements, the rate of decarbonization of the energy use (autonomous carbon efficiency improvements), and emissions of carbon dioxide from land use change, methane and nitrous oxide.

The scenarios of economic and population growth are perturbed by the impact of climate change. Population falls with climate change deaths, resulting from changes in heat stress, cold stress, malaria and tropical cyclones. Heat and cold stress are assumed to affect only the elderly, non-reproductive population. The other sources of mortality do affect the number of births. Heat stress only affects urban population. The share of urban in total population is, up to 2025, based on the World Resources Databases; after 2025, urban population slowly converges to 95% of total population (comparable to present-day Belgium or Kuwait). Population also changes with climate-induced migration between the regions. Immigrants are assumed to assimilate immediately and completely with the host population.

The tangible impacts of climate change are dead-weight losses to the economy. Consumption and investment are reduced, without changing the saving's rate. Climate change thus reduces long-term economic growth, although, in the short term, consumption takes a deeper cut. Economic growth is also reduced by carbon dioxide emissions abatement.

The energy intensity of the economy and the carbon intensity of the energy

Table 1. Parameters of equation (1)

Gas	α[a]	β[b]	Pre-industrial concentration
Methane (CH_4)	0.3597	1/8.6	790 ppbv
Nitrous oxide (N_2O)	0.2079	1/120	285 ppbv

[a] The parameter α translates emissions (in million metric tonnes of CH_4 or N_2O) into concentrations (in parts per billion by volume).
[b] The parameter β determines how rapidly concentrations return to their pre-industrial (and assumedly equilibrium) concentrations; $1/\beta$ is the atmospheric lifetime (in years) of the gases.
Source: After Schimel *et al.* (1996).

supply autonomously decrease over time. This process can be sped up by abatement policies.

The endogenous parts of *FUND* consist of the atmospheric concentrations of carbon dioxide, methane and nitrous oxide, the global mean temperature, the impact of carbon dioxide emissions reductions on economy and emissions, and the impact of the damages of climate change on the economy and the population.

Methane and nitrous oxide are taken up in the atmosphere, and then geometrically depleted:

$$C_t = C_{t-1} + \alpha E_t - \beta(C_{t-1} - C_{pre})1 \tag{1}$$

where C denotes concentration, E emissions, t year, and *pre* pre-industrial. Table 1 displays the parameters for both gases.

The atmospheric concentration of carbon dioxide follows from a five-box model:

$$\mathrm{Box}_{i,t} = \rho_i \mathrm{Box}_{i,t} + 0.000471\alpha_i E_t 2 \tag{2a}$$

with

$$C_t = \sum_{i=1}^{5} \alpha_i \mathrm{Box}_{i,t} 3 \tag{2b}$$

where α_i denotes the fraction of emissions E (in million metric tonnes of carbon) that is allocated to box i (0.13, 0.20, 0.32, 0.25 and 0.10, respectively) and ρ the decay-rate of the boxes ($\rho = \exp(-1/\text{lifetime})$, with lifetimes infinity, 363, 74, 17 and 2 years, respectively). Thus, 13% of total emissions remain forever in the atmosphere, while 10% is – on average – removed in two years (after Hammitt *et al.*, 1992). Carbon dioxide concentrations are measured in parts per million by volume.

Radiative forcing for carbon dioxide, methane and nitrous oxide are based on Shine *et al.* (1990). The global mean temperature T is governed by a geometric build-up to its equilibrium (determined by radiative forcing RF), with a half-time of 50 years. In the base case, global mean temperature rises in equilibrium by 2.5°C for a doubling of carbon dioxide equivalents, so:

$$T_t = \left(1 - \frac{1}{50}\right) T_{t-1} + \frac{1}{50} \frac{2.5}{6.3 \ln(2)} RF_t 4 \tag{3}$$

Global mean sea level is also geometric, with its equilibrium level determined by the temperature and a life-time of 50 years. Temperature and sea level are calibrated to the best-guess temperature and sea level for the IS92a scenario of Kattenberg *et al.* (1996).

The climate impact module is based on Tol (1995, 1996). A limited number of categories of the impact of climate change are considered: agriculture, sea level rise, heat and cold stress, malaria, tropical and extratropical storms, river floods, and unmanaged ecosystems. The damage module has two units of measurement: people and money.

People can die (heat stress, malaria, tropical cyclones), not die (cold stress), or migrate. These effects, like all impacts, are monetized. The value of a statistical life is set at US$250 000 plus 175 times the per capita income. The value of emigration is set at three times the per capita income, the value of immigration at 40% of the per capita income in the host region.

Other impact categories are directly expressed in money, without an intermediate layer of impacts measured in their 'natural' units.

Damage can be due to either the rate of change (benchmarked at 0.04°C/yr) or the level of change (benchmarked at 2.5°C). Benchmark estimates are displayed in Table 2. Damage in the rate of temperature change slowly fades at a speed indicated in Table 3. Damage is calculated through a second-order polynomial in climatic change. Thus, damage D_t in year t is either

$$D_t = \alpha_t W_t + \beta_t W_t^2 \tag{4a}$$

or

$$D_t = \alpha_t \Delta W_t + \beta_t \Delta W_t^2 + \rho D_{t-1} \tag{4b}$$

with W the appropriate climate variable (temperature, sea level, hurricane activity, etc.) and α, β and ρ parameters.

Damage is distinguished between tangible (market) and intangible (non-market) effects. Tangible damages affect investment and consumption; through investment, economic growth is affected; through consumption, welfare is affected. Intangible damages affect welfare.

Relative vulnerability to climate change – α and β in (4) – is a function of economic development in many ways. The importance of agriculture falls with economic growth. The share of agriculture in total output is assumed to change with per capita income with an elasticity of −0.31, which corresponds to the per capita income elasticity across *FUND*'s nine regions in 1990. Malaria incidence and the inclination to migrate are assumed to fall logistically with increases in per capita income. Heat stress is assumed to increase linearly with

Table 2. Monetized estimates of the impact of global warming (in 10^9 US$ per year)

Region	Species	Life	Agric.	Sea	Extreme	Total
Level (temperature: +2.5°C; sea level: +50 cm; hurricane activity: +25%; winter precipitation: +10%; extratropical storm intensity: +10%)						
OECD-A	0.0	−1.0	−5.3	0.9	2.5	−2.9
OECD-E	0.0	−1.1	−6.0	0.3	0.3	−6.5
OECD-P	0.0	−0.5	−6.1	1.5	5.5	0.3
CHEE&fSU	0.0	3.7	−23.2	0.1	0.2	−19.1
ME	0.0	3.5	3.1	0.1	0.0	6.6
LA	0.0	67.0	7.3	0.2	0.0	74.5
S&SEA	0.0	81.4	15.8	0.2	0.6	98.8
CPA	0.0	58.4	−22.2	0.0	0.1	36.3
AFR	0.0	22.5	5.4	0.1	0.0	28.0
Rate (temperature: 0.04°C/year; other variables follow)						
OECD-A	0.3	0.2	0.3	0.2	0.2	1.2
OECD-E	0.3	0.2	0.0	0.2	0.0	0.7
OECD-P	0.2	0.1	0.0	0.3	0.4	1.0
CEE&fSU	0.1	0.1	0.0	0.0	0.0	0.2
ME	0.0	0.0	0.1	0.0	0.0	0.2
LA	0.0	0.4	0.1	0.1	0.0	0.6
S&SEA	0.0	0.3	0.1	0.1	0.0	0.6
CPA	0.0	0.2	0.3	0.0	0.0	0.5
AFR	0.0	0.0	0.1	0.0	0.0	0.2

Source: Tol (1996).

Table 3. Duration of damage memory per category.[a]

Category	Years	Category	Years
Species loss	100	Immigration	5
Agriculture	10	Emigation	5
Coastal protection	50	Wetland (tangible)	10
Life loss	15	Wetland (intangible)	50
Tropical cyclones	5	Dryland	50

[a] Damage is assumed to decline geometrically at a rate of 1–1/life-time.
Source: Tol (1996).

urbanization. The valuation of intangible impacts is assumed to increase logistically with per capita income.

Emissions abatement is restricted to industrial sources of carbon dioxide. The costs of carbon dioxide emissions reduction are calibrated to the survey results of Hourcade *et al.* (1996), supplemented with results of Rose and Stevens (1993) for developing countries. Regional and global average cost estimates, and their standard deviations result. Regional relative costs are shrunk to the global average; that is, the weighted average of the regional and global average

Table 4. Parameters of the CO_2 emissions reduction cost function.[a]

OECD-A	2.0789	CHEE&fSU	2.0488	S&SEA	2.1268
OECD-E	2.3153	ME	2.1041	CPA	1.9544
OECD-P	2.2171	LA	2.1253	AFR	2.0931

[a] The proportional loss of GDP C in year t of proportional emissions reduction R in year t follows: $C_t = aR_t^2$. The costs to GDP are modelled as a dead-weight loss to the economy. Emissions reduction is brought about by a permanent shift in intensity of energy and carbon.
Source: After Hourcade *et al.* (1996) and Rose and Stevens (1993).

is taken, with the inverse variances as weights. This reduces the influence of a single study. It particularly influences the developing regions, for which much less information on emissions abatement costs is available. Costs are represented by a quadratic function. Table 4 presents the parameters. Roughly, a 1% cut in emissions costs 0.02% of GDP; a 10% cut costs 2%.

In *FUND*, each region has its own decision maker, nine in total. *FUND* also distinguishes generations of decision makers, 20 in total. Thus, there are $9 \times 20 = 180$ decision makers in the model. Each decision maker has control over a ten-year period only. Each decision maker maximizes the net present welfare of his/her region (in the non-co-operative cases) from the start of the control period up to 2200. In the case of global co-operation, the unweighted sum of the net present regional welfares is maximized (see Section 5). Each decision maker knows the emissions reduction efforts of all decision makers in all regions at all times. The equilibrium is found iteratively. That is, in the first iteration, each decision maker controls emissions so as to maximize net present welfare, assuming that the other decision makers do nothing. In the second iteration, each decision maker acts assuming that the other decision makers do as in the first iteration. And so on, until convergence.

3. Kant

Do not to others what you do not want to happen to you. It is simple, appealing and restraining. It does take a number of additional considerations, though, to make Kant operational in a climate change context. Firstly, there are costs of emissions reduction as well as costs of climate change. However, because of discounting and the slow workings of the climate system, *current* costs of climate change are likely to exceed *current* costs of emissions reduction. Therefore, it seems reasonable (and is found to be reasonable in the experiments below) to restrict the attention to the costs of climate change. Secondly, there are a great number of others whose potential discomfort should be internalized. The costs of climate change to various regions are strongly linked, however. If the costs of the most vulnerable are reduced to acceptable levels, the costs of the less vulnerable are likely to have fallen (and indeed are in the experiments

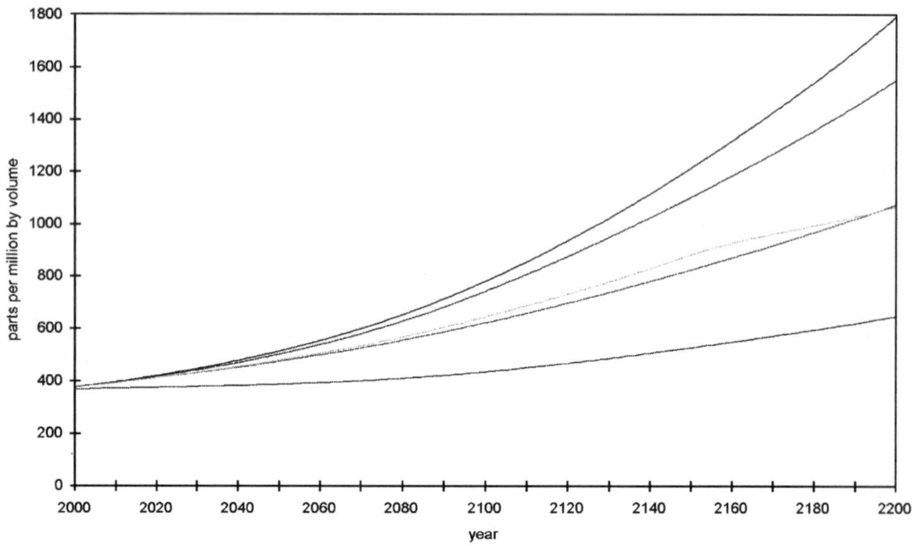

Figure 1. The atmospheric concentration of carbon dioxide for five scenarios, from top to bottom, business as usual (and non-co-operative optimal emissions control), non-co-operative average Kantian emissions control, non-co-operative maximum Kantian emissions control, co-operative optimal emissions control, and no-envy emissions control.

below) below acceptable levels. Thirdly, the costs of climate change to the most vulnerable regions are not reduced to a pre-ordained level. Instead, less vulnerable regions treat the relative costs of the most vulnerable region as if these were their own, and perform a cost–benefit analysis on that basis. That is, the climate change impacts of each region in each decade are multiplied by a factor $A_{t,r} = \max(I_{t,r})/I_{t,r}$, where $I_{t,r}$ denotes the impact at time t in region r. Fourthly, by focusing on the costs to the most vulnerable to climate change, the analysis is sensitive to scale. For instance, in *FUND*, the Maldives and India are grouped in one region. The impact on moderately vulnerable India dominates the impact on the highly vulnerable Maldives. The aggregation in *FUND* is such that little can be done about this. Therefore, the effect is demonstrated by looking at vulnerability per region averaged over time – $\max_r(\text{avg}_t(I_{t,r}))$ – and vulnerability per region per time period – $\max_{t,r}(I_{t,r})$. Fifthly, optimal emissions reductions are calculated in *FUND*'s non-co-operative mode.

Figure 1 displays some results. Using the time-averaged vulnerability adjustment, 'Kantian' emission abatement reduces concentrations substantially below the business as usual scenario. The common non-co-operative emissions abatement policy leads to concentrations which are indistinguishable from the business-as-usual case. Using the time-specific vulnerability adjustment, 'Kantian' abatement reduces concentrations even further, in fact close to the optimal co-operative solution. The latter may seem strange, since the 'Kantian' alternative has much higher 'damages'. However, the vulnerability adjustment is based on

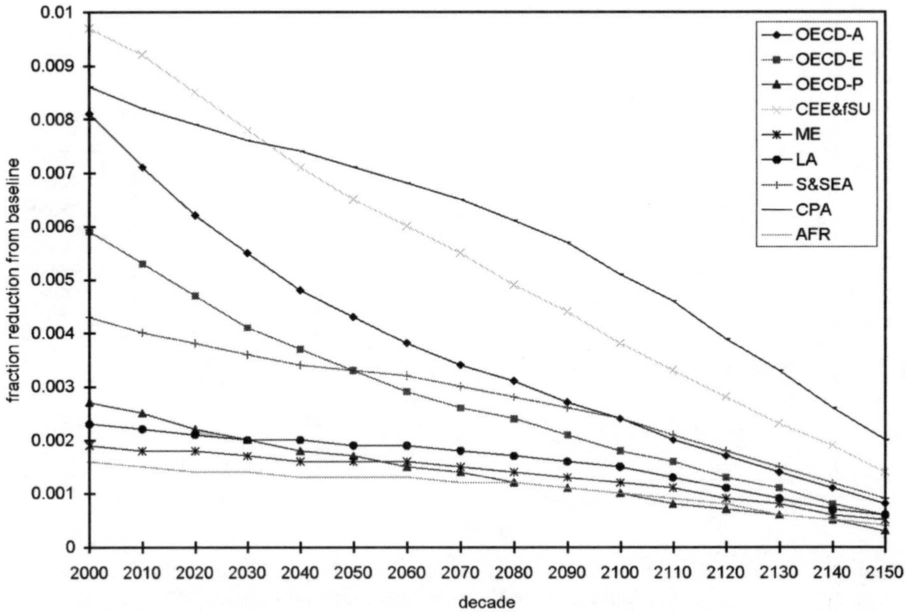

Figure 2. The regional and temporal emissions reduction efforts for co-operative optimal control.

impacts as a percentage of GDP. An equal percentage impact on, say, the European Union and Africa would still leave the latter worse off in terms of utility. In the 'Kantian' case, Africa's *relative* impacts on European utility count.[2] In the co-operative case, African impacts on African utility count.

Although the outcome in terms of concentrations is similar, the distribution of emissions reductions over regions and over time is not. Figure 2 displays the regional and temporal distribution of emissions reductions for the co-operative optimum, Figure 3 for the average Kant case, and Figure 4 for the maximum Kant case. Average Kantian emissions control is generally lower than emissions abatement in the other two cases. The main qualitative difference between co-operative and Kantian abatement is for Centrally Planned Asia, which starts poor but is little vulnerable to climate change, and is projected to emit large amounts of carbon dioxide.

4. No-envy

Climate change invokes additional inequities, as its impacts are unevenly distributed and disproportionally affect the poor. Greenhouse gas emissions reduction invokes other inequities. Consider the following thought experiment. The leaders of all countries and all generations meet to share the joint burden of climate change and emissions reduction. All are committed and no one is

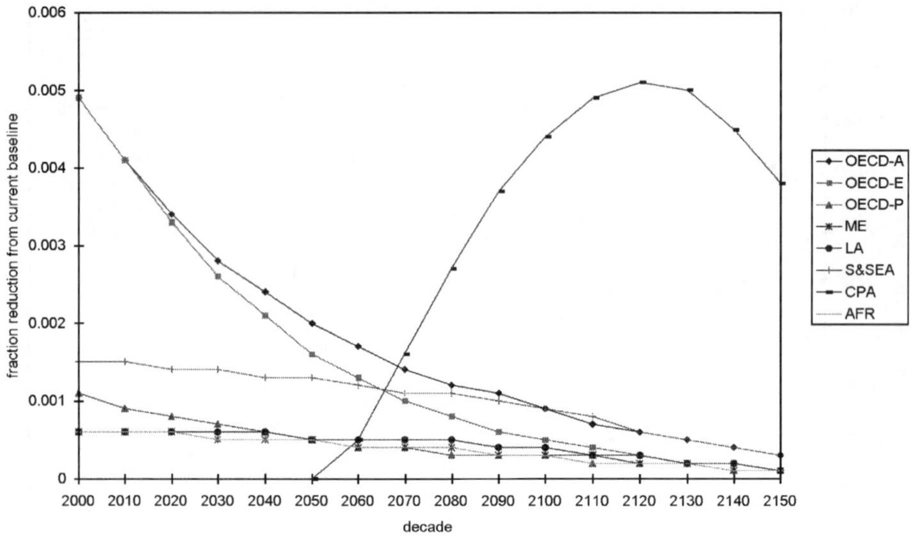

Figure 3. The regional and temporal distribution of emissions reduction efforts for the non-co-operative average Kantian emissions control case.

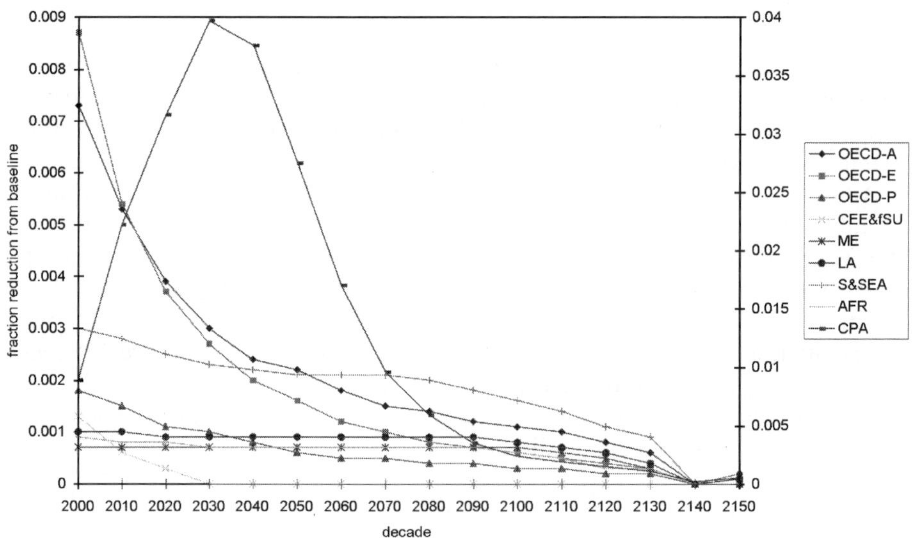

Figure 4. The regional and temporal distribution of emissions reduction efforts for the non-co-operative maximum Kantian emissions control case. CPA is on the right axis.

inclined to cheat. In real life, such meetings often agree on an equal effort for all (e.g., an equal percentage emissions reduction). This is not necessarily equitable, but, if the equal effort is in a proper metric, it does not introduce a

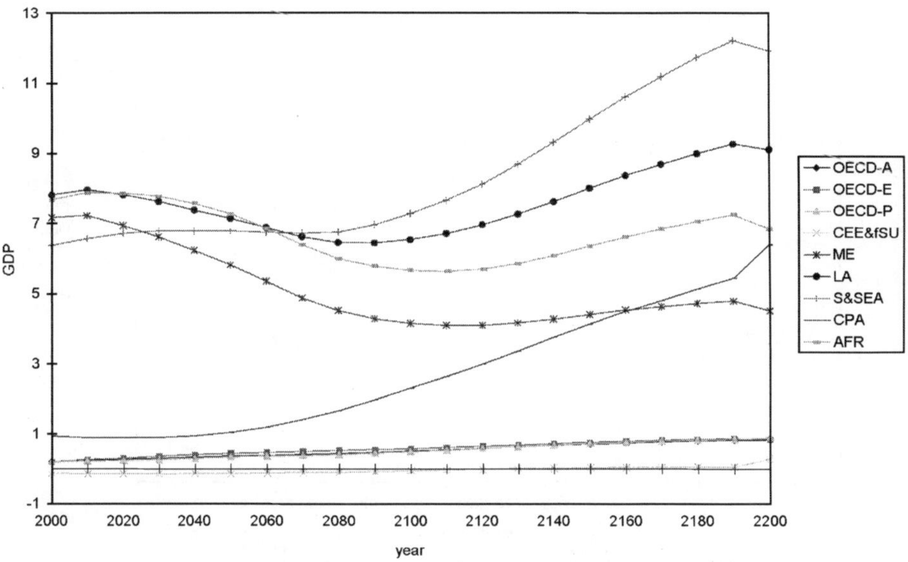

Figure 5. The regional and temporal distribution of the relative impacts of climate change in the business-as-usual scenario.

lot of new inequities either. The injustice of the status quo is by and large maintained.[3] Varian (1974) coined the term 'no-envy' for such a solution, and explores its implications.

In the hypothetical meeting of countries and generations, however, the situation without climate change is taken as the reference case. The sum of the costs of emissions reduction and the costs of climate change, relative to income, is equalized. This implies that the inequities of the no-climate-change scenario are more or less maintained. Greenhouse gas emissions reduction policy is used to counteract the inequities of climate change, but no more than that.

Figure 1 displays the results. The business-as-usual scenario is taken as the starting point. Its climate change impacts are displayed in Figure 5 for four regions. Emissions are reduced so that for each region for each time period the costs of emissions reduction minus the costs of climate change are equal to the average costs of climate change in the no-control scenario. Costs and benefits are expressed relative to total regional income.[5] Regions and time with higher than average impacts do not reduce emissions. Resulting emissions reductions are fed into *FUND*, new climate change impacts are calculated, and new emissions reductions are determined. Convergence is rapid.

The resulting emissions reductions are substantial. Atmospheric concentrations of carbon dioxide are kept below 650 ppmv. Figure 6 displays emissions abatement, expressed as the annual reduction from the current baseline as a fraction of that baseline. The OECD should reduce emissions by about 9% per year, Central and Eastern Europe and the former Soviet Union even more

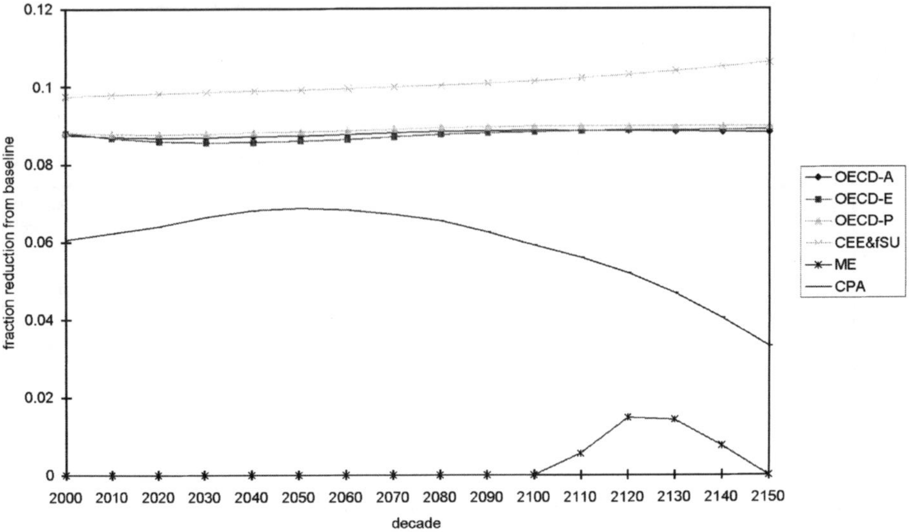

Figure 6. The regional and temporal distribution of emissions reduction efforts for the no-envy control case.

than 10%. Interestingly, China should start reducing emissions immediately, by about 5% per year. China's emissions reductions gradually rise to a peak of about 7%, and then fall to zero in the late 22nd century. By the year 2200, these five regions have moved to a almost carbon-free economy. Other regions do not face emissions cuts, except the Middle East for a short while.

The costs of this scenario are high. Figure 7 shows the net present consumption losses. These amount to US$162 trillion for the world. For comparison, the costs of the Wigley, Richels and Edmonds (1996) trajectory towards a stabilization of the carbon dioxide concentration at 650 ppmv costs only US$10 trillion, without international trade in emissions permits. The costs of the no-envy scenario can be reduced by allowing international emissions permit trade within each period. Costs then fall to US$51 trillion. This is in the self-interest of all trading parties. The Wigley *et al.* (1996) emissions reduction trajectory is cheaper because costs are minimized over time – the intertemporal cost distribution of Wigley *et al.* (1996) is not equitable and unlikely to be acceptable to all generations (Tol, 1998a).

5. Inequity aversion

Usually, co-operative solutions maximize the sum of the welfares of the actors in the game. There is no reason for this other than convenience. Alternatively,

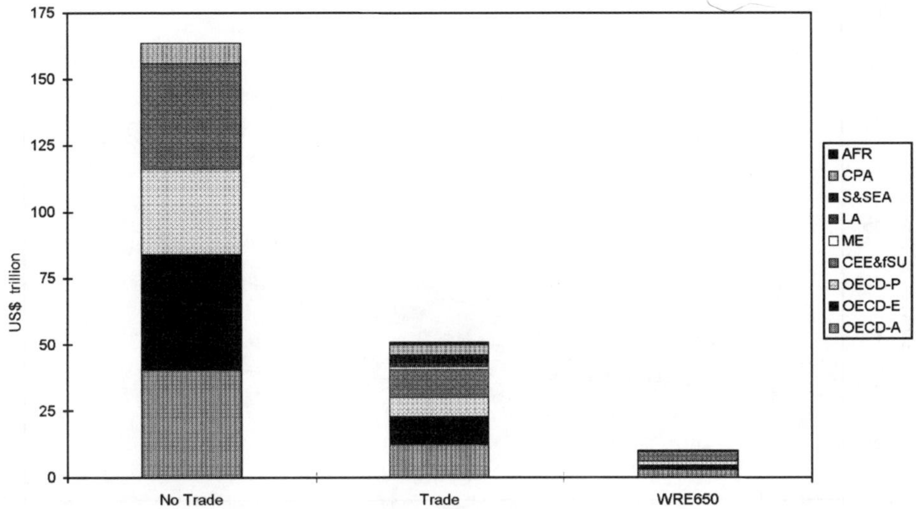

Figure 7. Consumption losses in the period 1990–2200, discounted to 1990 at 5% per year, due to emissions reduction for two no-envy emissions control scenarios – without and with international trade in emissions permits – and the WRE scenario – without international trade – that leads to the same CO_2 concentration in the long run.

one could maximize

$$W = \sum_r \frac{U_r^{1-\gamma}}{1-\gamma}$$

where U_r denotes the welfare of actor r, and γ is a parameter, denoting 'inequity aversion'. For $\gamma = 0$, W equals the conventional sum of welfare. The higher γ, the more W is determined by the welfare U of the poorer actors. This is easily seen since $\gamma \uparrow \infty$ implies that $W = \min(U_r)$ – the Rawlsian maximin approach – and $\gamma \downarrow \infty$ implies that $W = \max(U_r)$ – the Nietzschean maximax approach. If γ is unity, W is replaced by the – equivalent – product of the actors' welfares, a Bernouilli–Nash type of welfare function. Fankhauser *et al.* (1997) discuss the implications of alternative welfare specifications for the impact of climate change. Tol (1998b) treats of the implications for optimal control of greenhouse gas emissions.

In our experiment, inequity aversion γ assumes five different values: 0, 1, 2, 5 and 10. Figure 8 displays the results for the atmospheric concentration of carbon dioxide. Emissions are reduced more for higher inequity aversion. This is because of the implicit wealth transfer of climate change and emissions reduction. That is, the costs of greenhouse gas emissions reduction by the richer regions counts less and less for higher γ, while the avoided damages of climate change to the poorer regions count more and more. Poorer regions are thought to be more vulnerable to climate change, while welfare maximization concentrates emissions reductions in the richer regions. Co-operative emissions control

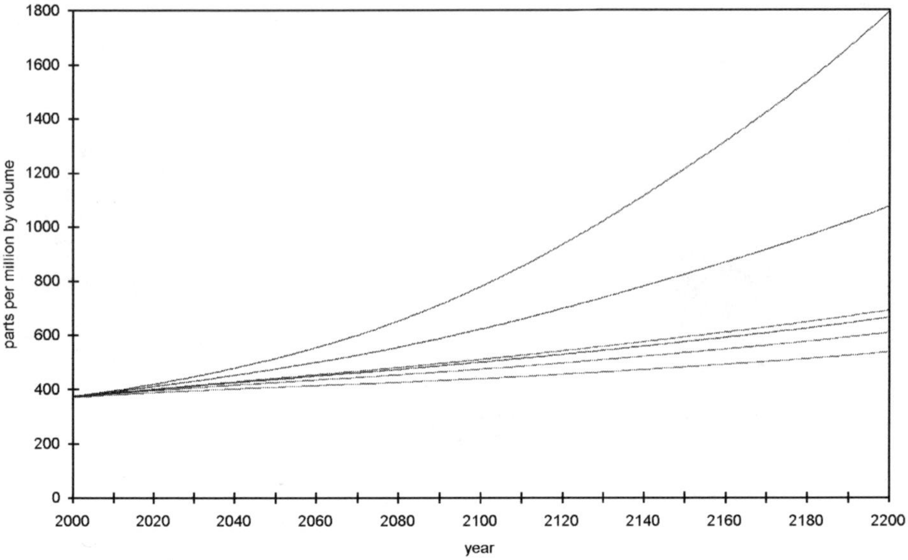

Figure 8. The atmospheric concentration of carbon dioxide for six scenarios, from top to bottom, business as usual, and co-operative optimal emissions control with inequity aversion of 0, 1, 2, 5, and 10.

leads to an atmospheric CO_2 concentration of over 1000 ppmv in 2200, and rising. If $\gamma = 10$, concentrations are kept below 550 ppmv.

6. Conclusions

This paper explores welfare-maximizing carbon dioxide emissions reductions that better adhere to equity issues than does conventional optimal control. If countries do not co-operate but do act as if the most vulnerable region's relative damage is their own, emissions are not substantially reduced, much less indeed than in the co-operative optimal control case. If, instead, the relative damage of the most vulnerable period of the most vulnerable region is considered, emissions are reduced about as much as in the co-operative optimal control case. This suggests two things. First, like any extremum operator, the Kant–Rawls framework is very sensitive to its temporal and regional resolution. If the grid is refined, emissions are likely to be reduced further. Second, the analysis demonstrates the power of co-operation over non-co-operation. (A similar conclusion can be drawn with respect to discounting (Tol, 1999a) and uncertainty (Tol, 1999d).)

In the case where the sum of relative carbon dioxide emissions reduction costs and relative climate change damage costs is equalized over regions and over time, emissions abatement is substantial, keeping CO_2 concentrations below 650 ppmv. This no-envy solution, however, is very expensive compared to other strategies aiming at the same concentration target.

If, in a co-operative setting, a premium is put on an equal distribution of per capita income, emissions abatement is stricter than in the case where premium is nil. For a high aversion of inequity, CO_2 concentrations may even be kept below 550 ppmv.

The numbers presented in this paper should be treated with great caution, as they depend on a single parameterization of a single model. The climate change impact estimates are particularly uncertain, but do drive the numerical results to a substantial extent. The qualitative results are more important. If one takes the climate-change-induced inequities into account, emissions abatement should be intensified. International co-operation in emissions control is crucial. Strong cuts in emissions may well be justifiable on grounds of equity.

The qualitative results need some caveats as well. Two of the three methods presented here (Kant, inequity aversion) are sensitive to resolution. It does matter whether one looks at groups of countries, countries, or sectors or regions within countries. This opens the door to differences of interpretation. All three methods are dependent on the baseline, and on the metric of expressing costs and benefits, both of which are open to dispute.

The main caveat, however, is that it is hard to observe concern for equity issues with the world's governments (Schelling, 1995). The paper presents academic constructs, not descriptions of the real world. These thought experiments may, however, help to inform further thinking about how to handle the enhanced greenhouse effect.

Acknowledgements

Seminar participants at the Center for Integrated Study of the Human Dimensions of Global Change, Carnegie Mellon University, and the Department of Energy, Environmental and Mineral Economics, Pennsylvania State University, provided helpful comments. Financial support from the Netherlands National Research Programme on Global Air Pollution and Climate Change, the European Commission DG12 Fourth Framework Programme, and the United States National Science Foundation is gratefully acknowledged. All errors and opinions are mine.

Notes

1. The source code of the model can be found at http://hdgc.epp.cmu.edu/people/fund/fund.html.
2. The Kantian case thus leads to more stringent emissions reductions than the case in which Europe would compensate Africa for the damage done by climate change, because in the latter case Africa's *absolute* impacts count.
3. This is much like the Pareto criterion, which takes resource endowments as given.

4. One could take changes in utility rather than changes in income. However, the income changes are relatively small, and so a reasonably good approximation to utility changes.

References

Alcamo, J. and G.J.J. Kreileman (1996). Emission scenarios and global climate protection. *Global Environmental Change*. 6(4):305–34.

Azar, C. and H. Rodhe (1997). Targets for stabilization of atmospheric CO_2. *Science*. 276: 1818–19.

Banuri, T., K.-G. Maeler, M.J. Grubb, H.K. Jacobson and F. Yamin (1996). Equity and social considerations. In: J.P. Bruce, H. Lee and E.F. Haites (eds.), *Climate change 1995: economic and social dimensions – contribution of Working Group III to the Second Assessment Report of the Intergovernmental Panel on Climate Change*. 79–124. Cambridge: Cambridge University Press.

Batjes, J.J. and C.G.M. Goldewijk (1994). *The IMAGE 2 Hundred Year (1890–1990) Database of the Global Environment (HYDE)*, 410100082, Netherlands National Institute of Public Health and the Environment (RIVM), Bilthoven.

Fankhauser, S., R.S.J. Tol and D.W. Pearce (1997). The aggregation of climate change damages: a welfare theoretic approach. *Environmental and Resource Economics*. 10: 249–66.

Farrow, S. (1998). Environmental equity and sustainability: rejecting the Kaldor–Hicks criteria. *Ecological Economics*. 27:183–88.

German Advisory Council on Global Change (1995). *Scenario for the derivation of global CO_2 reduction targets and implementation strategies*. Bremerhaven: Alfred Wegener Institute for Polar and Marine Research.

Ha-Duong, M., M.J. Grubb and J.-C. Hourcade (1997). Influence of socioeconomic inertia and uncertainty on optimal CO_2-emission abatement. *Nature*. 389:270–73.

Hammitt, J.K., R.J. Lempert and M.E. Schlesinger (1992). A sequential-decision strategy for abating climate change. *Nature*. 357:315–18.

Hourcade, J.-C., K. Halsneas, M. Jaccard, *et al.* (1996). A review of mitigation cost studies. In: J.P. Bruce, H. Lee and E.F. Haites (eds.), *Climate change 1995: economic and social dimensions – contribution of Working Group III to the Second Assessment Report of the Intergovernmental Panel on Climate Change*. 297–366. Cambridge: Cambridge University Press.

Kattenberg, A., F. Giorgi, H. Grassl, *et al.* (1996). Climate models – projections of future climate. In: J.T. Houghton, L.G. Meiro Filho, B.A. Callander, N. Harris, A. Kattenberg and K. Maskell (eds.), *Climate vhange 1995: the science of climate change – contribution of Working Group I to the Second Assessment Report of the Intergovernmental Panel on Climate Change*. 285–357. Cambridge: Cambridge University Press.

Leggett, J., W.J. Pepper and R.J. Swart (1992). Emissions scenarios for the IPCC: an update. In: J.T. Houghton, B.A. Callander and S.K. Varney (eds.), *Climate change 1992 – the supplementary report to the IPCC Scientific Assessment*. 71–95. Cambridge: Cambridge University Press.

Maddison, D.J. (1995). A cost–benefit analysis of slowing climate change. *Energy Policy*. 23(4/5):337–46.

Manne, A.S. and R.G. Richels (1995). The greenhouse debate: economic efficiency, burden sharing and hedging strategies. *Energy Journal*. 16(4):1–37.

Manne, A.S., R.O. Mendelsohn and R.G. Richels (1995). MERGE – a model for evaluating regional and global effects of GHG reduction policies. *Energy Policy*. 23(1):17–34.

Manne, A.S. and R.G. Richels (1996). The Berlin Mandate: the costs of meeting post-2000 targets and timetables. *Energy Policy*. 24(3):205–10.

Manne, A.S. and R.G. Richels (1998). On stabilizing CO_2 concentrations – cost-effective emission reduction strategies. *Environmental Modeling and Assessment*. 2:251–65.

Nordhaus, W.D. (1991). To slow or not to slow: the economics of the greenhouse effect. *Economic Journal*. 101:920–37.

Nordhaus, W.D. (1992). An optimal transition path for controlling greenhouse gases. *Science.* 258:1315–19.
Nordhaus, W.D. (1993). Rolling the 'DICE': an optimal transition path for controlling greenhouse gases. *Resource and Energy Economics.* 15:27–50.
Nordhaus, W.D. (1994). *Managing the global commons: the economics of climate change.* Cambridge: The MIT Press.
Nordhaus, W.D. and Z. Yang (1996). RICE: a regional dynamic general equilibrium model of optimal climate-change policy. *American Economic Review.* 86(4):741–65.
Peck, S.C. and T.J. Teisberg (1991). CETA: a model for carbon emissions trajectory assessment. *Energy Journal.* 13(1):55–77.
Peck, S.C. and T.J. Teisberg (1994). Optimal carbon emissions trajectories when damages depend on the rate or level of global warming. *Climatic Change.* 28:289–314.
Peck, S.C. and T.J. Teisberg (1995). Optimal CO_2 control policy with stochastic losses from temperature rise. *Climatic Change.* 31:19–34.
Peck, S.C. and T.J. Teisberg (1996). International CO_2 emissions targets and timetables: an analysis of the AOSIS proposal. *Environmental Modeling and Assessment.* 1(4):219–27.
Roemer, J.E. (1996). *Theories of Distributive Justice.* Cambridge: Harvard University Press.
Rose, A. and B. Stevens (1993). The efficiency and equity of marketable permits for CO_2 emissions. *Resource and Energy Economics.* 15:117–46.
Rose, A., B. Stevens, J.A. Edmonds and M.A. Wise (1998). International equity and differentiation in global warming policy – an application to tradeable emission permits. *Environmental and Resource Economics.* 12:25–51.
Schelling, T.C. (1995). Intergenerational discounting. *Energy Policy.* 23(4/5):395–401.
Schimel, D., D. Alves, I. Enting, *et al.* (1996). Radiative forcing of climate change. In: J.T. Houghton, L.G. Meiro Filho, B.A. Callander, N. Harris, A. Kattenberg and K. Maskell (eds.), *Climate change 1995: the science of climate change – contribution of Working Group I to the Second Assessment Report of the Intergovernmental Panel on Climate Change.* 65–131. Cambridge: Cambridge University Press.
Sen, A. (1982). *Choice, welfare, and measurement.* Cambridge: Harvard University Press.
Sen, A. (1987). *On ethics and economics.* Oxford: Blackwell.
Shine, K.P., R.G. Derwent, D.J. Wuebbles and J.-J. Morcrette (1990). Radiative forcing of climate. In: J.T. Houghton, G.J. Jenkins and J.J. Ephraums (eds.), *Climate change – the IPCC scientific assessment.* 41–68. Cambridge: Cambridge University Press.
Swart, R.J., H. De Boois and J. Rotmans (1989). Targetting climate change. *International Environmental Affairs.* 1:222–34.
Swart, R.J. and M.J.M. Hootsmans (1991). A sound footing for controlling climate change. *International Environmental Affairs.* 3:124–36.
Tol, R.S.J. (1995). The damage costs of climate change – towards more comprehensive calculations. *Environmental and Resource Economics.* 5:353–74.
Tol, R.S.J. (1996). The damage costs of climate change: towards a dynamic representation. *Ecological Economics.* 19:67–90.
Tol, R.S.J. (1997). On the optimal control of carbon dioxide emissions – an application of *FUND. Environmental Modelling and Assessment.* 2:151–63.
Tol, R.S.J. (1998a). *The optimal timing of greenhouse gas emission abatement, individual rationality and intergenerational equity.* Nota di Lavoro 3.98, Fondazione Eni Enrico Mattei. Milan. 20 pp.
Tol, R.S.J. (1998b). *Welfare specification and optimal control of climate change: an application of* FUND. Institute for Environmental Studies D98/04. Amsterdam: Vrije Universiteit. 8 pp.
Tol, R.S.J. (1999a). Time discounting and optimal control of climate change: an application of *FUND. Climatic Change.* (Forthcoming.)
Tol, R.S.J. (1999b). Kyoto, efficiency, and cost-effectiveness: applications of *FUND. Energy Journal.* (Forthcoming.)
Tol, R.S.J. (1999c). Spatial and temporal efficiency in climate policy: applications of *FUND. Environmental and Resource Economics.* (Forthcoming.)

Tol, R.S.J. (1999d). Safe policies in an uncertain climate: an application of *FUND*. *Global Environmental Change*. (Forthcoming.)

Toth, F.L., T. Bruckner, H.-M. Fussel, M. Leimbach, G. Petschel-Held and H.-J. Schellnhuber (1997). The tolerable windows approach to integrated assessments. In: O.K. Cameron, K. Fukuwatari and T. Morita (eds.), *Climate change and integrated assessment models: bridging the gaps – proceedings of the IPCC Asia-Pacific Workshop on Integrated Assessment Models, United Nations University, Tokyo, Japan, 10–12 March 1997*. 403–30. Tsukuba: Center for Global Environmental Research, National Institute for Environmental Studies.

Varian, H.R. (1974). Equity, envy and efficiency. *Journal of Economic Theory*. 9(1):63–91.

Weyant, J.P., O. Davidson, H. Dowlatabadi, *et al.* (1996). Integrated assessment of climate change: an overview and comparison of approaches and results. In: J.P. Bruce, H. Lee and E.F. Haites (eds.), *Climate change 1995: economic and social dimensions – contribution of Working Group III to the Second Assessment Report of the Intergovernmental Panel on Climate Change*. 367–96. Cambridge: Cambridge University Press.

Wigley, T.M.L., R.G. Richels and J.A. Edmonds (1996). Economic and environmental choices in the stabilization of atmospheric CO_2 concentrations. *Nature*. 379:240–43.

CHAPTER 13

Distributional Consequences of Alternative Emissions Trading Schemes*

Francesco Bosello and Roberto Roson

1. Introduction

Issues of equity and fairness have been much debated in the context of the international agreements on the reduction of greenhouse gases, especially before, during and after the negotiations for the Kyoto Protocol. This relates to the fact that a high proportion of the current CO_2 emissions are produced by developed countries, and to the fact that the costs of abatement, technology change and control are often too high to be afforded by Third World nations. By consequence, emissions limits have been set for only a limited number of industrialized countries.

However, it is also recognized that: (1) negative impacts of the global warming will likely be felt more in developing regions and (2) future emissions of some fast-growing countries, which have not signed the Kyoto agreement (e.g., China), will account for a large part of total world CO_2 emissions in the 21st century. The critical point is how to make the collaboration of these countries possible.

Co-operative actions between developed and developing nations can already been undertaken within the framework of the Kyoto Protocol. Beyond this, it is debated if, when, and under what conditions emissions constraints could also be set for some non-signatory countries in the future. Clearly, an important aspect that must be assessed here is the income and welfare distribution consequences of the proposed policy schemes.

These effects have been mostly analysed in terms of allocations of pollution rights and, to this end, axiomatic criteria for policy evaluation have been identified (Helm, 1999). These criteria differ in terms of the relative emphasis given to two conflicting principles: fairness in the distribution of rights and fairness in the distribution of costs.

In general, the issue of market efficiency has been regarded as totally separated from the equity issue. This is because efficiency is supposed to be achieved through a market for emissions permits, and the Paretian (or Coasian) conditions for the separability are assumed to hold (Rose and Stevens, 1999).

C. Carraro (ed.), Efficiency and Equity of Climate Change Policy, 291–304.

In an emissions trading system, instead of fixing individual abatement targets at the country level an aggregate level of emissions reduction is obtained by allowing the buying and selling of permits for those units emitting more or less than their individual reference target. It is well known (e.g. Baumol and Oates, 1975) that such a trading system allows to reduce the pollution at a lower cost, at any level, than in the case of command-and-control policies, especially when agents in the system are heterogeneous in terms of technology or preferences. Although command-and-control policies are still the standard approach in environmental regulation, recent experiences with emissions trading systems (Schmalesee *et al.*, 1998; Stavins, 1998) have reinforced the position of those advocating such a flexible market solution.

Since the aggregate abatement level would in this case be constant, and the greenhouse effect is a kind of 'global' externality, an ideal trading system (that is, without transaction and enforcement costs) could bring about a Pareto improvement, meaning that no country would be made worse off (in comparison with a situation of individual constraints) by the introduction of a market for emissions permits. In other words, the 'first theorem of welfare economics' can be applied here: a free market equilibrium amounts to a maximum of total welfare, defined as a weighted sum of individual welfare levels.

Nonetheless, there are many possible ways in which a market for emissions permits can be implemented, depending on the number of countries involved, on the volume of trade allowed, on the market clearing mechanisms, and so on. For example, it is sometimes claimed that only those countries bearing the costs of emissions reduction should be allowed to reap the benefits of a market, or that non-signatory countries should be allowed to trade only when they simultaneously agree to achieve some stabilization targets.

Since trading is a way of allocating resources, different trading systems of course produce different effects in terms of wealth distribution. As a consequence, *the possibility of pursuing distributional objectives through the allocation of rights is conditional upon the adoption of a specific emissions trading scheme.* Any change in the way the market operates would require an offsetting change in the allocation of rights.

It is therefore not sufficient to demonstrate that a market can make everybody better off. Other relevant questions are: who will gain more and who will gain less? What determines the 'weights' of the total welfare function which is implicitly maximized? Is the income inequality between countries in the world made larger by the introduction of an emissions trading system?

We shall explore these issues in this paper by means of an integrated assessment model, which is a variant of the model used by Nordhaus and Yang (1996). We shall consider here a set of idealized trading systems. Although data used in the model refer to real economies, the numerical results should be interpreted, because of the simplifying assumptions adopted, as examples providing an order of magnitude of the effects, rather than as precise forecasts.

2. Equivalence

Consider the position of a country which has signed the Kyoto agreement, and has thereby agreed to reduce its emissions below a certain level by the year 2010. Suppose that a perfectly competitive market for emissions permits is introduced, so that each country can choose, for each emissions unit exceeding the threshold level, between the alternatives of paying for the abatement and paying to purchase an emissions permit from another country. If cost minimization is the guiding principle in the choice, the market equilibrium will be achieved at a point in which marginal abatement costs in all countries are equal, and equal to the price of permits as well. Indeed, if two marginal costs are different, the country with the lower cost could sell permits to the higher-cost country, and both parties would benefit from the transaction. Furthermore, each country must be indifferent between abatement and trade.

An allocation in which marginal costs are equalized coincides with an allocation that could be chosen by a central planner who faces a constraint on total emissions, but can distribute the abatement costs among a set of 'technologies' or countries i, like in the case of a multi-plant producer:

$$\min \sum_i c_i(e_i) \quad \text{s.t.} \sum_i c_i \leq \bar{E} \tag{1}$$

First order conditions of the above problem clearly bring about the equalization of marginal abatement costs $c'(e)$. In the case of international agreements, the abatement costs can be stated in terms of foregone net revenue y (per capita) in each country (or consumption), so that (1) can be also written, equivalently:

$$\max \sum_i y_i(e_i) \quad \text{s.t.} \sum_i e_i \leq \bar{E} \tag{2}$$

where $y'(e) > 0$. Although higher aggregate emissions may lower income and consumption levels in all countries, the reduction of a country's own emissions is expected to reduce the country's income, because the positive effects of the reduction are shared, while the abatement costs are entirely borne by the country under consideration.

The optimization problem (2) is also equivalent to the following one:

$$\begin{gathered} \max \sum_i w_i U[y_i(e_i)] \\ w_i = U'[y_i(e_i)]^{-1} \quad \text{s.t.} \sum_i e_i \leq \bar{E} \end{gathered} \tag{3}$$

where $U(.)$ are utility functions with standard properties. Again the first order conditions are the same. However, (3) is a problem of total welfare maximization, as it could emerge in the case of a co-operative agreement with the weights w expressing the 'bargaining' power of each country.

This illustrates an interesting but well-known result from general equilibrium theory, stating that a market equilibrium in a perfectly competitive exchange economy can be replicated by a single optimization problem with appropriate weights. This principle turns out to be very useful for applied general equilibrium models, because it allows the avoidance of the computation of the equilibrium as a solution of several simultaneous optimization problems (a Nash equilibrium), often involving lengthy iterations and convergence problems in numerical simulations.

It is worth stressing that, because of the concavity of the utility functions (implying decreasing marginal utility of income), the richer a country is, the stronger it is in this game. Also, the solution of (3) depends neither on the initial income distribution[1] nor on the emissions limit set for each country. The latter can be used to verify ex-post, by comparing it with the emissions level obtained in (3), whether a country is an importer or an exporter of permits. In turn, the market for tradable permits must be in equilibrium if the constraint on total emissions coincides with the sum of the individual emissions limits.

This can also be seen by rewriting the problem in a fourth equivalent way:

$$\begin{aligned} &\max \sum_i w_i U[y_i(e_i)] \qquad w_i = U'[y_i(e_i)]^{-1} \\ &\text{s.t.} \quad e_i \leq \bar{e}_i + m_i \; \forall i \left(\sum_i \bar{e}_i = \bar{E} \right) \qquad \sum_i m_i = 0 \end{aligned} \tag{4}$$

where m stands for net imports of permits, a decision variable emerging in the optimization problem faced by each country:

$$\max U[y_i(e_i) - pm_i] \qquad \text{s.t.} \quad e_i \leq \bar{e}_i + m_i \tag{5}$$

and p is the price of permits, taken here as a given.

Observe that the price of permits is not computed directly in (4), but it can be obtained indirectly from the solution, on the basis of the marginal abatement cost. The price level is needed in order to estimate net income levels in each country. Notice also how the individual welfare is affected by the amount of trade: the lower the country emissions limit, the higher the volume of net imports of permits (and/or abatement effort) and the lower the available income.

3. Simulating an emissions trading system for Kyoto

The simulation experiments illustrated in this section have been carried out with a simplified version of the RICE'96 model (Nordhaus and Yang, 1996). In common with RICE our model maintains the regional disaggregation (the world is divided into six macroregions) and the environmental sub-model. CO_2 emissions are a by-product of the economic activity and their impact on global temperature is determined by the Schneider–Thompson model, where increases in the temperature translate in terms of GDP losses. The interested reader may refer to the original Nordhaus–Yang for a description of the model structure.

Exactly as in RICE the link between the economic and the environmental part is given by a particular function accounting for the fact that income is negatively affected by both the environmental damage and the environmental protection effort. This function transforms *potential* income, which is exogenously given here,[2] in net income or consumption, which in turns affects the utility of the representative consumer in each region:

$$Y(t) = \bar{Y}(t)\frac{1-\beta_1\mu(t)^{\beta_2}}{1+\alpha T(t)^{\sigma}} \tag{6}$$

where $Y(t)$ is the *net* income of a generic region at time t, m is a control variable representing the rate of emissions control, μ stands for world average temperature, and all remaining symbols are parameters whose values vary between regions and time periods.

Parameter values have been borrowed directly from the RICE model, but potential income and emissions levels has been estimated on the basis of the IPCC 'A1 Marker Scenario Family' (Morita *et al.*, 1998) and a report of the White House.[3]

The main characteristics of the reference business-as-usual scenario (that is, without the imposition of emissions limits and trading) are:

- average GDP of so-called 'non-Annex I countries' (countries which have not signed the Kyoto agreement) is expected to reach average 1990 GDP levels of signatory countries by the year 2030;
- non-Annex I countries' CO_2 emissions are expected to become higher than signatory countries' CO_2 emissions between 2010 and 2020;
- Former Soviet Union countries will reach emissions levels higher than the limits imposed by the Kyoto agreement only after the year 2020;
- The average temperature is expected to increase by 2.7°C in 2090, whereas with the implementation of the Kyoto agreement this increase is reduced to 2.5°C.

Other scenarios that have been considered here are:

- The imposition of constraints on emissions for signatory countries without the possibility of emissions trading[4] (scenario NOT). Non-Annex I countries are assumed to act here as in the BAU scenario, without emissions abatement, although in this case their net income turns out to be higher because of the reduced climate change. This constraint operates from the year 2010 onwards.
- The introduction of a perfect emissions trading system including all countries which have signed the Kyoto agreement (scenario EMT). This amounts to setting only an aggregate emissions limit for this group of countries, whereas other countries are modelled exactly as in the scenario NOT.
- The trading system is extended to include all countries in the world (scenario CDM). Non-Annex I countries must not exceed the emissions levels corresponding to the BAU trend, but they can trade on the basis of possible

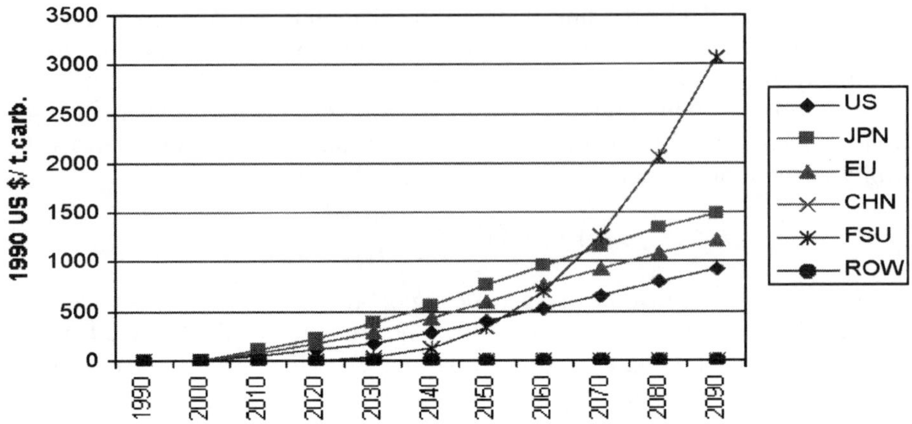

Figure 1. Marginal abatement costs in the NOT case.

emissions reductions. This scheme is sometimes termed 'Clean Development Mechanism'.

The latter scenarios consider three alternative ways of implementing the Kyoto agreement. However, it is worth to notice that aggregate emissions are actually higher, but only for the first period, when an emissions trading is introduced. This is because one region (former Soviet Union, FSU) would be 'naturally' below its threshold initially, but it could still sell – under a trading system – the difference between imposed and actual emissions levels.

Figure 1 shows the marginal abatement costs faced by each country when the Kyoto agreement is applied without trading. Substantial differences, due to both assumptions on economic growth and on control costs, indicate that significant gains should be expected from the introduction of a trading mechanism.

As it has been pointed out in the previous section, trade brings about the equalization of the marginal costs to the market clearing price of emissions permits. Figure 2 shows the evolution over time of the world price under the two schemes considered (EMT and CDM)

When the Kyoto agreement comes into play, the price is US$32.72 per tonne of carbon in the EMT case, whereas in the CDM case it is only US$9.2. Afterwards the prices diverge sharply in the two scenarios, with the price in CDM actually falling in the last periods. The latter effect is basically due to some 'optimistic' hypotheses about technical progress and new abatement technologies.

As the price can be interpreted as a sort of average of the initial marginal abatement costs, the much lower price level obtained in the case of worldwide trading is essentially due to the fact that the marginal costs are significantly below the average for non-Annex I countries. This depends both on the initial

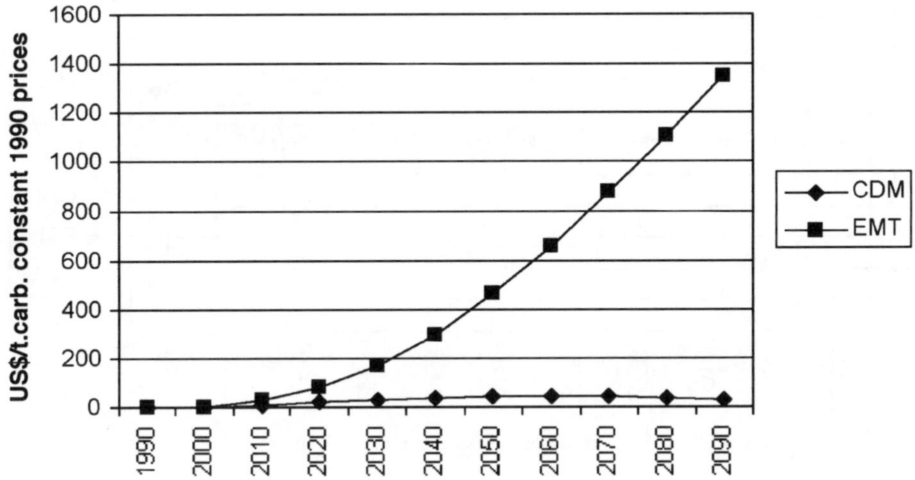

Figure 2. Prices of permits EMT and CDM cases.

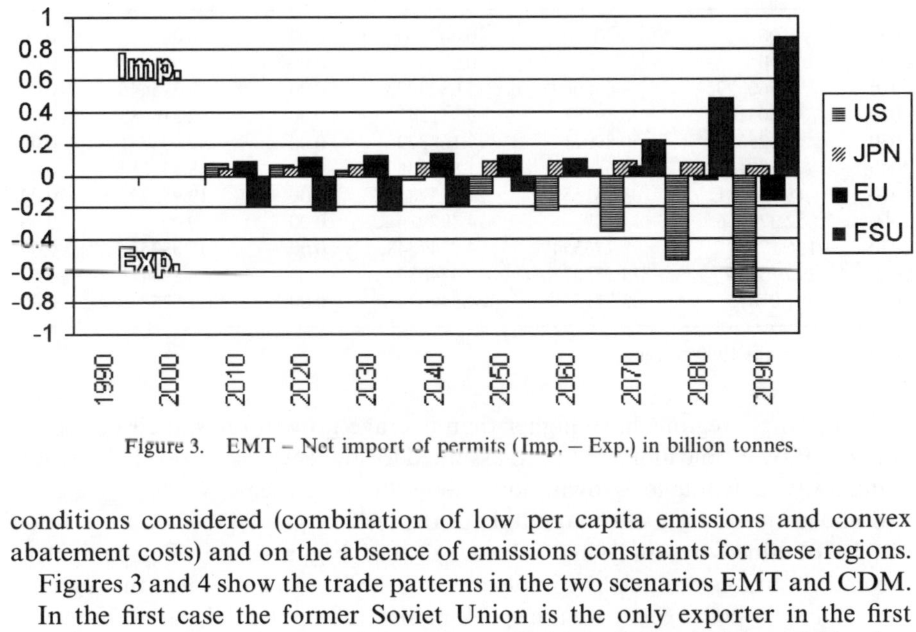

Figure 3. EMT – Net import of permits (Imp. – Exp.) in billion tonnes.

conditions considered (combination of low per capita emissions and convex abatement costs) and on the absence of emissions constraints for these regions.

Figures 3 and 4 show the trade patterns in the two scenarios EMT and CDM. In the first case the former Soviet Union is the only exporter in the first periods, but later the (assumed) strong economic growth transforms this region in a net importer, whereas the US and the European Union (EU) replace FSU as supplier of permits. Although technical efficiency does play a role here, the main driving force is the relative growth rate of each country.

Because of the much lower abatement costs, China (CHN) and Rest of the World (ROW) are and remain exporters of permits in the CDM world market.

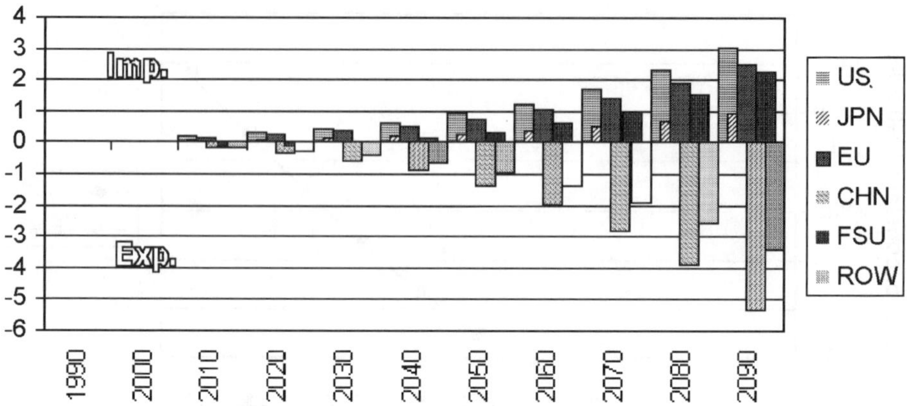

Figure 4. CDM – Net import of permits (Imp. – Exp.) in billion tonnes.

Table 1. Equivalent variation of per capita net income (1): comparison NOT-EMT

	US	JPN	EU	CHN	FSU	ROW
1990	0.00	0.00	0.00	0.00	0.00	0.00
2000	0.00	0.00	0.00	0.00	0.00	0.00
2010	2.5608292	9.6566693	5.0829093	0.00	13.395645	0.00
2020	2.5031847	26.59181	9.9512475	0.00	32.216683	0.00
2030	3.5266809	50.890714	17.168379	0.00	49.906031	0.00
2040	3.5001011	78.633727	21.878038	0.00	48.456053	0.00
2050	13.960081	101.78155	21.774784	0.00	19.982792	0.00
2060	52.374166	114.587	14.492916	0.00	0.00	0.00
2070	136.57839	98.775554	4.8347348	0.00	117.08755	0.00
2080	299.15713	74.555077	2.4240843	0.00	636.89841	0.00
2090	581.07875	33.37152	24.349624	0.00	2041.5262	0.00

(1) US$, constant 1990 prices.

Although these regions have higher than average growth rates, their emissions limits shift over time along the path assumed for the BAU scenario. This creates comparative advantages growing over time for these areas.

Some insights about the magnitude of the welfare gains expected from trade can be obtained by analysing Tables 1–3 and Figure 5, showing equivalent income variations obtained by implementing the Kyoto agreement through the two trading systems. Since we are speaking here of a potential Pareto improvement, every region is expected to gain in every period from the possibility of trade.

The amount of gain is broadly proportional to the volume of trade, independently of the direction of the trade flow. This explains, for example, why the FSU obtains insignificant benefits when it switches from the position of net exporter to that of net importer.

Table 2. Equivalent variation of per capita net income in absolute terms (1): comparison NOT-CDM

	US	JPN	EU	CHN	FSU	ROW
1990	0.00	0.00	0.00	0.00	0.00	0.00
2000	0.00	0.00	0.00	0.00	0.00	0.00
2010	13.901644	17.703894	12.707273	0.8138674	2.1806864	0.2371279
2020	41.123749	62.047558	37.317178	2.7999079	1.806543	0.8510994
2030	105.80043	160.75003	98.105025	7.3251041	0.00	1.5621094
2040	231.00668	351.82513	213.91859	15.133666	15.863589	3.101584
2050	471.15273	677.45801	416.14032	25.535829	118.18394	5.1888127
2060	872.90276	1186.7939	736.72321	37.722556	469.5409	6.6355794
2070	1512.8683	1934.3546	1215.9358	48.532704	1390.7717	8.0332905
2080	2463.647	2982.2031	1895.6339	56.2641	3426.5711	9.4136818
2090	3798.2709	4371.6691	2809.9467	61.661568	7459.3102	15.433215

(1) US$, constant 1990 prices.

Table 3. Equivalent variation of per capita net income in absolute terms (1): comparison EMT-CDM

	US	JPN	EU	CHN	FSU	ROW
1990	0.00	0.00	0.00	0.00	0.00	0.00
2000	0.00	0.00	0.00	0.00	0.00	0.00
2010	11.339839	8.0450117	7.6228634	0.8138674	−11.18996	0.2371279
2020	38.617838	35.433763	27.357097	2.7999079	−30.29791	0.8510994
2030	102.26528	109.75258	80.899118	7.3251041	−49.71096	1.5621094
2040	227.49109	272.85559	191.94649	15.133666	−32.50829	3.101584
2050	457.09028	574.92634	394.20664	25.535829	98.130283	5.1888127
2060	819.96151	1070.9187	722.07059	37.722556	469.5409	6.6355794
2070	1374.2532	1834.0214	1211.0275	48.532704	1271.248	8.0332905
2080	2158.7375	2906.1222	1893.1625	56.2641	2770.3066	9.4136818
2090	3203.5919	4337.4628	2785.0223	61.661568	5337.6996	15.433215

(1) US$, constant 1990 prices.

Which trading system performs better in terms of welfare gains? Since the trade pattern obtained in the EMT scenario can always be replicated in the case of an enlarged trading system, but not vice versa, aggregate equivalent income variations must be always larger in the CDM case. In other words, the global payoff must be higher when there are more degrees of freedom (Chander *et al.*, 1999).

It is not true, however, that every country would prefer an enlarged market for emissions permits. This is the case, for example, in the former Soviet Union which is made worse off in the first four periods because of a worsening of its terms of trade. This can be easily understood by considering the situation occurring in the first period, where this region sells the difference between allowances and emissions obtained without abatement effort. In this case, any

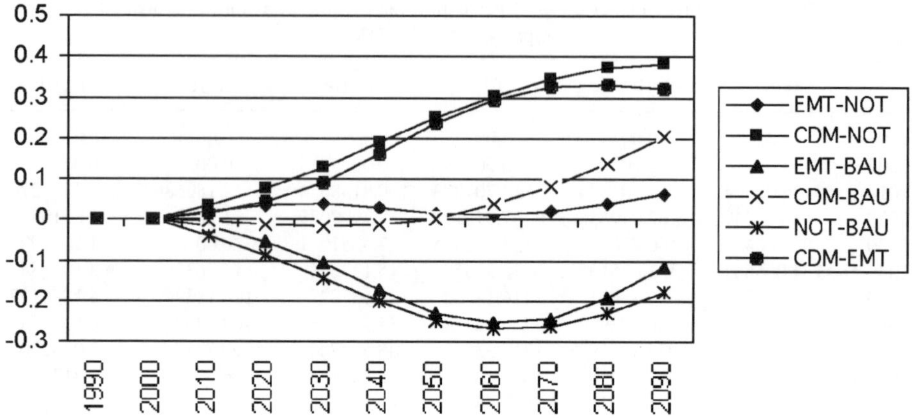

Figure 5. Differences in world per capita average net income (percent).

price reduction, like that induced by the entry of new low-cost competitors, immediately translates into a reduction of income. In the subsequent periods, the FSU can only partly offset this effect by reducing the export volume. Eventually, however, the FSU becomes a net importer and start to gain more in the CDM scenario.

4. Distributional effects

Income inequality can be measured in several different ways. For example descriptive statistics, like the Gini coefficient, can be computed for each distribution considered. Usually, the indices used to this purpose obey to the so-called 'principle of transfers', meaning that if income is transferred from a person to another person with a lower income, then the index of inequality must decrease. However, a transfer may occur at different levels of the distribution: it may involve high- and middle-income persons, or middle- and low-income persons. Even if the principle of transfers is satisfied, the variation of the index could depend on the point of the scale of income at which the transfer occurs.

As Atkinson (1970) pointed out, weighting income transfers in different ways implicitly defines a social welfare function. Since the most popular descriptive statistics imply quite implausible assumptions about the shape of the implicit social welfare function, Atkinson suggests starting by making explicit the social function, and deriving an index conditional on this.

In the context of the issues addressed in this paper, the most natural candidate is an 'egalitarian' social welfare function, that is a function in which the welfare of all individuals is considered in the same way. In this case, it is possible to compute an 'equally distributed equivalent level of income', which is the level of income per head j that, if equally distributed, would give the same level of

Table 4. Variations in the equity index due to different implementation mechanisms ($\times 10^5$)

	NOT-BAU	EMT-BAU	CDM-BAU
2010	19	14	16
2020	44	35	33
2030	81	69	50
2040	119	108	61
2050	140	132	56
2060	134	129	42
2070	94	90	22
2080	25	25	7
2090	−67	−56	−5

social welfare as the distribution under consideration, where the social welfare is defined here as a simple sum of utility levels:

$$U(j)=\sum_i s_i U(y_i) \tag{8}$$

where s is the population shares of individuals having the same income level.

Applying (8) to the case of emissions trading, and assuming a logarithmic utility function, the equivalent income would be given by:

$$j=\exp\left(\sum_i s_i \log[y_i(e_i)-pm_i]\right) \tag{9}$$

It is to be noticed that:

- the index j never exceeds the average income, and the ratio between equivalent and average income provides an index of income equity, ranging between zero and one;
- a Pareto improvement, like that generated by the introduction of a competitive market, increases the equivalent income j. The index of equity does not change if the gains are distributed so as to make the relative incomes unaltered;
- the market equilibrium is defined by an implicit social welfare function maximization in which the function which is maximized is different from that used here in constructing the index j. In particular, the maximized function has been defined so as to make the total welfare invariant to income transfers between any pair of income classes.

Distributional implications of the scenarios considered in the simulation exercises can be drawn by looking at Table 4, presenting the differences observed in the equity index (ratio between equivalent and average income) for the three implementation schemes (NOT, EMT, BAU), in comparison with the BAU baseline.

Three mechanisms determine the equity effects here. First, the Kyoto constraints are applied only to a subset of countries which have (initially) the higher income per capita levels. Of course this produces a positive distributional effect which is especially evident when emissions trading is not allowed (NOT). Second, trading creates a welfare improvement which is, however, 'inequitable', because the social welfare function which is implicitly maximized is not an egalitarian one. Third, emissions limits are set in such a way that the poorer countries are exporters of permits, thereby creating a trade revenue redistribution effect (growing over time because non-Annex I countries have no limits), with positive distributional consequences.

If the impact of the Kyoto Protocol on the world income distribution were to be mainly determined by the allocation of emissions rights, one would see quite similar values in the three columns. In other words, if trading is only a matter of efficiency, with distributional effects due to cost allocation and trade revenue cancelling out, the relative incomes would be affected only by the reduction targets.

However, this is not the case. Trade does have distributional consequences which are often comparable to those due to the initial allocation of rights, especially when the market is a global one. Richer countries are expected to gain more, despite the fact that in this exercise non-Annex I countries have emissions targets growing along the BAU trend. As a consequence, it is in the interest of developed nations to trade *always* with developing countries, even when these countries have not agreed to reduce their carbon emissions.

5. Conclusions

The simulation exercises presented in this paper have highlighted the following points: first, although the Kyoto Protocol is costly in terms of GDP, (ideal) emissions trading schemes would produce improvements for all countries (both trading and not). The cost reductions are substantial when countries that have not signed the Kyoto agreement are allowed to trade. Second, in a global trading system non-signatory countries are always exporters of permits, whereas when trading is restricted the former Soviet Union is initially the only exporter, becoming an importer in later periods. These findings are remarkably similar to those which have found with other modelling approaches (e.g. Nordhaus and Boyer, 1999).

In contrast with other simulations and theoretical analyses, however, we have illustrated in this paper the distributional effects associated with a number of different emissions trading schemes, stressing that equity and efficiency issues should not be addressed separately. Indeed, although each trading system is superior to a no-trade situation, the countries do experience different impacts in the different cases. FSU countries, for example, would be made worse off in the first periods by an enlargement of the trade area.

The simulation exercises have been carried out by assuming the so-called 'Kyoto forever' hypothesis: that is, that the emissions level targets fixed by the

Kyoto agreement remain in place for the signatory countries from the year 2010 onwards. Since non-Annex I countries are not affected by these limits, this hypothesis is quite advantageous for the non-signatory nations, especially when these are allowed to trade on permits.

It is sometimes argued that non-signatory countries should be allowed to participate in a world market for emissions permits only when they will agree to set domestic constraints on CO_2 emissions, possible in the future and conditional on their level of economic development. In this work we have made clear, however, that from the signing nations' point of view this argument has no economic rationale.

Notes

* This work has been realized within the FEEM working group on Integrated Assessment Modelling. Members of this group, including Paolo Buonanno, Carlo Carraro, Efrem Castelnuovo, Marzio Galeotti, Michele Moretto and the authors, have contributed with useful comments and suggestions. Remaining errors, however, are the sole responsibility of the authors of the paper. We are grateful to Z. Yang who kindly provided the RICE model.

1. If weights in (3) were to be set in a different way, emissions would be allocated in order to influence the distribution of income.
2. This is different from RICE, where investments are determined by intertemporal optimization. The FEEM research group has, however, made some experiments by introducing emissions trading in an intertemporal setting, and found that there are not significant differences in the results, whereas the solution algorithm of the model becomes substantially more complicated and less accurate.
3. The Kyoto Protocol and the President's policies to address climate change. July 1998.
4. We consider that the US, Japan, Europe, and the former Soviet Union (our Annex I) in 2010, as an approximation of the period 2008–2012, have to reduce their emissions by 7%, 6%, 8% and 0% below 1990 level respectively.

References

Administration Economic Analysis (1998). The Kyoto Protocol and the President's policies to address climate change. Published report.

Atkinson, A.B. (1970). On the measurement of inequality. *Journal of Economic Theory*. 2:244–63.

Baumol, W.J. and Oates, W.E. (1975). *The theory of environmental policy: externalities, public outlays and the quality of life*. New York: Prentice-Hall.

Chander, P., Tulkens, H., van Ypersele, J.P. and Willems, S. (1999). *The Kyoto Protocol: an economic and game theoretic interpretation*. CLIMNEG working paper, n. 12, CORE. Université Catholique de Louvain, Belgium.

Helm, C. (1999). *Axiomatic fairness criteria and the allocation of climate protection burdens*. Paper presented at the EFIEA workshop on Integrated climate policies in the European environment, costs and opportunities. Milan, March 1999.

Morita, T., Matsuoka, Y., Jiang, K., *et al.* (1998). A1 marker scenario. Mimeo.

Nordhaus, W.D. and Boyer, J.G. (1999). *Requiem for Kyoto: an economic analysis of the Kyoto Protocol*. Mimeo.

Nordhaus, W.D. and Yang, Z. (1996). A regional dynamic general equilibrium model of alternative climate change strategies. *The American Economic Review.* 86(4):741–65.

Rose, A. and Stevens, B. (1999). *A dynamic analysis of fairness in global warming policy:* Kyoto, *Buenos Aires, and beyond.* Paper presented at the EFIEA workshop on Integrated climate policies in the European environment, costs and opportunities. Milan, March 1999.

Schmalensee, R., Joskow, P.L., Ellerman, A.D., Montero, J.P. and Bailey, E.M. (1998). An interim evaluation of sulfur dioxide emission trading. *The Journal of Economic Perspectives.* 12(3):53–68.

Stavins, R.N. (1998). What can we learn from the grand policy experiment? Lessons from SO_2 allowance trading. *The Journal of Economic Perspectives.* 12(3):69–88.

CHAPTER 14

A Different Perspective for Global Climate Policy: Combining Burden Sharing and Climate Protection

Janina Onigkeit and Joseph Alcamo

1. Introduction

The climate summits in Kyoto and Buenos Aires achieved some tentative first steps for international climate protection. However, an important question that was left open by both summits was the issue of strategies for long-term climate protection strategies and their consequences on emissions reduction commitments for both industrialized and developing countries. This question was later given high priority at the 6th International Workshop on 'Using Global Models to Support Climate Negotiations', in Kassel, Germany (see Onigkeit *et al.*, 1998) and is addressed by the authors in this paper. The purpose of this paper is to provide an approach that combines the question of stabilization targets with the question of allocation of greenhouse gas (GHG) emissions. We use this approach to evaluate the implications of different long-term climate protection targets on the allocation of emissions reductions in non-Annex B[1] and Annex B countries. This allocation is based on two indicators that reflect considerations of capability and equity. These are (1) per capita income to select the point in time when developing countries can afford to take measures in a climate protection regime and (2) a convergence of per capita emissions with the aim of overcoming the current disparity in per capita emissions between industrialized and developing countries. Why is the allocation of emissions an international issue? First of all, according to the Berlin Mandate and the Kyoto Protocol, most industrialized countries are required to begin reducing their greenhouse gas emissions. Some have argued that this is justified because of their high level of per capita emissions and their historical contribution to climate change. On the other hand, results of scenario analysis have shown that many climate targets[2] require very significant reductions in global emissions that would be virtually impossible for Annex B countries to achieve alone. One reason is that greenhouse gas emissions from developing countries are expected to increase substantially (Alcamo *et al.*, 1995). In response to this situation, we present here an approach that can help identify strategies for

C. Carraro (ed.), Efficiency and Equity of Climate Change Policy, 305–314.

both long-term climate protection and sharing the burden of emissions reductions between Annex B and non-Annex B parties. This approach is one of the first attempts to combine climate protection and burden sharing with indicators for equity and capability in a single analysis. In this paper we apply this approach to two CO_2 stabilization targets, taking into account CO_2, CH_4 and N_2O emissions from the energy/industry sector as well as land-use emissions.

2. The burden-sharing concept

The main idea behind the proposed presented burden-sharing scheme is that the per capita emissions of non-Annex B parties, which are currently substantially lower than those of Annex B countries, are allowed to increase uncontrolled until they reach a specified income level (the so-called graduation criterion). Above this level, developing countries are expected to participate in international emissions regimes. In principle, the graduation income level must be set high enough so that developing countries will have a high enough national income to afford to control their emissions. The first step in participating in international regimes is to freeze per capita emissions, and the second is to reduce emissions (see Figure 1a). Reductions start when non-Annex B per capita emissions equal Annex B per capita emissions aiming at an equitable distribution of emissions rights.

In the following paragraphs we specify the rules for allocating global emissions between Annex B and non-Annex B countries. For this allocation, a baseline ('no policy') emissions pathway and an economic growth scenario are

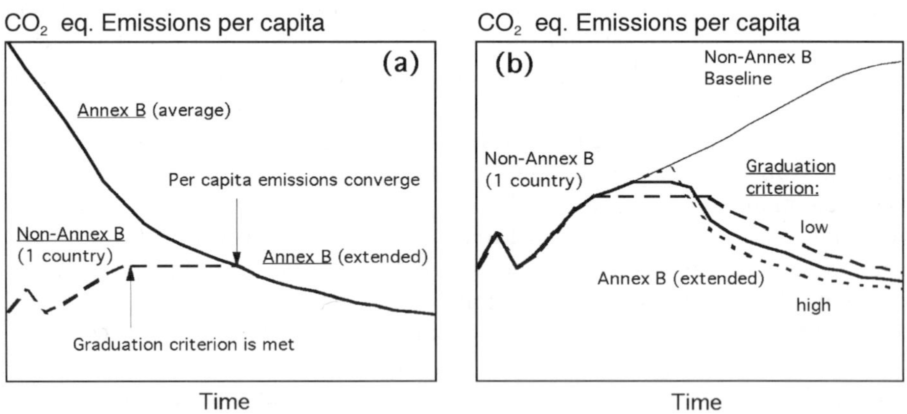

Figure 1. Illustration of the burden sharing concept. (a) Per capita emissions pathway of one non-Annex B country and averaged Annex B per capita emissions. (b) Illustration of the implications of different graduation criteria. A high graduation criterion stands for a high per capita income and vice versa.

needed for each non-Annex B country or group of countries. A population scenario is required for both Annex B and non-Annex B countries.

2.1. Procedure for setting climate protection goals and allocating emissions

1. *Pathway of global emissions:* A stabilization target for the atmospheric CO_2 concentration is first specified, and then the global emissions that comply with this stabilization target are computed. If the analysis is based on CO_2 equivalent emissions, additional assumptions must be made for non-CO_2 greenhouse gas emissions.
2. *Non-Annex B emissions up to and after the graduation income level.* Emissions from non-Annex B regions are not controlled until their income reaches the graduation income level. Up to this point they follow their baseline emissions pathway. After the graduation level is reached, emissions are frozen until non-Annex B per capita emissions are equal to the average of Annex B emissions.
3. *Non-Annex B emissions equal Annex B emissions*: When the per capita emissions of the non-Annex B region converge with the average per capita emissions of Annex B countries, they then both follow the same per capita emissions pathway (see Figure 1a). The non-Annex B party then joins an 'Extended Annex B' group.
4. *Annex B total emissions.* The emissions from Annex B are computed to be the global emissions from step 1 minus the total emissions of all non-Annex B emissions. After this step, the total emissions of non-Annex B countries consist partly of baseline emissions (for those non-Annex B countries that did not yet meet the graduation criterion) and partly of controlled non-Annex B emissions.

These calculations are performed iteratively for each of the non-Annex B countries and the average of all Annex B countries for each time step. Since the emissions from all Annex B countries are summed into a single region, their total emissions have to be distributed between the countries of this group in a later step (e.g. all have to carry the same reduction burden or allocate emissions in the most cost-effective way by emissions trading). This allocation is not a subject of this paper.

Main characteristics

1. Allowable emissions of Annex B depend on the total amount of non-Annex B emissions. Since this analysis is based on per capita emissions, it is advantageous to Annex B countries for non-Annex B countries with a large and fast-growing population to join the Annex B group as early as possible.
2. If a high graduation criterion is chosen, i.e. a non-Annex B region has to reach a high per capita income before taking action, it is likely to follow

the baseline emissions pathway for a longer time compared to the case of a lower graduation criterion. The price this region has to pay is the need for more rapid reductions in per capita emissions after convergence (see Figure 1b). The decision for a low criterion leads to earlier participation in emissions controls but to less stringent annual reduction rates.

3. A stricter stabilization target results in a lower global emissions pathway. In this case stricter emissions reductions are required for Annex B countries, since baseline emissions profiles of non-Annex B parties remain the same. The consequence for non-Annex B parties is that their per capita emissions sometimes converge with the per capita emissions of Annex B countries before they (the non-Annex B parties) reach the graduation income level. Hence, the 'grace period' of constant per capita emissions is omitted for these countries and they must reduce emissions as their first action.

3. Application of the burden-sharing concept

The implications of two stabilization targets on burden sharing were evaluated: (1) Stabilization of the atmospheric CO_2 concentration at 550 ppmv, a target which is under discussion in the European Union; and (2) a stricter target of 450 ppmv. The following were underlying scenarios and assumptions for the application:

1. The analysis was performed for the six non-Annex B regions of the IMAGE 2.1 model (Alcamo *et al.*, 1998a). The seven Annex B regions of the IMAGE 2.1 model were aggregated to one region.
2. Population for both Annex B and non-Annex B regions increase according to the IPCC medium scenario IS92a (Alcamo *et al.*, 1998b).
3. Economic growth of the non-Annex B regions was based on IPCC scenario IS92a assumptions.
4. The baseline emissions of the non-Annex B regions originated from the Baseline A scenario of the IMAGE model. In this scenario, population growth and economic growth assumptions of IS92a have been implemented (Alcamo *et al.*, 1998b). We included CO_2, CH_4 and N_2O emissions from land use, industry and energy production.
5. As a graduation criterion we used a specified percentage of the average per capita income of Annex B regions in 1990. Around this year Annex B countries declared their intention to become active in a climate protection regime (Earth Summit in Rio, 1992). It is therefore appropriate to use the economic strength (in GDP per capita) of Annex B countries at this time as a measure for the capability of non-Annex B regions to initiate emissions controls.

3.1. Stabilization targets and global emissions

For this analysis both a long-term concentration target, and the pathway to reach this target, had to be specified. For both targets we used the pathways of the Intergovernmental Panel on Climate Change (IPCC) described in Enting *et al.* (1994). The targets will be reached in 2100 and 2150 for a CO_2 concentration of 450 ppmv and 550 ppmv, respectively. The IMAGE model was used to back-calculate allowable global CO_2 emissions between 1990 and 2100. Non-CO_2 emissions (CH_4 and N_2O) from the energy/industry sector were assumed to be reduced proportionally to CO_2 emissions whereas future non-CO_2 emissions from land use were allowed to increase according to the IMAGE 2.1 Baseline A scenario. The resulting CO_2 equivalent concentrations and global emissions are shown in Figures 2a and 2b, respectively. To illustrate the necessary global emissions reductions we also present the so-called 'Kyoto' scenario as a reference scenario. For this scenario we assumed that Annex B regions fulfil the Kyoto commitments until year 2010 and then freeze their total emissions up to 2100. Non-Annex B emissions follow the IMAGE Baseline A emissions pathway without further restrictions.

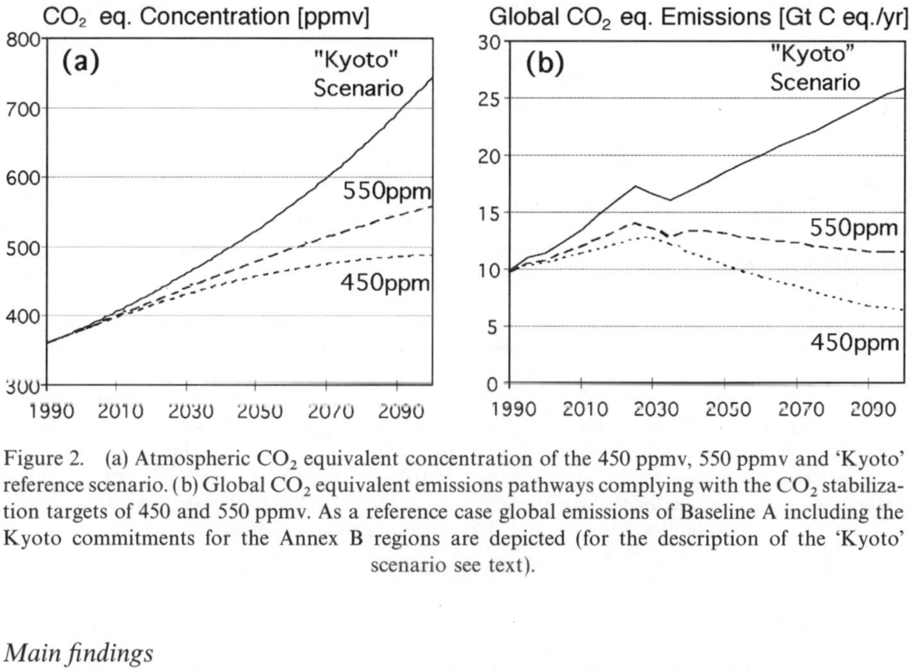

Figure 2. (a) Atmospheric CO_2 equivalent concentration of the 450 ppmv, 550 ppmv and 'Kyoto' reference scenario. (b) Global CO_2 equivalent emissions pathways complying with the CO_2 stabilization targets of 450 and 550 ppmv. As a reference case global emissions of Baseline A including the Kyoto commitments for the Annex B regions are depicted (for the description of the 'Kyoto' scenario see text).

Main findings

1. Under the specified assumptions for non-CO_2 greenhouse gases, the 550 ppmv and 450 ppmv CO_2 stabilization scenarios result in an atmospheric CO_2 equivalent concentration of 560 ppmv and 490 ppmv, respectively. For the 'Kyoto' scenario the CO_2 equivalent concentration increases to 744 ppmv in 2100.
2. Global CO_2 equivalent emissions for both stabilization scenarios may increase until 2030, then emissions have to decrease to a level slightly above the 1990 level for the 550 ppmv scenario and to about half the 1990 emissions

for the 450 ppmv scenario. Taking no further steps in emissions controls than those agreed on in Kyoto could lead to more than a doubling of CO_2 equivalent emissions between 1990 and 2100 ('Kyoto' reference scenario).

3.2. Per capita emissions resulting from the burden-sharing concept

The calculated allowable global emissions were distributed year by year on a per capita basis between Annex B and non-Annex B regions following the specified rules (see Figure 3). We present the implications of two different graduation income level criteria for non-Annex B regions: They must freeze their emissions when their per capita income equals (1) 10% and (2) 100% of the Annex B average per capita income in 1990.

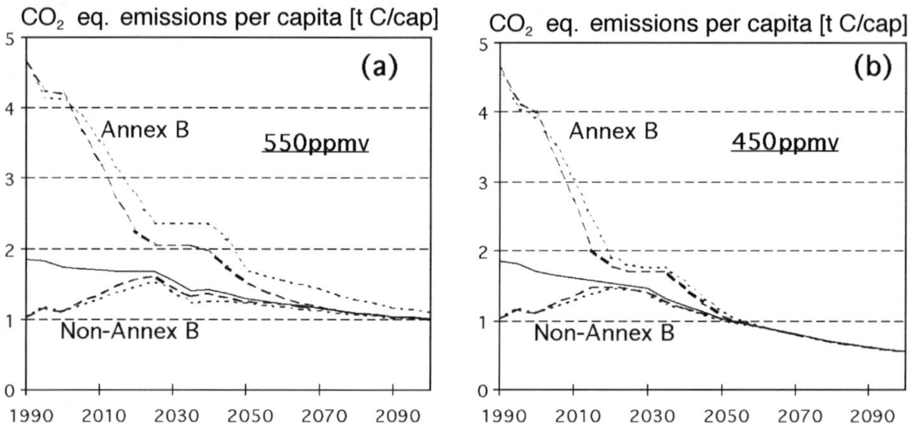

Figure 3. Application of the burden sharing concept to on two CO_2 stabilization targets: (a) 550 ppmv and (b) 450 ppmv. Depicted are the pathways of average per capita emissions (CO_2 equivalent) of Annex B and non-Annex B countries. The short-dashed line shows results for the specified graduation gives a graduation criterion of 10% of Annex B average GDP/cap. The long-dashed line shows results for the a higher criterion – 100% of Annex B average GDP/cap. The solid line gives the global average of per capita emissions.

Main findings

1. Average Annex B per capita emissions start at an emissions level that is more than double the global average per capita emissions. As a consequence their emissions have to be reduced by 78% and 88% between 1990 and 2100 for the 550 ppmv and 450 ppmv stabilization scenarios, respectively (Figure 3). A reduction rate of about 2–3% per year is needed between 1990 and 2030 for the 550 and 450 ppmv target, respectively. After 2030, reduction rates decline to about 0.75 and 1% per year, but reductions must continue. The stringent reductions required for Annex B per capita emissions is caused by three factors: (1) the concentration targets, (2) the growing population

and per capita emissions in non-Annex B countries, regions and (3) the growing population in Annex B regions.

2. Average non-Annex B per capita emissions start at half of the globally averaged emissions level and may therefore increase by 50% up to 2030 (Figure 3). After 2030, however, non-Annex B per capita emissions must decline to their 1990 levels by 2100 for the 550 ppmv scenario. For the 450 ppmv case, per capita emissions must decrease below this level by 2100. Although Annex B regions substantially reduce their per capita emissions, non-Annex B emissions must stay below the decreasing globally averaged emissions.
3. Varying the graduation criterion has a bigger impact on Annex B per capita emissions reductions under the 550 ppmv stabilization scenario than under the 450 ppmv scenario. This is due to the slower reduction of Annex B emissions between 1990 and 2020 in the 550 ppmv scenario. In the 450 ppmv scenario it makes no difference for many of the non-Annex B regions whether the graduation criterion is high or low. In any case, their emissions equal Annex B emissions before they meet the graduation criterion, i.e. they are not allowed to go through a stage of stabilized emissions but have to reduce emissions as their first action. It should be noted that for graduation criteria between 10% and 100%, emissions pathways do not necessarily lie between the two presented profiles. Instead, these pathways will be equally influenced by the assumed GDP per capita growth rate for non-Annex B countries; these growth rates will determine when the different non-Annex B countries reach the graduation income level, and when they will freeze their emissions.

3.3. *Total emissions of Annex B and non-Annex B*

Total CO_2 equivalent emissions of Annex B and non-Annex B resulting from the application of the burden-sharing concept are depicted in Figure 4.

Main findings

1. Annex B total CO_2 equivalent emissions have to be reduced by 76% and 86% up to 2100 for the 550 ppmv and 450 ppmv scenarios, respectively (Figure 4).
2. Non-Annex B total CO_2 equivalent emissions are allowed to more than double up to 2030 for both stabilization scenarios (Figure 4). Afterwards, emissions slowly decrease under the 550 ppmv scenario, and sharply decrease under the 450 ppmv scenario.
3. Varying the graduation criterion makes a difference of about 0.1 Gt C and 0.4 Gt C in total emissions per year for the 450 ppmv and 550 ppmv scenarios, respectively. The reason for the difference between the two stabilization scenarios has already been described in section 3.2.

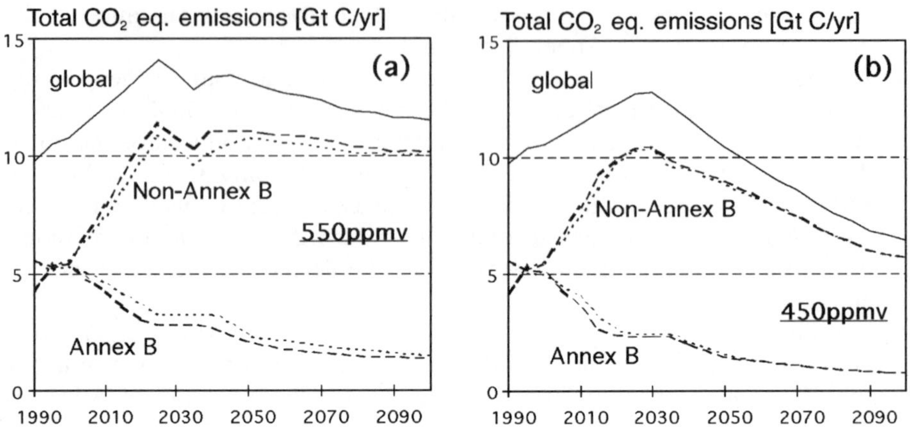

Figure 4. Total CO_2 equivalent emissions of Annex B and non-Annex B compared to allowable global emissions for the two stabilization targets of (a) 550 ppmv and (b) 450 ppmv CO_2. The short-dashed line represents a graduation criterion of 10% of Annex B average GDP/cap, and the long-dashed line a higher criterion of 100%.

4. Concluding remarks

The first intention of this paper is to present a method that allows easy evaluation of the implications of long-term climate protection targets for necessary emission controls in industrialized as well as developing countries. The second point is to present the results for two stabilization targets. These results, however, are based on only one medium scenario for population growth and economic growth. As a follow-up to the work presented in this paper, it would therefore be valuable to carry out further sensitivity analyses using the described approach. In particular, the effects of assumptions about population and economic growth on emissions pathways should be identified. In addition, the sensitivity of results to different graduation criteria should be investigated.

The question arises, can the burden-sharing approach presented in this paper be used in climate policy making? In answering this, an important consideration is that this approach requires the specification of only two factors: a graduation criterion and a climate target and pathway. It also uses a widely accepted parameter as a graduation criterion (GDP per capita) and considers the economic heterogeneity within the group of non-Annex B countries. This simplicity and transparency could be a valuable asset in its use for climate policymaking. On the other hand, the simplicity of the method can also be viewed as a drawback, because policy makers might prefer to specify a wider set of goals and policy options than are included in this approach. Nevertheless, it is not intended to cover all options for assessing and developing climate policies.

Despite its simplicity the approach presented in this paper takes into account both concrete climate goals and important equity and capability considerations in determining future emissions pathways. As such it can provide useful input to the climate policy-making process.

Acknowledgements

The analyses for this paper were funded by the German Federal Environment Agency. The authors are indebted to Paul Reuter for his assistance in this evaluation and to the participants of the 6th and 7th 'International Workshop on Using Global Models to Support Climate Negotiations' for the fruitful discussions during the workshop.

Notes

1. Annex B countries are those countries that agreed as part of the Kyoto Protocol to control their emissions. These countries are listed in 'Annex B' of the Kyoto Protocol, and include all industrialized countries. Countries not listed in Appendix B have assumed no official commitments to control their emissions. These include all developing countries.
2. For example, stabilization of atmospheric CO_2 at 450 ppmv, or limiting global temperature increase to 1.5°C between 1990 and 2100 require global CO_2 emissions reductions of at least 50% relative to 1990 by the year 2100.

References

Alcamo, J., Bouwman, A., Edmonds, J. Grübler, A., Morita, T. and Sugandhy, A. (1995). *An evaluation of the IPCC IS92 emission scenarios.* In: Houghton, J.T., Meira Filho, L.G., Bruce, J., Lewe, H., Callander, B.A., Haites, E., Harris, N., Maskell, K. (eds). Climate change 1994: Radiative forcing of climate change and an evaluation of the IPCC IS92 emission scenarios. Cambridge: Cambridge University Press. 247–304.

Alcamo, J., Kreileman, G.J.J., Krol, M., *et al.* (1998a). *Global modelling of environmental change: an overview of IMAGE 2.1.* In: J. Alcamo, R. Leemans and G.J.J. Kreileman (eds). Global change scenarios of the 21st century. Results from the IMAGE 2.1 model. Oxford, UK: Pergamon/Elsevier Sci. 3–94.

Alcamo, J., Kreileman, G.J.J., Bollen, J., *et al.* (1998b). *Baseline scenarios of global environmental change.* In: Alcamo, J., Leemans, R., Kreileman, G.J.J. (eds). Global change scenarios of the 21st century. Results from the IMAGE 2.1 model. Oxford, UK: Pergamon/Elsevier Sci. 97–139.

Enting, I.G., Wigley, T.M.L. and Heiman, M. (1994). *Future emissions and concentrations of carbon dioxide.* Research technical paper no. 31, CSIRO Division of Atmospheric Research, Australia.

Onigkeit, J., Berk, M. and Alcamo, J. (1998). *Report on the Sixth International Workshop on Using Global Models to Support Climate Negotiations.* Center for Environmental Systems Research, University of Kassel, Germany. The report is available as a pdf file at http://www.usf.uni-kassel.de/service/bibliothek/kassel-1-report.pdf

CHAPTER 15

Discounting and Sustainability in Applied IAMs

Reyer Gerlagh[1]

1. Introduction

Since the potential climate change caused by antropogenic greenhouse gas (GHG) emissions has been recognized as a major environmental problem, a number of applied Integrated Assessment Models (IAMs) have been developed to advise policy makers as to the optimal GHG emissions reduction levels and the supporting prices of present GHG emissions, e.g. CETA (Peck and Teisberg, 1992), DICE (Nordhaus, 1994), MERGE (Manne and Richels, 1995), MERGE-II (Manne *et al.*, 1995) and RICE (Nordhaus and Yang, 1996). Most of the IAMs are of the dynastic, or Ramsey (1928), type, in which there is a central planner who maximizes a welfare measure that aggregates utilities over generations. The calculated allocation of goods thus follows from the rules for the welfare distribution over generations. In an alternative so-called overlapping generations (OLG) framework, the dynamic allocation is primarily determined by the distribution of property rights over generations. There has been much analysis of the dynastic model, and the analysis has led to much debate on the use of different welfare functions (Azar and Sterner, 1996). In this paper, we will argue that the dynastic mechanism of welfare distribution is unrealistic and misleading. Its prevalent use in IAMs has led to unnecessary controversies between researchers who advocate efficient and those who advocate sustainable resource use.

The dynastic IAMs commonly use a constant Rate of Pure Time Preference (RPTP) to discount future welfare levels before aggregation in a net present welfare measure. The discount rate substantially affects the calculated optimal emissions and their supporting prices (Azar and Sterner, 1996). This finding is confirmed by the theory in which it is shown that 'discounting advances the doomsday' (Koopmans, 1974), or, in other words, where it is shown that the choice for the RPTP determines whether a sustainable allocation is optimal or not (Pezzey, 1992) (Pezzey and Withagen, 1998). The importance of discounting is intuitively clear. The environmental resource can produce an indefinite stream of valuable services, the aggregate net present value of which directly depends on the discount rate. This potential future value has to be balanced with the immediate returns of present exhaustive resource use that decreases future output. The

C. Carraro (ed.), Efficiency and Equity of Climate Change Policy, 315–326.

discount rate determines which value is maximal, and thus, which use is most profitable.

Given the importance of the RPTP in the sustainability debate, many papers have been devoted to argument about the proper choice (e.g. Cline, 1992). Some economists suggest that historic data on interest rates can be used to calibrate the level of the RPTP in IAMs, while these in turn determine the optimal environmental resource use in applied models (Nordhaus, 1994, Ch. 6). In other words, they assume that historic data determine whether a sustainable environmental resource use is desirable. Others strongly oppose this so-called 'descriptive' view, and argue for a 'prescriptive' approach in which one should use a low, or even zero, discount rate, since future generations should not be discriminated by lower welfare weights (Broome, 1992). Between these two views, no agreement seems possible. In this paper, we will argue that part of the discord is due to the use of a dynastic framework, which gives a convenient simplification of the dynamic competitive equilibrium, but which cannot be used to calculate whether sustainable environmental resource use is optimal or not. Stated in other words, dynastic models are not fit for calculating optimal emissions reductions and supporting prices if these emissions affect welfare levels in the very long run.

The argument developed in this paper is straightforward. First, let us note that the RPTP as used in a dynastic framework represents a fictitious parameter that cannot be observed in a competitive dynamic equilibrium. The observed discount rate – say, the interest rate – is a variable that changes over time. (We assume that there is one aggregate consumer good price the depreciation of which defines the interest rate.) This variable is not correctly treated in the dynastic model, especially if the RPTP is assumed to be constant so that the interest rate is rigidly linked to economic growth. Use of an overlapping generations (OLG) model reveals that there need not be such a fixed relation between economic growth and the interest rate. In fact, various factors determine the change in the interest rate over time, including the expected demographic change in the next century, and most importantly, including social security policies and potential sustainability policies (Howarth and Norgaard, 1992). Thus, discounting is endogenous, and an overall trend of historic interest rates cannot be extrapolated to the future; the dynastic assumption is inappropriate. As an alternative, we discuss the OLG framework in which it is possible to distribute a kind of 'climate endowments' over generations. Although this is, given the present set of goods over which property rights are currently established, also fictional, a political debate about it is more straightforward than one on the abstract concept of pure discounting.

2. Numerical illustration

To illustrate the arguments given above, we use an OLG-IAM model, ALICE 2.0, that is applied to the issue of climate change, a comprehensive description

Table 1. Scenario specification

	Environmental resource endowed to first generation	Environmental resource shared with future generations
No ageing included	Victims pay 1	Polluters pay 1
Ageing included	Victims pay 2	Polluters pay 2

of which can be found in (Gerlagh, 1998, Section 5.2). A brief specification is given in the Appendix. The model extends the existing OLG-IAMs (Stephan *et al.*, 1997; Howarth, 1998) in two ways. First, it includes a rise in life expectancy ('ageing') that is supposed to take place during the 21st century (the first five periods), represented by a transition from generations that live two adult periods to generations that live three adult periods. Second, it includes a redistribution mechanism for the entire value of the environmental resources at hand; redistribution is not restricted to sharing the revenues of emissions taxes between the currently old and young. The extensions in the model presented here alter the results from the existing studies mentioned above, which suggest that the OLG model does not add much insight to the findings of a typical dynastic model. The conclusions from the other OLG-IAMs can be explained by the assumed stationarity for the structural economic variables such as for demography. If a realistic transition of structural parameters is taken into account, such as in ALICE 2.0, the OLG model is found to show substantially different resource allocations compared to a model in which the transition is absent. The conclusion of our analysis is therefore that the two elements, the expected demographic transition and the intergenerational distribution of property rights over natural resources, have far-reaching consequences for optimal emissions reductions.

ALICE 2 distinguishes periods of 20 years each, the first of which corresponds to the interval 2000–2020. In every period, a new generation is born. In the figures that follow, state variables, such as CO_2 concentrations, are defined for the years between two periods – 2000, 2020, 2040, etc. – while flow variables, such as consumption and emissions, describe a flow during a period, e.g. 2000–2020, and these periods are identified with the central year, e.g. 2010. ALICE 2 includes a simple production sector for man-made commodities, using labour, man-made capital and emissions units as production factors.

We specify four scenarios to show how the demographic change and the intergenerational distribution of property rights over the environmental resource affect the results (Table 1). The first scenario assumes that the environmental resource is grandfathered to the first generation, and abstracts from the increase in life expectancy. Grandfathering implies that the future generations will have to pay if they want to enjoy a clean environment, it is therefore labelled *Victims pay 1*. The second scenario, labelled *Victims pay 2*, includes the demographic change. The third scenario assumes that the property rights

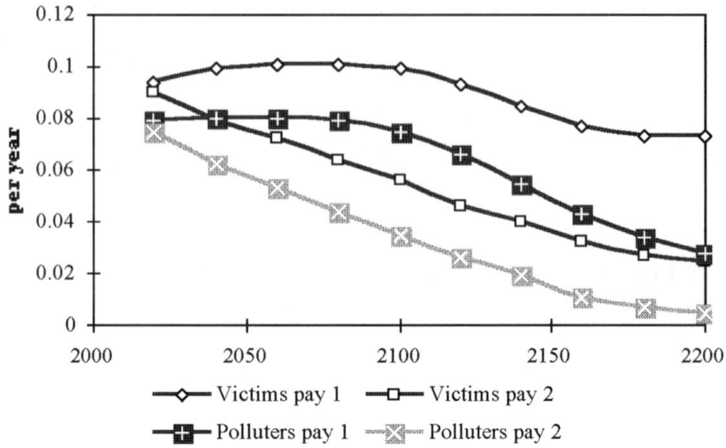

Figure 1. Interest rates.

over the environmental resource are equally distributed over all generations by use of a trust fund (Gerlagh, 1998, Section 4.3.3), and abstracts from the demographic change. Sharing the property rights over the resources implies that the polluters will have to buy the permits to emit; the scenario is therefore labelled *Polluters pay 1*. The fourth scenario, labelled *Polluters pay 2*, includes the demographic change.

Figure 1 shows the interest rates for the four scenarios. We notice that all scenarios have significantly different interest rate paths, indicating that, in constrast with the dynastic model, there is not a unique extrapolation of past interest rates.[2] There are two outstanding findings. First, the intergenerational distribution of the property rights over the environmental resource has a major impact on future interest rates. Sharing the property rights with future generations decreases the interest rate from 7.3 to 2.9% per year in 2200, if we omit ageing. Including the demographic change, we find that sharing the property rights decreases the interest rate from 2.5 to 0.5% in 2200. Second, comparing the scenarios with and without ageing, it is immediately clear that aging decreases the future interest rate substantially, from 7.3 to 2.5% in 2200 under grandfathering, and from 2.9 to 0.5% in 2200 if property rights are shared with future generations. The figure thus undoubtedly reveals the importance of both demographic changes and environmental policies for long-term interest rates.

Now, we turn our attention to the consequences of the different interest rates for climate change itself. We have assumed that CO_2 emissions decrease by one percentage point for each US\$4/tC increase in the emissions price, and this implies that we abstract from endogenous technological improvement and transition costs associated with a shift towards a 'backstop' energy technology.

Omitting ageing, grandfathering leads to a very low emissions price that reaches US\$100/tC in 2200. If we include ageing, the discount rate decreases,

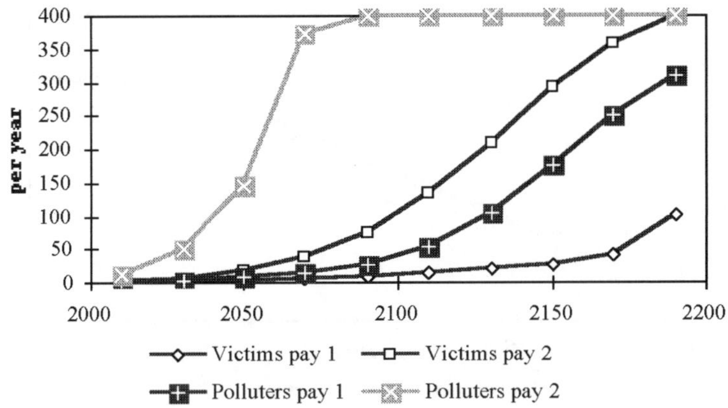

Figure 2. CO_2 emissions prices.

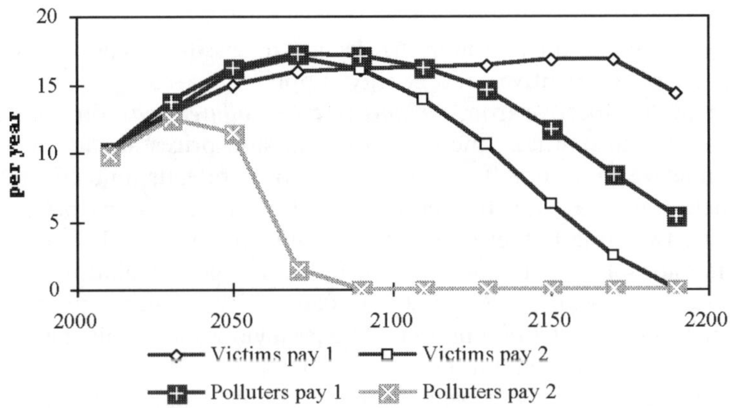

Figure 3. Net CO_2 emissions.

and the carbon emissions price slowly increases from nearly zero in 2000 to US\$100/tC in 2100 and US\$400/tC in 2200 (Figure 2). Consequently, in the first scenario, the reduction in emissions is mainly explained by an assumed autonomous increase in energy efficiency (Figure 3). In the second scenario, in the early periods, emissions exceed the levels of the first scenario, because of the increased capital stock that is related to the decreasing interest rate. In the 22nd century, the rise in emissions prices outweighs the autonomous increase in emissions due to economic growth. This leads to a steady decrease of emissions, and by 2200, net emissions are zero. In the *Victims pay 2* scenario, in 2100, cumulative emissions amount to 1450 GtC, which is comparable with the IPCC IS92a scenario (1992), but it should be understood that these estimates are highly dependent on the assumptions made. Chakravorty *et al.* (1997)

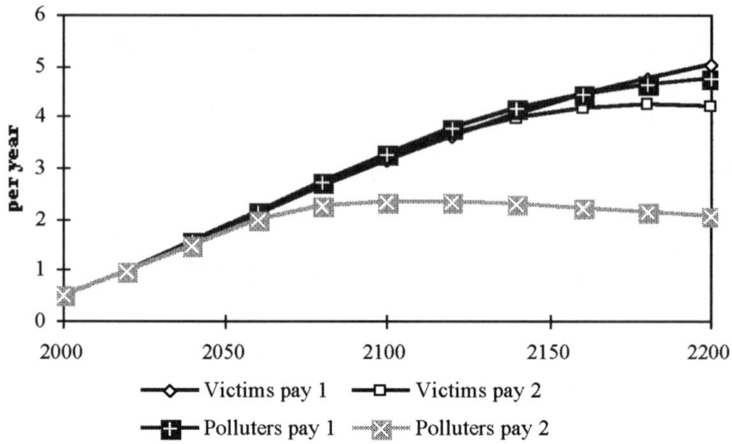

Figure 4. Global mean temperature.

for example estimate an autonomous decline in fossil-fuel energy use before 2100 because of competitive solar energy supply.

Shifting the perspective from *Victims pay* to *Polluters pay* through the use of a trust fund substantially increases the emission prices already in the first period, but as we have already seen for the discount rate, its total effect depends on the inclusion of ageing. If ageing is omitted, the emissions price remains between the two grandfathering scenarios, and this more or less also applies to the emissions levels. If ageing is included in the calculations, the price increases rapidly towards US$400/tC in 2080, which by assumption implies a 100% reduction in net CO_2 emissions. The policy does not limit the expansion of net emissions in the medium term, leaving present generations the possibility of adapting and developing alternative energy sources. After 2050, net emissions decrease rapidly, and by 2100 a complete substitution of fossil-fuel energy carriers by fossil-free energy carriers has taken place.

CO_2 concentration levels in the atmosphere lag behind CO_2 emissions patterns, and in turn the global mean temperatures lag behind the CO_2 concentration levels. Figure 4 shows temperatures over the interval 2000–2200; the delay becomes visible as the first three scenarios nearly coincide. Only the fourth scenario, in which both future generations are given a share of the environmental resource value and in which account is taken of the demographic change succeeds in reducing future temperature change before the 22nd century. We may say that the environmental policy of the fourth scenario, which consists in sharing property rights over environmental resources with future generations, endogenously adjusts the interest rate to a sustainable level.

Let us reiterate that the figures presented are meant to illustrate the importance of taking into account various exogenous changes that may affect future discounting, but that are easily neglected in dynastic models. It is not our purpose to present a new set of numbers that can be used for applied greenhouse gas policies. Instead, we want to emphasize that a cautious use of the results is justified.

Conclusion

Summarizing the findings, discounting is of major importance to the sustainability issue, since too high a discount rate makes it unprofitable to sustain a high-quality environment, whereas a low discount rate makes sustainable use the most profitable venture. In the standard dynastic model with additive discounted welfare functions (a constant RPTP), the future discount rates depend on a limited number of parameters only. Since these parameters are calibrated on historic data, it seems possible to determine whether sustainability is optimal or not. It is undeniable that the dynastic model simplifies the analysis. However, it makes no sense to use the dynastic model for predicting future interest rates, since there is no reason to assume that welfare is distributed as if there was a central planner.

In OLG models, on the other hand, the discounting is related to many other variables in the economy that may change in the future. Demographic changes, changes in social security, and the specification of property rights over environmental resources all affect the discounting. These changes may lead to future interest rates that are much lower than past levels. Probably, the OLG model better reflects the actual budgets in the dynamic competitive economy than the dynastic model. Even more important, the OLG model shows that various public policies affect the interest rate. It is even possible to specify property rights over environmental resources and to distribute these over generations in such a way that, if necessary, the discount rate will endogenously adjust to its sustainable level.

Stated in other terms, the way in which applied dynastic IAMs are used for environmental policy calculations suggests that the historic data on interest rates determine whether sustainability is desirable or not. The dynastic models sometimes suggest that a choice for sustainability is at a contrast with efficiency. Clearly, this thinking mistakes the responsibility of present policy makers. An OLG analysis reveals that there can exist instruments that bring forward an efficient and sustainable environmental resource use. If a policy maker prefers to combine efficiency and sustainability, it is preferable not to use a dynastic model that is not capable of doing so.

Acknowledgements

The author thanks Richard Tol, Bob van der Zwaan, and two anonymous referees for their comments.

Appendix. Condensed model description for ALICE 2.0

The ALICE 2.0 model distinguishes discrete time steps, $t \in \mathbf{T} = \{1, \ldots, \infty\}$, each representing periods of 20 years' duration, the first of which corresponds to 2000–2020. In every period, a new generation is born, denoted by the date of birth, using the label t or i if convenient. The model includes a rise in life expectancy that is supposed to take place during the 21st century (the first five periods), represented by a transition from generations that live two (adult) periods to generations that live three (adult) periods. The first and third scenarios abstract from the demographic change, assuming that all generations live only two (adult) periods. See Gerlagh (1998, Table 3.5) for a comparison of modelled life expectancy with data from the World Bank (WB, 1994).

Generations maximize their lifetime utility $\mathrm{U}(C_i, B_i)$ derived from rival consumption of the consumer good during the life-cycle, $C_i = (C_i, C_{i,i+1}, C_{i,i+2})$, and non-rival consumption of the resource amenity, for convenience referred to as 'environmental services', $B_i = (B_t, B_{t+1}, B_{t+2})$ where we use the subscript $t = i$ to stress that the amenity level is the same for all consumers. Their utility function U(.) is a nested CES function with a constant expenditure share of 90% for the consumer good and 10% for the resource amenity, and an intertemporal elasticity of substitution of 0.67. The generations maximize utility subject to the budget constraint:

$$\max \left\{ \mathrm{U}(C_i, B_i) \middle| \sum_{t=i,\ldots,i+2} p_t C_{i,t} + \varphi_{i,t} B_t \leq \sum_{t=i,\ldots,i+2} w_t L_{i,t} + H_{i,t} \right\} \tag{1}$$

where p_t, w_t denote the prices of the (rival) consumer good and labour, respectively, $\varphi_{i,t}$ are the Lindahl prices for non-rival consumption of the resource amenity of generation i in period t, $L_{i,t}$ denotes the labour endowment and $H_{i,t}$ is the income generation i receives in period t as its share in the value of the environmental resource. The first generations not living three periods have adjusted utility functions and budget constraints in the obvious way.

There is one private firm that uses 'emissions units', E_t, and 'labour', $L_t = L_{t-2,t} + L_{t-1,t} + L_{t,t}$, for the production of the consumer good, and also a capital stock, K_t, which is itself produced by the same sector. For convenience, we assume that the capital stock is made up of the consumer good, and that it has to be replaced after use in one period. The production structure can thus be represented by

$$Y_t + K_{t+1} = \mathrm{h}_t(K_t, E_t, L_t) \tag{2}$$

where $\mathrm{h}_t(.)$ is a continuous, differentiable, and concave production function, linearly homogeneous, strictly increasing in the first and third arguments, and with bounded derivative for the second argument, $0 < \mathrm{h}_{t,2}(., 0, .) < \infty$, which implies that, as in the original economy, the emissions units are valuable, but not essential for production. The subscript t is maintained to allow for technological innovation. More specifically, capital and labour are combined in a

Table A1. Parameter values for the linear Maier-Reimer and Hasselman 5-box model

e-folding time (τ_i) (yr)	∞	313.8	79.8	18.8	1.7
share (a_i)	0.14	0.24	0.32	0.21	0.09

Cobb–Douglas aggregate, and the productivity of this aggregate good is supposed to be a quadratic function of the emissions intensity, such that the optimal emissions levels decrease by 1% for every US$4/tC price increase for the emissions units, and such that the maximum emissions levels follow the IS92a scenario (IPCC 1992).

The climate change issue is represented through an 'environmental firm'. Peck and Teisberg (1992) and Nordhaus (1994) have contributed much to the development of stylized economic IAMs by providing highly simplified representations of geophysical interactions to make these applicable in macroeconomic models. The typical simplified aggregate representation that has thereafter evolved links emissions to concentrations, concentrations to temperatures, and temperatures to damages. We follow this literature.

As the emissions of CO_2 account for the main antropogenic contribution to the greenhouse effect, we focus on the carbon cycle and its relation to climate change. Let us assume that the atmospheric GHG accumulation can be represented by a 'linear box' model in which each of the boxes $i = 1, \ldots, 5$ is well mixed as described by Maier-Reimer and Hasselman (1987). If CO_2 is emitted, it is distributed over the boxes for given shares a_i. Within a box, CO_2 concentrations exponentially adjust to their 'natural' levels at annual adjustment rate $1/\tau_t$, where τ_i is the so called 'e-folding time' or turnover time, which is the expected period that a gas particle will remain in the box. Let M_t^i denote the accumulated antropogenic emissions in box i at the beginning of period t and E_t the emissions during period t, we then have:

$$M_{t+1}^i = e^{-N\tau_i} M_t^i + a_i E_t \tag{3}$$

for $\sum_i a_i = 1$, where N is the period length of the discrete time model. Maier-Reimer and Hasselman (1987) have estimated the following parameters a and τ for a five-box model.

Remarkably, 14% of emissions remain in the atmosphere for the infinite horizon, which implies that the absorption capacity of the biogeochemical system is exhaustible.

The accumulation of GHGs causes an increase in the equilibrium global mean temperature (see IPCC (1992, Table 2.2) for a list of GHGs and their potential contribution). For CO_2, the temperature increase is expected to be of approximate logarithmic nature,

$$T_t^{eq} - \bar{T}\,{}^2\log\left(1 + \sum_i M_t^i \Big/ \bar{M}\right)^3 \tag{4}$$

where T_t^{eq} is the long-term equilibrium temperature for given concentrations

$\sum_i M^i$, and $\bar{T}$ is the benchmark equilibrium temperature associated with a doubling of total accumulated antropogenic GHGs in the atmosphere compared with the 'natural' level $\bar{M}$.

The earth surface and atmosphere have a warmth capacity, and therefore the temperature slowly adjusts to the long-term equilibrium level:

$$T_{t+1} = e^{-Ne} T_t + (1 - e^{-Ne}) T_{t+1}^{eq} \tag{5}$$

where e is the annual adjustment rate, which we set to 2% per year following Peck and Teisberg (1992).

So far, the above equations relate to complex geophysical relationships, and though present knowledge of these relationships is limited, the approximations are based on a reasonable understanding of physical processes. However, when calculating impacts of climate change, the scientific understanding is grossly insufficient to warrant even something like a 'best guess'. In general, it is assumed that damages caused by climate change will outweigh the benefits, but the lack of knowledge is unmistakably revealed by many sensitivity analyses that are carried out with a variety of so-called damage functions. These damage functions are supposed to provide a reduced form for the many complex damages associated with climate change, such as loss of coastal zones due to sea level rise, loss of biodiversity, vector-borne diseases, and extreme events. Some damage functions take the global temperature as arguments, others take the rate of increase of global temperature as an argument, some damage functions are quadratic, other are of higher or lower order, cf. Tol (1995). The lack of understanding is also recognized by the IPCC (1996a, Section 6.2.13). However, in the same report, several damage estimates are listed (IPCC 1996a, Table 6.4). The use of these figures for our analysis does not mean that we consider them reliable, but reflects our wish to maintain compatibility with the prevalent IAMs. Typically, the IAMs include a reduced damage function h(.):

$$D_t = \mathrm{h}(T_t) \tag{6}$$

where D_t is the damage, usually a function of actual temperature increase, expressed in monetary units or as a percentage of GDP.

The IAMs have different ways of incorporating the damage functions, either by subtracting damages from production or consumption, or directly from utility. We find this practice misleading, because environmental damage is better understood as a decrease in the quality or quantity of environmental functions, than as a reduction in a flow of man-made goods. Environmental degeneration can lead to a reduction both in biomass and in biodiversity, with the former measuring the quantity and the latter the quality of the environmental functions (IPCC 1996b, Section 1.3.2). A decrease in biodiversity constitutes a change in environmental quality, as opposed to the quantity, and in a reduced model this can best be described via a non-rival amenity value (Gerlagh, 1998, Section 5.2.3). This explains our use of a non-rival good labelled 'environmental services', which level is given by:

$$B_{t+1} \leq e^{-Nd} B_t + (1 - e^{-Nd})[1 - \mathrm{f}(T_{t+1})] \tag{7}$$

where d is the annual adjustment rate, and f(.) is the reduced form function that describes long-term losses of environmental services which we take to be increasing on its positive domain, and for which $\mathrm{f}(0) = 0$ (see Tol, 1995) for various functions. To summarize, the biogeochemical system produces emissions units E_t, and environmental services B_t. It is a renewable resource that, however, deteriorates because of emissions, and this is the core of the sustainability problem (Clark, 1997).

Now, let us return to the resource income shares which, by definition, should sum to the total value of the resource, that is:

$$\sum_{i=1,\ldots,\infty} H_i = \sum_{i=1,\ldots,\infty} (p_t^e E_t + p_t^b B_t) \tag{8}$$

where p_t^e and $p_t^b = \varphi_{t-2,t} + \varphi_{t-1,t} + \varphi_{t,t}$ represent the price of the emissions units and the environmental services level respectively, and $H_i = H_{i,i} + H_{i,i+1} + H_{i,i+2}$ is the aggregate resource share received by generation i over the life cycle.

In the grandfathering scenario, the environmental resource is given to the first generation, $t = 0$, who receive all present and future revenues as income:

$$H_0 = \sum_{t=1,\ldots,\infty} (p_t^e E_t + p_t^b B_t) \tag{9}$$

$$H_t = 0 \quad \text{for } t = 1, 2, \ldots, \tag{10}$$

where p^e and p^b represent the price of the emissions units and the environmental services respectively.

In the trust fund scenario, all generations receive a claim for the maximum environmental services level:

$$H_{i,t} = \varphi_{i,t} \tag{11}$$

and the trust fund implements the required redistribution of revenues over the generations; see Gerlagh (1998, Section 5.2.4) for a detailed description.

Notes

1. Correspondence: Reyer Gerlagh, IVM/VU, De Boelelaan 1115, 1081 HV Amsterdam, the Netherlands, tel. +31 20 44 49502, fax. +31 20 44 49553, E-mail: reyer.gerlagh@ivm.vu.nl
2. Consistency with observed past interest rates follows from the calibration of the model based on historic data.
3. We use the notation that ${}^2\log x = y$ is equivalent to $2^y = x$.

References

Azar C. and T. Sterner (1996). Discounting and distributional considerations in the context of global warming. In *Ecological Economics*. 19:169–84.

Broome J. (1992). *Counting the costs of global warming*. Cambridge: White Horse Press.
Chakravorty U., J. Roumasset, and K. Tse (1997). Endogenous substitution among energy resources and global warming. *Journal of Political Economy*. 105:1201–34.
Clark C.W. (1997). Renewable resources and economic growth. *Ecological Economics*, 22:275–76.
Cline W.R. (1992). *The economics of global warming*. Washington, DC: Institute for International Economics.
Gerlagh, R. (1998). *The efficient and sustainable use of environmental resource systems*. Amsterdam: Thela Thesis.
Howarth, R.B. (1998). An overlapping generations model for climate-economy interactions. *Scandinavian Journal of Economics*. 100:575–91.
Howarth, R.B. and R.B. Norgaard (1992). Environmental valuation under sustainable development. *American Economic Review*. 82:473–77.
IPCC (Intergovernmental Panel on Climate Change) (1992). *Climate change 1992; the supplementary report to the IPCC scientific assessment*. Cambridge: Cambridge University Press.
IPCC (Intergovernmental Panel on Climate Change) (1996a). *Climate change 1995; impacts, adaptations and mitigation of climate change: scientific-technical analyses*. Cambridge: Cambridge University Press.
IPCC (Intergovernmental Panel on Climate Change) (1996b). *Climate change 1995; economic and social dimensions of climate change*. Cambridge: Cambridge University Press.
Koopmans T.C. (1974). Proof for a case where discounting advances the doomsday. *Review of Economic Studies*. 41:117–22.
Maier-Reimer, E. and K. Hasselman (1987). Transport and storage of CO_2 in the ocean – an inorganic ocean-circulation carbon cycle model. *Climate Dynamics*. 2:63–90.
Manne, A.S., R. Mendelsohn and R. Richels (1995). MERGE, a model for evaluating regional and global effects of GHG reduction policies. *Energy Policy*. 23:17–34.
Manne, A.S. and R. Richels (1995). The greenhouse debate; economic efficiency, burden sharing and hedging strategies. *Energy Journal*. 16:1–37.
Nordhaus, W.D. (1994). *Managing the global commons*. Cambridge, Massachusetts: MIT Press.
Nordhaus, W.D. and Z. Yang (1996). A regional dynamic general equilibrium model of alternative climate-change strategies. *American Economic Review*. 86:741–65.
Peck, S.C. and T.J. Teisberg (1992). CETA: a model for carbon emissions trajectory assessment. *Energy Journal*. 13:55–77.
Pezzey, J. (1992). *Sustainable development concepts, an economic analysis*. Washington, D.C.: World Bank.
Pezzey, J. and C. Withagen (1998). The rise, fall and sustainability of capital-resource economies. *Scandinavian Journal of Economics*. 100:513–27.
Ramsey, F.P. (1928). A mathematical theory of saving. *Economic Journal*. December: 543–59.
Stephan, G., G. Müller-Fürstenberger and P. Previdoli (1997). Overlapping generations or infinitely-lived agents: intergenerational altruism and the economics of global warming. *Environmental and Resource Economics*. 10:27–40.
Tol, R.S. (1995). The damage costs of climate change toward more comprehensive calculations. *Environmental and Resource Economics*. 5:353–74.
WB (World Bank). Bos, E., M.T. Vu, E. Massiah and R.A. Bulatao (eds.) (1994). *World population projections 1994–95 edition; estimates and projections with related demographic statistics*. Baltimore, Maryland, USA: Johns Hopkins University Press.

PART 4

Policy Conclusions

CHAPTER 16

Integrating Climate Policies in the European Environment: A Policy Report*

Carlo Carraro and Alessandra Goria

1. Introduction

The discussion on climate policy is still quite open, and many issues remain unsolved. Previous chapters have shown many advances and new useful results. However, the debate that took place during the Second EFIEA Policy Workshop on 'Integrating Climate Policies in the European Environment',[1] where the above chapters have been circulated and discussed, is far richer than the results that they contain.[2] This is why we would like to summarize in this concluding chapter the many ideas, proposals, doubts, conflicting and consensus views that were debated during the conference, that were discussed within the panel sessions, that are contained in the many papers that have been prepared for the conference, and of which only a small part is gathered in this book.

The main goal of this chapter is therefore to organize the lively discussions among the workshop panellists and participants within a framework that can help the reader to understand, on the one hand, the main issues that were raised during the workshop, and, on the other hand, on which issues a consensus could be found and on which other issues conflicting views still prevail.

Let us recall that the EFIEA workshop was attended by policy makers at the EU and national levels, editors of main environmental journals, environmentalists and other NGO members, scientists and economists, etc. Many of them did not provide written contributions. Hence, this chapter is partly based on the recording of the workshop. One of the workshop main goals was to improve communication between science and policy. This is also why it is important to summarize in this chapter the outcomes of the participants' communication effort.

The workshop was organized according to the following steps:

1. Policy makers and scientists identified relevant policy questions (demand-driven workshop) one year before the date of the workshop.
2. Economists and scientists then carried out research which provided results adequate to answer those questions.[3]
3. This research was circulated among the workshop participants, who compared questions and answers, raised new questions, and created interaction

C. Carraro (ed.), *Efficiency and Equity of Climate Change Policy*, 329–362.

between policy making and science on the basis of sound, high-quality research.

This chapter reflects the process that led to the workshop. First, the main questions that were circulated among scientists and economists are summarized and then, for each of these questions, the chapter analyses and compares different viewpoints and results, in order to identify those issues on which a consensus could be found. Indeed, it is worth noticing that during the workshop, despite the presence of different, sometimes conflicting stakeholders, some general ideas were supported by all parties. For example, the link between equity and efficiency of climate agreements was clearly identified and emphasized. On the one hand, an equitable agreement induces more countries to commit to emissions control, thus enhancing its effectiveness; on the other hand, a cost-effective agreement reduces the conflict on distributional issues. Another example is the importance of using the so-called Kyoto flexibility mechanisms to reduce the costs of compliance. These ideas, and the related policy perspectives, are discussed at length also in this chapter. In particular, in the last part of the chapter, some policy guidelines that could help improve the chances of a successful control of climate changes are proposed.

The structure of the chapter is as follows. The next section will present the main policy questions. Section 3 will identify a few issues on which consensus emerged during the long and lively debates that took place during the workshop. Section 4 will focus instead on the issues on which conflicting views could not be resolved and for which further research efforts and deeper interactions between science and policy making are necessary. Finally, Section 5 proposes a few policy guidelines that reflect the main ideas debated on the last day of the workshop. Let us stress again that the papers quoted in the following sections were all submitted for circulation and discussion at the workshop. The final section summarizes the main ideas that emerged during the five sessions of the EFIEA policy workshop.

2. Relevant policy questions

The most relevant policy questions addressed at the EFIEA workshop are grouped into a few broad classes, reflecting policy questions relating to equity, policy integration and policies by sector.

2.1. Equity

Equity issues were at the core of the workshop. The terms of reference of first the workshop, and then the scientific surveys and debates, identified some key equity issues for global warming policies. The following main broad policy questions were addressed:

- *Which equity dimensions should be addressed by international climate policies?*

When equity objectives are concerned, international policies involve several dimensions of equity: equity between countries, referred to as *international equity*, equity between different social groups and stakeholders, i.e. *national/social equity*, and equity between generations, mostly referred to as *intergenerational equity*. It was acknowledged that climate policies should not aim at a unique definition of equity, or to the achievement of a unique equity principle. Instead, given the multiple facets of equity, climate policies should be developed with a better understanding of their equity dimension and the related socio-economic implications. The multiple equity dimensions apply to equity on impacts, adaptation and mitigation of climate change. Current climate change policies mainly address international equity issues, which are crucial to the implementation and evolution of the UNFCCC and of the Kyoto Protocol. So far international equity in climate change mitigation has been central to the climate change debate; the international equity aspects of climate change impacts and adaptation have instead received little attention.

- *Does equity in climate-change impacts and adaptation deserve more policy attention?*

This question relates to the multiple facets of the equity problem among which the following ones emerge in the literature:

1. The precautionary principle should be applied with equity issues in mind: what counts is the risk of the most vulnerable (Martin, 1997).
2. Climate-change impacts are likely to vary considerably between countries; in hotter climates and natural-resource-dependent countries, the impacts are widely expected to be adverse (IPCC, 1996).
3. Institutional and social structures in developing countries tend to be weaker and these countries lack financial resources: present inequalities are likely to increase (Shukla, 1999).
4. Small islands, low-lying areas, and very dry regions are exceptionally vulnerable; adaptation options vary widely and may not exist at all (Banuri *et al.*, 1996).

- *How does equity relate to efficiency, i.e. the achievement of emissions reduction targets at the lowest cost? Is there a trade-off between equity and efficiency objectives in climate-change policies?*

There is a widespread opinion that, through emissions trading, efficiency in emissions reduction can be made independent of equity (the initial distribution of permits). In other words, a system of tradable permits may allow for the separability of equity and efficiency: equity objectives could be pursued in line with efficiency criteria, provided that the distribution of permits is equitable. Rose *et al.* (1998) note that an equitable distribution of permits will enhance efficiency: 'Interestingly, this pursuit will actually promote efficiency as well. We have taken for granted the issue of the comprehensiveness of the number of parties to an agreement. If more parties can be enticed into it by appealing to their equity concerns, environmental quality will be further enhanced. If these parties have relatively lower abatement costs than the original signatories, overall economic efficiency will be improved'. Rose *et al.* (1998) note that the concept of separability between efficiency and equity applies to both a static and a dynamic framework, allowing intergenerational equity issues to be addressed along with inter-temporal efficiency: 'Permit trading over time promotes even higher levels of efficiency by capitalizing on abatement cost differentials in different periods'.

However, are these conclusions really sound? Do they refer to a specific definition of efficiency and of equity? Do equitable and efficient outcomes need to reflect a common equity principle? Can the amount of resources that a system of emissions trading distributes affect other economic variables, in addition to the environmental ones, thus affecting economic efficiency and equity? Is there a social dimension of equity which is neglected by economic analysis? If the above conclusions are correct, how can an equitable distribution of permits be achieved? Is the distribution of emissions allowances implicit in the Kyoto Protocol equitable?

- *How can climate-change policies be made consistent with other public policies? Are Costs–Benefits Analysis techniques applicable to the climate-change problem? And what would be the most appropriate discounting procedure to compare costs and benefits over time?*

The choice of the appropriate discounting technique is crucial to weight the costs and benefits of integrated climate policies in an inter-temporal dimension. Ferenc Toth, from PIK, Germany, provided a review of the various discounting techniques that have been proposed and applied in integrated climate-change models. Dealing with the effects of climate change, which imply long-time horizons, possibly irreversible changes, the threat of environmental catastrophes, leads to Toth's dilemma: 'One can attempt to be consistent with economic theory and empirical observations, but in this case the derived discount rate will be in the order of 5 to 8%. As a result, even possibly significant damages from climate change turn out to be negligible when considered at their present value. The artificially low discount rate based on ethical reasoning, on the other hand, makes climate-related decisions and resource allocation inconsistent with the majority of other public decisions'.

Toth concludes that three major lines of thought regarding the choice of costs–benefits techniques can be identified in the literature and in the current debate:

1. A first view, based on the consideration that impacts and benefits related to climate change will occur very far in the future, believes that standard

cost–benefit analysis (CBA) is appropriate even for policies applied to climate change. Therefore the discounting technique should not differ from other policies standards, provided they are revised on a regular basis.
2. A second view 'recognizes that the technique of CBA is appropriate to address climate policy but tries to bring distant economic losses due to global warming closer to the attention of today's decision makers', using lower discount rates for the evaluation of far future impacts. Nordhaus (1997), however, points out that this procedure does not help to achieve efficient abatement policies, nor to save unique environments.
3. A third view sustains that, if the environment is highly at risk and irreversible changes are foreseen, CBA may have only a limited validity. The best strategy would be to define specific long-term environmental goals and work out the optimal cost-effective policy to reach them.

Is there a way to compare the above proposals? Is there a consensus on how to compare environmental benefits and abatement costs? How should the intertemporal discount rate be fixed when evaluating the net benefits of climate policies?

- *Which main equity issues should be addressed by climate-change mitigation policies?*

Some policy issues are often indicated as the most relevant in terms of their implications for equity: the definition of the emissions reduction quota for Annex I parties, the criteria for 'burden' sharing, the implementation of the sinks provision, the design of the Kyoto flexibility mechanisms, the share of proceeds for the Clean Development Mechanism project activities, the consequences of participation in the Kyoto mechanisms for both Annex I and non Annex I parties, the future participation of parties in mitigation.

However, are there other issues that are relevant for equity but are not sufficiently analysed? What is the link between equity issues addressed by mitigation policies and those addressed by adaptation policies?

- *Which equity criteria could be reflected in 'burden' sharing?*

The evaluation of costs and opportunities of alternative climate policies should fully address their implications for equity. Berk *et al.* (1999) emphasize that there is no common accepted definition of equity in climate-change policies. Rose *et al.* (1997) distinguish three types of alternative equity criteria for global warming regimes:

1. *Allocation-based criteria,* defining equitable burden sharing in terms of principles for the distribution of emission rights.
2. *Outcome-based criteria,* defining equitable burden sharing in terms of the consequences of mitigation policies.

3. *Process-based criteria,* defining equitable burden sharing in terms of the process for arriving at a distribution of emissions burdens.

This distinction raised a broad fundamental question: *should equitable policies imply equitable outcomes and/or equitable procedures*? Rose *et al.* (1998) write: 'Fairness in one area of the policy may have quite different implications in another, i.e. a principle that gives all countries an equal proportion of permits in relation to their initial emissions will not result in equal percentage emissions reductions or equal percentage changes following trading'.

Ringius *et al.* (1998) propose the following equity principles to address climate change policies:

1. *Egalitarian*: people have equal rights to use the atmosphere.
2. *Sovereignty*: current emissions constitute a status quo right now.
3. *Horizontal*: actors under similar (economic) conditions have similar emissions rights and burden-sharing responsibilities.
4. *Vertical*: the greater the capacity to act/ability to pay, the greater the economic burden.
5. *Polluter pays*: the greater the contribution to the problem, the greater the burden.

Ringius *et al.* (1998) also suggest that during the negotiations on the Kyoto Protocol the principle of horizontal equity has been dominant amongst Annex I countries, whereas the relationship between the developed and the developing countries, even in the broader context of the UNFCCC, can be better described by a vertical-equity and polluter-pays approach.

Banuri *et al.* (1996) assert that, with regard to emissions allocation criteria for burden sharing, the following *concrete proposals* exist:

1. Allocation based on *equal per capita entitlements*, reflecting an egalitarian approach.
2. Allocation based on *current emissions*, reflecting a sovereignty approach.
3. Allocation based on the *polluter-pays principle*, as reflected in the Brazilian proposal presented at Kyoto, whereby the emissions reduction efforts should be proportional to each party's contribution to the global temperature increase.
4. Allocation based on a *mixed system*, rooted both in population and in GDP, whereby proportional reduction efforts should lead to per capita convergence.

It is therefore necessary to clarify how equity is defined, which equity criteria are reflected by climate change policies, and how equitable policies can be implemented.

- *How should the sink provision be implemented?*

The definition and implementation of the sink provision will have major implications for equity: among Annex I parties due to further definitions and retro-active inclusions of additional sink categories, for developing countries depending on the inclusion of sinks in the Clean Development Mechanism.

As stressed at the EFIEA workshop by Tom Heller from Stanford University, this is an issue which is emerging as one of the most relevant and therefore deserves additional careful research. Tom Heller noticed that the US has the highest potential for carbon sinks, and that the definition of rules with regard to the sink provision will strongly influence future political games.

- *How should the guidelines and the operational modalities of the Kyoto flexibility mechanisms be effectively designed to pursue equitable objectives?*

The design of the Kyoto flexibility mechanisms will have equity implications for any restriction imposed in the form of caps (definition of the 'supplementarity' criterion), participation, types of projects, share of proceeds or fees. Banking of emissions and transaction costs for the project-based mechanisms will carry out equity implications as well. As stressed by Sujata Gupta, from the Tata Institute, India, developing countries are particularly concerned about higher future costs of emissions control after the Clean Development Mechanism phase.

- *Should country participation in emissions reduction commitments be enlarged? Which commitments and which timing would be most appropriate?*

With regard to countries' participation, even if the Kyoto Protocol starts from a 'no harm to developing countries' approach, based on a sense of historic responsibility and allowing for efficient reductions through global trading, increased participation of countries in emissions reduction will be necessary over time in order to be able to stabilize concentrations of greenhouse gases (IPCC, 1997; Berk *et al.*, 1999). Berk *et al.* (1999) suggest that for the evolution of the UNFCCC two options could be foreseen:

1. A gradual extension of the group of Annex I countries, taking on binding emissions levels under the Convention (sometimes referred to as 'graduation').
2. Defining the emissions rights of all parties over a longer period.

The first regime would mean a gradual extension of differentiated commitments, as in the Kyoto Protocol, based on rules for participation and burden sharing which are part of an incremental decision-making process. This 'increasing participation regime' is combined with a long-term perspective by using a global emissions constraint.

The second regime would be a major shift away from the Kyoto Protocol approach and have a long-term perspective with respect to the distribution of rights and their evolution over time. One example would be the so-called 'contraction and convergence' scenario of the Global Commons Institute, which defines emissions permits on the basis of a convergence of per capita emissions under a contracting global emissions profile. Another example is provided by J. Onigkeit and J. Alcamo (1999), who develop long-term scenarios combining burden sharing and climate protection. Based on a specific burden-sharing scheme, their methodology allows the definition of two stabilization targets (at years 2100 and 2150 respectively) that can be achieved even if total emissions from non-Annex I countries increase until around 2030.

The equity impacts of the two regimes for the world regions should be addressed, and the socio-economic implications in terms of costs and opportunities of alternative policy scenarios should be explored.

- *Which other crucial elements underlie the policy debate on equity?*

The surveys and the subsequent debates identified some additional variables affecting equity that need to be assessed by climate research and addressed by policy making:

1. The patterns of GHGs emissions, in terms of their temporal, geographical and sectoral distribution. These patterns are needed to address their impact on present and future generations, to assess the nature of current and future emissions by source and sector, as well as in terms of their geographical concentration, and to address their distributional impact at both the national and the international levels.
2. The inclusion of the different greenhouse gases, the methods to weight the contributions of the different gases to overall emissions (Gottinger, 1999), emissions sources and sinks.
3. The inventory methodology (consumption versus production-based inventories).
4. The relationship between equity and technology, identifying which forces among which social groups may enhance technological change.
5. The choice of indicators (Berk *et al.*, 1999).
6. The role of nuclear energy with respect to intergenerational equity issues: it was claimed that policies should address the social legitimacy of nuclear waste for present and future generations.

Does existing research address the above issues? Is there enough information on the role played by the above variables for policy makers to undertake reliable actions?

- *How could weaker countries be empowered to correct or compensate for inequities in the contractual power of climate-change policy negotiators?*

Joyeeta Gupta, from IVM, Amsterdam, claimed that developing countries' representatives are in a weaker position to negotiate with respect to developed countries' negotiators; appropriate mechanisms should be developed to empower them to address the potential net benefits of climate policies for the South. Adequate incentives should be directed for instance towards the design and implementation of national and regional climate policies in the South. One other suggestion was to get China and India, as leading G-77 representatives, involved in the G-7 meetings. Another one is the creation of an international arbitrator for climate conflicts. Which institutions are most appropriate to give countries a balanced and therefore equitable bargaining power in climate negotiations?

2.2. *Sectoral policies*

2.2.1. *Transport policies*

The scientific survey on 'Policies in the Transport Sector' by Stef Proost highlighted that in the EU the transport sector represents approximately 25% of all CO_2 emissions; the major part of emissions comes from the use of fossil fuels for road transport, i.e. cars and trucks. Emissions in the transport sector in Europe have been growing extremely fast, faster than GDP growth, and are expected to grow further. This makes transport a priority in climate-change policies. The relevant policy questions addressed at the workshop are:

- *Which instruments would be most appropriate to meet climate-change policy goals in the transport sector?*

Stef Proost (1999) notes that at present the *fuel efficiency standard* for new cars is the only policy decision which has been taken to reduce CO_2 emissions from cars in the EU. He also suggests that 'the emphasis of current European policies on fuel efficiency of cars (via standards, eco-bonus, etc.) is wrong. This is not a cost-effective policy. The main reason is that the fuel efficiency of present cars has already been designed in function of the present excise on fuels that represent a disguised CO_2 tax of 300% or more'. Does this imply that further efficiency gains cannot be achieved by increasing energy taxation? Or should energy prices be further increased?

Moreover, energy taxation (or other instruments) should also be used to internalize the external costs generated in the transport sector, such as air pollution, accidents, noise, and congestion. As regards air pollution, for instance, major externalities generate from the diesel, which has a strong impact on health. Policy measures should be balanced taking such considerations into account. In reality transport pricing appears to be highly inefficient, where inefficiencies are dominated by external congestion costs. Addressing external costs correctly, with measures which may generate additional CO_2 emissions reductions, is a challenge for policy making. Is this type of policy politically and economically feasible? What could impacts on income distribution be?

- *How could transport policies be integrated with other sectoral policies to meet climate-change policy objectives?*

Stef Proost (1999) underlines that 'in certain urban areas air quality problems can be addressed by a combination of local and global emission measures; due to the strict emission standards that are in place now for the transport sector, a cost-effective solution could require important efforts from other emissions sources'. This may require the integration of policy instruments and measures from different sectors towards a common policy goal. For instance, transport policies could be integrated with land-use planning policies, and integrated policies should address which patterns of land-use planning would enhance energy efficiency towards climate change mitigation. Transport policies could also enhance fuel switching from traditional fossil fuels to clean fuels such as NLG and bio-gas.

2.2.2. Energy policies

The following relevant policy questions were raised:

- *Which key elements should be taken into account in the design of future energy policies?*

Alessandro Lanza (1999) highlights three key features which are likely to shape energy policies in Europe: the increasing concern about the environment, the ongoing process of market diversification and deregulation, and the increasing role of non-Annex I parties.

The increasing concern about the environment is reflected by the focus of policy making on greenhouse gases emissions and concentrations, and on the economic and modelling analysis, which accounts for emissions trading and the other Kyoto flexibility mechanisms. The process of market diversification and deregulation is characterized by the following facts:

1. Oil is no longer the dominant fuel, and grid-based energy sources are increasing.
2. In developing countries there is a substitution of non-commercial energy source (i.e. bio-mass) with commercial bio-mass and other energy sources.
3. Public ownership and state monopolies in the electricity and gas sectors are being privatized, opening up to competitive markets.

The increasing role of non-OECD, i.e. non-Annex I parties is testified by the world energy balance projections contained in the 1998 World Energy Outlook: by the year 2020 the current group of OECD countries will use 42% of the world energy, compared to a 55% share in 1995.

Which other elements should be taken into account when designing future energy policies? How can climate-unrelated energy policies help to address climate problems?

- *What does the achievement of the Kyoto targets imply in terms of energy investments and infrastructures? Are the additional emissions reductions needed to reach the Kyoto targets feasible from an economic, energy and political perspective? Are financial transfers to economies in transition or developing economies, implied by the Kyoto mechanisms, realistic given the present situation?*

In Lanza (1999) great emphasis is given to the issue of financial transfers implicit in the implementation of the Kyoto Protocol through the flexibility mechanisms, and more generally to the problems of financial needs arising from the large investments necessary to control GHGs emissions. Are these transfers politically and economically feasible? Will they induce a 'credit crunch'? How do they affect income redistribution world-wide? Can these environmentally related financial transfers be linked to financial aid to economies in transition and developing countries?

2.3. *Policy integration*

- *Is policy integration a powerful tool to achieve climate change policy objectives? Which types of policy integration are more necessary? How should transport and energy policies be integrated with environmental, economic and fiscal policies?*

Alessandro Lanza from IEA, Paris, suggests that policy integration avoids sending contradictory signals to the economy, induces win-win policy options, and creates a self-reinforcing environment across policy measures. Environmental policies could be integrated with energy policies, welfare and labour market policies (Marsiliani *et al.*, 1998), technology transfer and co-operation (J. Janssen, 1998), policies to achieve side-effects and co-benefits of greenhouse gases emissions control (Markandya *et al.*, 1998), and transport and land-use planning policies (S. Proost, 1999). Energy and environmental policies could be harmonized with fiscal policies, and 'double dividend' opportunities could be exploited whenever they exist (cf. Bosello, Carraro and Galeotti, 1999; Marsiliani, 1999). It was emphasized by several participants that fiscal policies should aim at enhancing energy efficiency: if any form of taxation on energy were imposed, the money raised should be used towards energy efficiency and fuel switching. It was suggested that the contribution of other fiscal policy instruments towards energy efficiency both in a static and in a dynamic horizon should be further investigated.

How can these policy integrations be actually implemented? Are there political obstacles or economic costs of integration? Do countries need new institutions? Do they lack a long-term perspective when designing their optimal policy-mix?

3. Consensus issues

As said in the introduction, climate policies are often the subject of deep controversy and even during the workshop conflicting views were hotly debated.

Nonetheless, there are a few issues on which, at the end of the debate, consensus could be found among the different participants.

3.1. Equity

- *Equitable outcomes are as important as equitable procedures.*

Integrated policies should aim at both equitable outcomes and equitable procedures. At present in climate-change policies it would be unrealistic to expect to find an agreement on equitable procedures, whereas it may be possible to reach an agreement on equitable outcomes. To provide an example, with regard to emissions trading, equitable procedures could refer to the criteria according to which emissions permits are initially allocated, and to the process by which the policy decisions are made; equitable outcomes would instead refer to the post-trading welfare outcome. An equitable post-trading outcome should involve both developed and developing regions, and could be set in terms of a global emissions constraint, or in terms of per capita emissions 'entitlements'.

The criteria for the initial allocation of marketable emissions permits were discussed in depth. Berk *et al.* (1999) for instance found that participation and burden-sharing rules for emissions reduction based on indicators of a region's absolute contribution to climate change act to the disadvantage of large regions/countries, while disregarding differences in levels of economic development. Rose *et al.* (1999) noted that principles that allocate to all countries an equal proportion of permits in relation to their initial emissions will not necessarily result in equal levels of post-trading emissions reductions, thus highlighting a possible trade-off between equitable procedures and equitable outcomes.

Manfred Bergmann, from the European Commission, DGXII, stressed how hard it is to agree on equity procedures because of political reasons, but he expressed a positive view on the possibility of reaching agreements on equitable outcomes, helping developing countries to identify winning strategies and to undertake commitments. Generally, the need for a long-term focus on equitable objectives, inclusive of both Annex I and non-Annex I countries, was stressed by most participants.

Frances MacGuire, from Friends of the Earth, UK, claimed that the Kyoto Protocol does not establish equitable procedures and is not so equitable from a real-world point of view. Tom Heller noticed that in Buenos Aires, equity issues were raised in two ways: (a) it was questioned if trade could occur before the overall distribution of entitlements was decided, (b) it emerged that equity issues should be discussed with regard to the individual Kyoto mechanisms. It was agreed that making the Kyoto mechanisms operational and equitable cannot be a long-term policy objective: it should be a priority of short-term forward-looking climate policies. Consensus was expressed on the fact that 'per capita allocation' of emissions 'entitlements' will be the most equitable solution to pursue international climate-change policy objectives.

- *Emissions permits should be auctioned rather than grandfathered.*

Peter Bohm (1999) made a strong case for auctioning permits rather than grandfathering, i.e. free quota allocation, supporting his argument mainly because of revenue recycling options: 'Auctioning the whole volume of permits provides government revenue that allows a reduction of pre-existing distortionary taxes, a so-called double-dividend ... the auction price reflects this environmental concern and emerges as a corrective rather than distortionary levy'. Frank Convery (1999) noted that 'grandfathering allows benefiting firms to (1) remain in business for some non-economic reason, (2) have more funds for risky investments, and (3) have cheaper access to bank loans and capital markets ... giving away permits for free to existing firms can be expected to slow down productivity growth. Thus, the fear that countries using grandfathering will have a competitive edge is unlikely to be valid, at least in the medium term. And any advantage will be further undermined by revenue recycling, and neutrality towards new firms, which imply that auctioning of permits provides other important efficiency benefits'.

On the above remarks and conclusions there was a widespread consensus at the workshop. It was also proposed to look at the recent proposal by Resources for the Future (RFF) which provides an interesting step in the right direction, since it proposes a model of domestic trading with auctioning of emissions permits and a cap on prices, where equity is served by a recycling mechanism. The three key features of the RFF proposal are summarized by its broad coverage, its modest target, and its equitable burden sharing. Concerns were expressed, however, about the fact that the recycling process may re-propose the 'carbon-tax loop'.

- *Further research on the equity dimension of climate-change impacts is necessary.*

Little research has in fact been carried out on the equity dimension of impacts; instead, research has focused on mitigation policies, addressing the equity implications of mitigation mainly through the debate on procedures and criteria for emissions allocation and trading. Bill Hare, from Greenpeace International, emphasized the need for careful evaluation of climate-change impacts, particularly impacts on human life, in order to address equitable targets.

3.2. Operational modalities

With regard to the implementation of the Kyoto Protocol, general consensus was found around the following key issues:

- *Despite their high variability, all estimates show that the Kyoto flexibility mechanisms significantly reduce the costs of compliance.*

J. Shrogren noted that it is estimated that any agreement without the cost flexibility provided by trading will at least double the US costs, where flexibility can be measured as the ability to reduce carbon at the lowest cost, either domestically and internationally, including the so-called 'when and where' flexibility; the key is to distribute emissions internationally so as to minimize the costs of climate policy. The model proposed by Manne and Richel (1999) indicates that 'losses in 2010 are two and one-half times higher with the constraint on the purchase of carbon emission rights; international co-operation through trade is essential if we are to reduce mitigation costs'.

These are just a couple of examples of the many models that reach the same conclusion: emissions trading can reduce overall mitigation costs without reducing the effectiveness of the climate policy (see also Rose *et al.*, 1998; Bosello and Roson, 1999; Hourcade *et al.*, 1999). As noticed by Carlo Carraro, from the University of Venice and FEEM, the same conclusion can also be achieved independently of the specific climate model which is adopted. Tulkens *et al.* (1999) showed that the application of simple economic principles is sufficient to achieve the following conclusions:

1. Flexibility mechanisms reduce total compliance costs.
2. The largest cost reduction is achieved when no constraint is imposed on the trading system.
3. There exists a system of transfers such that this cost reduction benefits all countries.

- *Quota or permit trading markets can be made to work effectively.*

P. Bohm (1999) made the point that 'carbon trading involves a homogeneous good with no transport, implying that transaction costs per unit of trade will be small: therefore designing a market system should not be impossible. The units traded could be units of emissions reductions or units of national emissions quotas that can be used once during a five-year commitment period, or banked for a subsequent period, using an international emissions exchange which never closes'. Frank Convery from UCL, Dublin, noted that 'trading rules should be designed and defined to be effective and ensure competitive markets ... it would seem to be necessary to make all transactions subject to a *transparency requirement* where the prices are made commonly known to all traders ... anonymity makes market transactions more efficient'.

- *Monitoring and enforcement should be feasible.*

Monitoring of national emissions levels will be crucial to the implementation of mitigation policies. P. Bohm (1999) writes that 'it is in the interest of every fossil-fuel-exporting country to avoid underestimates of its export volume, and of every importing country to avoid over-estimates of its import volume'. Appropriate institutions should be put in place to initiate the monitoring process. At the same time special measures may be required for monitoring imports from non-signatory countries: measures such as sample unannounced inspections of fossil-fuel transports could be implemented.

- *Without strict pre-conditions, early crediting is likely to be ineffectual and to undermine the achievement of targets.*

Frank Convery noted that 'the Kyoto Protocol sets legally binding emission targets for industrialized countries, without accounting for reductions carried out prior to 2008, the beginning of the first commitment period'. Only the Clean Development Mechanism allows for a still-undefined pre-commitment crediting. However, 'an early credit Bill was introduced in the US Senate in October 1998 which allows crediting for future domestic emissions reductions, but also for past actions identified under par 1605 of the Energy Policy Act of 1992 – where companies can voluntarily report mitigation activities which will then be registered; no verification process exists and in some cases the same reductions are reported by different entities (double counting) – as well as for projects under the USA initiative on Joint Implementation'. Credits could be used to offset future emissions control obligations.

In the literature, however, this proposal has generated unanimous criticism: A. Michaelowa and M. Stronzik concluded that 'the following safeguards are necessary but not sufficient to guarantee an advantageous system for early crediting: strict baseline; incentives for innovation; incentives for rigid sectors to install climate-friendly capital replacement'. RFF suggests that 'the early credit Bill proposal risks distributing too many credits for questionable early reductions. The only way to reduce this risk is to thoroughly examine each project and evaluate the true reductions incurred: a cumbersome and potentially expensive administrative process. Further, the proposal still connects the incentive for early action to the future worth of emissions rights under an uncertain permit scheme, a speculative value that may fail to motivate sufficient activity'.

- *There is scope for experimental domestic trading markets.*

RFF provides a complete proposal for a US credible early action in climate policy from the research modelling side. On the other end, Michael Wriglesworth, from BP-Amoco UK, illustrated BP-Amoco action, which offers an interesting operative example from the industry side, having launched a domestic emissions trading scheme in 1998. Both schemes are introduced in more detail below in the section on 'new policy perspectives and recommendations'.

- *Sufficient arguments exist with respect to using discounting procedures in the evaluation of climate-change policy options.*

The discount rate is at the heart of any inter-temporal decision and plays a central role in models of economic growth: in integrated climate–economic models the choice of the appropriate discount rate is crucial. Various discounting techniques are, however, available and the optimal discounting choice is under debate.

3.3. Sectoral policies

- *An effective transport policy in Europe is crucially necessary, which would imply new policy reforms.*

General agreement was expressed on the fact that so far European transport policies have been a failure. De Pretto, from the International Road Transport Union, Switzerland, stated the need for a transport policy in Europe. From the panel debate, it emerged that *in the short run it will be relevant to ameliorate access to infrastructures*, and that *efficiency in the transport sector should be improved across transport modes*: Gert-Jan Koopman, from the European Commission, stressed the pressing need for a railway reform in Europe, and for shifting away from a modal transportation system towards an inter-modal system, improving efficiency across transport modes. Inter-modal systems should be made available and accessible. Further consensus among panellists and participants in the workshop was reached on the *need to integrate transport policies with other sectoral policies* in order to achieve climate-change policy objectives. It was agreed that correct and socially efficient transport pricing should be a complement to a mix of policies.

- *Energy policies should provide incentives to supply side actions on renewables.*

This view did not generate any conflicts at the workshop. It was noted that in Europe there is a great scope for cost-effective measures on renewable energy sources. Kevin Leydon, from the European Commission, DG XVII, stressed the obligation to increase energy from renewables in Europe, doubling it by 2010 mainly through the use of biomass.

- *Too little attention has been devoted to sinks; sinks provisions should be developed to evaluate sinks as sources of carbon sequestration as well as sources of energy.*

However, carbon sinks should be dealt with carefully: Frank Convery noted that ‘countries can claim carbon sequestered from 1995 as a contribution to meeting their targets, and some countries which have a comparative advantage in forest plantation growth, e.g. New Zealand, Ireland, the USA, will be able to avail of this option’. J. Shogren called carbon sinks a *wild card*: ‘It may cost as little as $25 to sequester a tonne of carbon in the US but both the base line and operational (sequestration rates etc.) data are poor’.

- *Climate-change policies should address not only the technology side, but also the behavioural side.*

As stressed by Kevin Leydon, technology improvements alone will not be able to drive parties to reach their emissions reduction targets. In specific sectors, consensus on enhancing a ‘joint responsibility’ approach was expressed. Philip

Daubeney, from the Electricity Association, UK, suggested that efforts should be made in order to induce consumers to buy 'cleaner' energy at higher prices, stressing that the electricity industry has a major role to play, not only as a source of the climate-change problem, but even as a source of possible solutions.

3.4 Policy integration

- *Energy and environmental/economic policies need to be closely integrated.*

It was underlined that policy integration may also lead to effective synergies across institutions. As explicitly addressed by Kevin Leydon, policy integration should be achieved across various sectors, including energy, transport and agriculture. It was agreed that policy integration may help to achieve climate-change targets at lower cost. At the same time, the achievement of climate goals may help to lower the impacts of economic activities on other environmental problems.

With regard to institutional integration, the consideration that the integration of environmental and competitiveness regulatory authorities may achieve synergies, in the form of lower costs or increased benefits, seemed acceptable. Frank Convery noted that 'market failures derive from the ability to exercise market power (competitiveness issues) and ability to impose environmental externalities (environmental issues)'. M. Acutt and C. Elliot, who presented a paper modelling the effects of a change of regulatory regime from the current system of separate simultaneous regulation by the two regulators to a co-operative regulatory regime, concluded that '... a potential welfare improvement is available as a result of a move to regulator co-operation. Under the co-operative regime, whilst welfare is reduced as a result of less investment in cleaner technology by firms, welfare is simultaneously increased as industry output and consumer surplus plus profit will be greater than under non-co-operation.'

3.5. Research, policy and market strategies

With regard to research, policy and market strategies there was agreement on the fact that:

- *Research should make it possible to assess the effectiveness of various policy instrument combinations, in terms of both their environmental effectiveness and their macro-economic impact.*

Frank Convery quoted the SPOT3 model for Belgium, which indicates that a carbon trading scheme designed to comply with Kyoto targets could be implemented with minimum recessive effects on the economy, incorporating the interactions of energy economy and environment for the Belgian economy. This approach should be pursued on a wider scale. In particular, research should focus on the analysis of the mechanisms driving technological change in the different sectors of the economy. It was agreed that *technological change represents the weakest area in energy and economic modelling and deserves special attention.*

- *One of the major problems in the design of international climate-change policies is the long-term involvement of developing countries; there was general consensus on the consideration that the overall UNFCCC process will not succeed unless effective globalization mechanisms are found.*

Priyadarshi Shukla from the Indian Institute of Management made the point that 'the whole venture will fail unless the South is productively engaged, and this will involve radical thinking'. A. Manne and R. Richels (1999) made the same point: 'The Kyoto forever scenario will fail to stabilize global emissions and concentrations'. Hence, the participation of developing countries is needed.

- *Kyoto was a win-win option for Annex I countries.*

At Kyoto, Annex I players were winners in the game. The US obtained the insertion of the option of International Emissions Trading in the Kyoto Protocol, giving up their opposition to the EU bubble. The EU was thus able to develop a framework that other countries could join. Eastern European countries became signatories thanks to hot air, i.e. loose stabilization targets with an allocation which exceeds their likely emissions in 2008–2012 under the BAU scenario. If hot air had not been provided, Russia and Ukraine would not have played the game and everyone would have lost. P. Bohm (1999) wrote: 'First, Russia and the Ukraine may not have accepted a reduction of their quotas to equal their most likely BAU levels, especially not given the time-pressed conclusion of the Kyoto negotiations. Instead, the risk would seem to have been significant that, confronted by such a quota offer, one or both of these countries would have dropped out of the agreement. And other Annex I countries would realize that their commitments then would have become more costly and therefore may have negotiated larger quotas themselves', with an increased risk of carbon leakage.

- *Efforts should also be made domestically.*

There was general agreement on the need for domestic policies in addition to the use of global flexibility mechanisms, and on the fact that *technology efficiency and its improvement should be a major issue and concern.* The US in particular should make domestic efforts to enhance technology efficiency. It was proposed that climate-change policy targets and measures should allow decoupling of greenhouse gases emissions from GDP growth.

- *Risk management strategies should be incorporated in the climate-change policy framework.*

Josef Janssen from the University of St. Gallen underlined the need for instruments to manage risks related to project-based mechanisms. These instruments may include integrated existing insurance institutions, such as the MIGA (Multilateral Investment Guarantee Agency), and mechanisms for projects' portfolio diversification, with a wide range of project types and host countries. Tom Heller suggested that, to promote national and international equity, this process might be guaranteed by a multilateral arrangement with very little overall regulation, in the form of an open multiple market, careful to avoid the formation of cartel prices.

- *Dialogue between actors such as fuel producers, governments, NGOs and consumers should be enhanced.*

The workshop showed that, despite the large amount of information which is circulating around the world on climate issues, little dialogue is taking place among different parties. Both at the research and at the policy level it is necessary to develop more careful comparisons of results and proposals. Alain Heilbrunn, from the Energy Policy Action Group (EUROPIA), suggested the need for 'more and better science', and greater acceptance by the general public, to promote the acceptance of the Kyoto Protocol. Michael Wriglesworth, representing the oil and chemical industry, also expressed the need for more dialogue and communication.

4. Conflicting issues

The lively debate that took place during the three days of the workshop highlighted that there is consensus on some issues, at least the ones described above, but that many issues remain controversial. Environmentalists, policy makers and industrialists have different views on crucial aspects of the climate debate and in particular on the priorities to be given to different policy goals. Here is a summary of the conflicting issues that emerged during the panel debates and the discussions.

4.1. Equity

As far as equity is concerned, conflicts emerged about:

- *The possibility of establishing equitable procedures in climate-change policies.*

Even if agreements on equitable outcomes could be reached, scepticism was raised about the possibility of agreement on equitable procedures, due to the very different economic and political interests of UNFCCC parties. With regard to carbon sinks, for instance, Tom Heller noted that 'the USA contributes to the climate change process as the biggest terrestrial sink: equity issues are something the world doesn't want to look into'. With regard to intergenerational equity objectives it was argued that, if present technology does not allow us to reach the Kyoto targets, it should be possible to accept that future generations will have less than we do. Focusing on international and national equity concepts, the achievement of equitable objectives in climate policies may occur through the application of different principles of equity at the national, regional and international levels. Hence, whereas on the one hand equity was considered to be an important step toward a successful stabilization of carbon emissions, on the other hand there was no consensus on the possibility of pursuing a unique equity principle. It was noted that the search for a common definition of equity should be de-emphasized; policies should instead be designed with a better understanding of their equity implications. As already stated, pessimistic views were raised about this issue, thus suggesting that an excessive focus on equity would be detrimental to effective protection of the environment.

- *The real meaning of long-term convergence.*

Assuming convergence as a long-term equitable objective, which convergence should countries aim at? Per capita income convergence? It was emphasized that income convergence between the North and the South is not likely to occur in the next hundred years. Furthermore: would convergence in income also imply convergence in structural patterns? Should countries aim at convergence by sector? What would be the linkage with sectoral technological change? Again, the claim was that excessive attention given to convergence issues would delay the implementation of effective climate policies. However, the opposite view was that without convergence no climate policy can be effective.

- *The possibility of identifying win-win options for developing countries.*

Northern and Southern positions appear to be quite divided. Paul Metz, from the European Business Council for a Sustainable Energy Future, was optimistic about the possibility of identifying win-win options in developing countries. However, no consensus was found on the acceptability of 'meaningful participation' of non-Annex I countries in the climate-change process. Frances MacGuire claimed that there is too much pressure from Annex I countries on developing countries with regard to 'meaningful participation'. Priyadarshi Shukla noted that 'meaningful participation' sounds highly inequitable given that India's per capita emissions are 5% of those in the US. Sujata Gupta emphasized that setting quantitative targets as 'meaningful participation' would not be enough, and recognized the need and the scope for energy efficiency improvements, particularly in non-Annex I countries.

- *The relevance of the social equity dimension in the context of the UNFCCC.*

Michael Grubb from RIIA, London, stressed the relevance of both international and national/social equity in the context of the UNFCCC, finding some opposing views; two developing countries' representatives, Sujata Gupta and Priyadarshi Shukla, claimed that social equity is a relevant concept, but that it should not be addressed in the UNFCCC: rather, national policies sovereignty and non-interference principles should be addressed. In general, full agreement was not reached on the proposal of using climate-change policies and instruments to address social equity, for instance through the creation of labour market opportunities in emerging energy markets, or through revenue-recycling mechanisms.

- *The possibility of attaining both equity and efficiency objectives, accepting the separability of equity and efficiency.*

Northern and Southern interpretations do exist with regard to the Coase[4] theorem. Some participants argued that equity should be dealt with through an initial world-wide allocation of emissions rights, and then markets would guarantee the achievement of equity. Whatever the initial allocation, efficiency would be attained. Others suggested that equity is a preliminary condition for any action and therefore only if the burden of reducing emissions is equitably shared can emissions abatement take place.

It was also noted that conflict on the equity of a climate agreement may lead to paradoxical situations. For example, it is clear that a generalized emissions trading system constitutes a Pareto improvement (all countries are going to gain, including developing countries), but nonetheless it is opposed by developing countries on the grounds that it would favour developed countries more than anyone else (cf. Bosello and Roson, 1999).

- *How options for mitigation should be implemented across countries.*

It was stressed that mitigation should be evaluated at its price, rather than at its costs. Costs are certainly lower in developing countries but there should be some reward. Fair competition would imply a surplus to suppliers of cheap mitigation. But this view obviously reduces the efficiency of global emissions trading systems and no consensus could be found on it.

- *Who should share the burden.*

Should it be the politicians, the producers, or the consumers? It was proposed that, rather than concentrating on 'burden sharing', the focus of climate-change policies should shift to both costs and opportunities, from 'costs sharing' to 'commitment sharing'. NGOs claimed that it would be unrealistic to expect people to act: rather, industry should act.

Andrew Warren, from the Association for the Conservation of Energy, UK, stressed the critical role of energy suppliers in taking action. Kenneth Ruffing, from the UN Division for Sustainable Development, illustrating the case of China, claimed that energy efficiency efforts from the industry side occur outside international climate change commitments, and questioned how much the present scenario could be changed by new instruments. Delia Villagrasa, from Climate Network Europe, Belgium, questioned the transport and energy industries' interest in international policies and measures, and she claimed that industry should move ahead, influencing behaviour and demand. With regard to transport, for instance, car companies should make cars attractive because of their energy efficiency, in the attempt to impact on the emotional and psychological component of demand. Frances MacGuire questioned the ability of consumers to pay for 'cleaner' energy, rather than their willingness to do so, raising equity considerations. Among the environmental groups' representatives therefore emerged the general view that industry should act and take responsibilities towards the improvement of energy efficiency and the mitigation of climate change. However, besides a few encouraging examples, such as the BP-Amoco and a few other experimental trading markets, as well as some voluntary agreements towards climate-change mitigation between industry and national governments, industry appeared not to be very proactive and expressed the need for a more favourable political and institutional environment. In this regard Stefano Micossi, former General Director of DGIII, European Commission, emphasized that the European and national public administration systems do not have the capability of efficiently managing voluntary agreements on climate change and expressed his doubts on the possibility to make the flexibility mechanisms work effectively, stressing that achieving the Kyoto targets is too costly. The different positions of environmentalists, industrialists and policy makers shed doubts on how and who should 'share the burden'.

4.2. Operational modalities

Conflicts on operational modalities focused on the definition of the 'supplementarity criterion' for the use of the flexibility mechanism, and on the nature of the restrictions to be imposed on global trading.

- *There was no agreement on the nature of the ceiling or on how much of the emissions reduction quota should be traded.*

Article 17 of the Kyoto Protocol calls for emissions trading to be only 'supplemental to domestic actions for the purpose of meeting quantified emissions limitation and reduction commitments under Article 3'. To make it operational, it has been suggested that quantitative constraints on imports of emissions reduction be introduced. Frank Convery, as well as many other participants in the workshop, stated that 'to the extent that constraints are placed on carbon trade, the costs of compliance will be increased .... the negative effects would spill over also to the countries likely to export carbon credits, since the volume of their sales and the price they get will be lower because of lower demand induced by higher costs of mitigation measures taken at home by the EU and the USA. All countries lose, and emissions reduction commitments in subsequent periods will be made more expensive and therefore less likely to be significant. Since the gains from trade experienced by Russia and Ukraine will be reduced, it will also reduce the prospects for trade gains from potential entrants from the developing countries in the future'.

P. Bohm (1999) asserted that 'this can be expected to increase marginal costs or shadow prices of emissions reductions in the importing countries', and 'forcing countries to produce more of the emissions reduction quantity commodity at home than it wants to is like forcing cold Nordic countries to grow some minimum share of bananas before it is allowed to import bananas from countries that have a comparative advantage in banana production .... Supplementarity will not only make present Annex B countries less likely to accept more stringent future commitments, it will also make it harder to get new countries to join the set of Annex B countries. Finally, in the more immediate time frame, the success of the Kyoto Protocol stands and falls with the US ratifying it', which will require maximum flexibility.

There were, however, arguments in favour of a ceiling, mainly justified by the fact that restrictions on trading could be an effective stimulus for technical innovation and dynamic efficiency, which in the long run would make future targets more affordable. Jean-Charles Hourcade, from CIRED, France, who supports this view, proposed a minimum level of emissions reduction from domestic actions. A similar proposal came from Stephan Schleicher, from the Austrian Climate Council, even if a well-defined quantification of the ceiling was not offered.

Priyadarshi Shukla claimed that the North and the South are quite divided on this issue. He said that perhaps less developed countries would like to see a better-defined supplementarity clause, to make Annex I countries' commitment more credible. In fact, although the global flexibility mechanisms may allow room for reducing mitigation costs and achieving a global policy goal, the 'hot air' issue has provided a bad example. Arguments in favour of a ceiling were also supported by environmental associations. Bill Hare, given the environmental consequences of the whole 'hot air' issue, and the need for technological development, made a stringent case for having a cap.

- *The relevance of the 'hot air' issue was debated.*

Frank Convery suggested that 'there is too much hot air about hot air', pointing out that hot air was a pre-requisite to secure an agreement in Kyoto. Others, in particular NGOs and environmental associations' representatives, see hot air as a threat, in terms of the future credibility of the agreement itself, and as a disincentive for non-Annex I parties to join the agreement with long-term commitments. Bill Hare stressed that the hot-air issue has strong equity and environmental implications, increasing emissions levels which otherwise would not have been reached. Political, economic and environmental arguments were expressed and confronted without achieving a consensus.

- *Different views were expressed with regard to having a buyer's or a seller's liability system.*

It was argued that a buyer liability system may be a more sound system. Bill Hare underlined that a compliance regime should work after the first commitment period, stressing that liability clauses could make the system more efficient. Since not all selling parties have a robust legal system, he proposed that a buyer, or a mixed buyer–seller liability scheme, would be preferable. P. Bohm (1999) instead suggested that in an ideal system of international emissions exchange 'the seller will be involved in a transaction with a neutral international exchange which, once the question of sanctions has been settled, holds the seller responsible for its sales of emissions reduction'.

4.3. Sectoral policies

- *The view that CO_2 emissions reduction from the transport sector is the most relevant issue to be addressed was challenged.*

Stef Proost suggested that the major problem in the transport sector is to address external costs such as congestion and accidents. A carbon tax should not be used to achieve higher efficiency in the transport sector, although revenues could be used as an instrument to support relevant energy policies. Other instruments, such as road pricing, or land use planning, could be used to achieve the climate policy goal. This view was challenged by several speakers. In particular Jean-Charles Hourcade expressed his view in favour of a carbon tax, noting that without a carbon tax there would be no way to achieve effective policies in the transport sector: road-pricing or permits for land use would not be feasible instruments to be implemented. A carbon tax would instead have the advantage of being simple, and of generating revenues which could be used to implement energy efficiency policies.

- *It was highly controversial whether transport should make a real contribution to reaching the Kyoto objectives.*

It was claimed that in Europe there is little marginal efficiency gain in the transport sector. Transport is a sector where significant emissions reduction cannot be achieved cost-effectively. Other views proposed that tradable permits or quotas be targeted at those sectors where there is likely to be most growth in carbon emissions if they are not constrained, such as the transport sector, to enhance their effectiveness. Focusing on the transport sector, Johan Albrecht, from the University of Gent, commented that 'current designs and proposals for tradable CO_2 emission systems do not provide incentives to stimulate cross-sectoral energy efficiency investments. Manufacturers should be "rewarded" for their products that make it possible for consumers to save energy during consumption'.

Supporting the need for measures to control emissions in the transport sector, it was argued that in urban areas the transport sector has been the most important source of emissions for traditional pollutants (CO, VOC, NO_2, PM), representing the most important source of emissions not only in quantitative terms, but also in terms of harmfulness: in fact, transport emissions, affecting mostly urban air quality, have much greater impact on local health than other emissions sources. The emphasis on CO_2 fuel efficiency standards in transport could be shifted towards the use of clean fuels, switching from CO_2 fuels to NLG or bio-gas.

- *Conflicting views on the possible future role of land-use planning.*

Despite the view that land-use planning would not be an effective instrument in attaining climate change policy targets, Carlo Jaeger, from the University of Darmstatd, argued that land-use planning will have to emerge as a relevant policy objective in the next 20–30 years because of the alarming growth of 'mega' cities in the developing world. Per Sorup, from the Institute for Perspective Technologies (IPTS), suggested that land-use planners and policy makers should be more involved in the climate-change debate. Richard Tol from IVM, Amsterdam, stressed that changing transport patterns is a long-term issue; he questioned how consistent signals on the effectiveness of such a policy over a long time period could be obtained now and how measures on transport policies taken today could be valid in the far future. An interesting view was proposed by Virginia Sonntag-O'Brien, from the International Council for Local Environmental Initiatives, Germany. She suggested that land-use planning policies should be targeted at intervening on the city structure, since cities were conceived for no-mobility, and instead they have become a place for automobiles; policies should aim to reduce demand for mobility, to create special structures, such as pedestrian departments, and generally to shift from the automobile car industry to a 'mobility service' industry. Marc Darras, from Gaz de France, stressed the need for land-use planning in particular with regard to job organization and location.

4.4. Research, policy and market strategies

- *There is disagreement between experts on the significance of short- versus longer-term action.*

A. Manne and R. Richels (1999) argued that 'rather than requiring sharp near-term reductions, it appears that a more sensible strategy would be to make the transition at the point of capital stock turnover – eliminate the need for premature retirement of existing plants'. On the other hand, many participants argued in favour of early action to send the right signals to investors. Among the environmentalists it was claimed that environmental irreversibilities make it impossible to delay action. The issue of the optimal timing of climate policy is still very controversial and little support is provided by models where uncertainty and technical change are unsatisfactorily modelled.

- *There is a lack of clarity or convergence on the origins of technological advance.*

Jason Shogren noted that 'engineers argue that the origins of technological advance are firmly rooted in non-price responses; but for economists it is the search for profits that can create R&D breakthroughs that reduce the costs of backstop technologies'. Furthermore, Vijay Vaitheeswaran, Environment and Energy Correspondent for *The Economist,* expressed some doubts on whether the technology option for climate-change mitigation could be played if oil prices have continuously fallen, and a decreasing trend is forecast. Again the issue of technological innovation, of the availability of low-cost climate-friendly technological options, and of the right policy to favour innovation and diffusion is very controversial and consensus could not be found.

- *No consensus could be found on adopting a unique discounting procedure and on how to use Cost–Benefit Analysis (CBA) to address climate-change policies.*

Reyer Gerlagh, from IVM, Amsterdam, underlined that in most dynastic integrated assessment models (IAMs) the debate on the proper level of discount rates in view of sustainability is based on an assumed relation between a 'pure time preference' economic growth and the interest rate. Quantitative results depend highly on future discounting. 'Consequently, dynastic IAMs cannot be used to provide policy makers with quantitative figures about the desirable emissions reduction levels and the supporting prices. Instead, economists should inform policy makers as to which instruments can be used to bring forth both efficient and sustainable resource use .... Using the overlapping generations model (OLG) reveals that there need not be such fixed relations between economic growth and the interest rate. In fact various factors determine the change in the interest rate over time, including the expected demographic change in the next century, and, most importantly, including social security policies and potential sustainability policies'.

Chris Hope, from Cambridge University, argued that specific objectives should be defined, and that discount rates could then be targeted towards those specific objectives in the long term. Note that these views are in contrast with the practice adopted by most empirical analysis of climate change policies and were not supported by Ferenc Toth in his survey paper and by other participants in the workshop.

Kenneth Ruffing, from the UN Division for Sustainable Development, suggested that, given the net benefits of climate-change mitigation options, Cost–Benefit Analysis (CBA) should be used to identify the least-cost options. However, CBA should be considered as an important source of information, but not as the sole base of social decision making. In fact it was argued that identifying the magnitude of mitigation costs is not the ultimate relevant question. The real issue should be how to mitigate, and *mitigation options should be compared to adaptation costs.*

- *No agreement could be found on the possibility of having a sound reference scenario for developing countries in the light of their future involvement in long-term climate change mitigation commitments.*

Furthermore, scepticism was manifested about the possibility of getting developing countries involved in the climate change policy process with a long-run commitment. It was questioned *how any voluntary target from developing countries could be credible.*

5. Conclusions: new policy perspectives and recommendations

The previous sections provided a summary of the main issues debated at the EFIEA policy workshop on climate policies. In this concluding section, we attempt to summarize a few policy guidelines that were proposed. We call them 'perspectives' because they are not normative recommendations but rather positive scenarios that suggest how climate policies could be implemented.

5.1. Perspectives on equity

- Practical and effective solutions in climate-change policies balance equity, efficiency, and political feasibility criteria.
- Policies focus on equitable procedures as well as on equitable outcomes. Long-term policies aim at defining a global cap, which will induce efficiency, identifying at the same time who would bear the costs and benefits of mitigation, and addressing equitable procedures.
- Before any agreement on a global cap, there is, however, the need to get developing countries on board: get them involved in the whole climate change debate, and make climate-change policies part of their national policies. To achieve this goal, equity and efficiency issues need to be carefully integrated. Increased equity would help increase the number of participating countries, thus enhancing the efficiency of the agreement. Increased efficiency would reduce each country's cost of controlling emissions, thus facilitating the adoption of transfer and co-operation policies that increase equity as well as the number of participating countries.

- Policy makers fully evaluate the socio-economic implications of climate-change impacts (damage to the environment, and human life) and adaptation, particularly addressing their equity aspects, confronting the costs and benefits of alternative policies. A better understanding of the linkage between technology (efficiency) and equity is therefore a prerequisite for policy actions.

5.2. *Perspectives on policy integration*

- Policy integration takes into account the linkages of environmental policy with energy policies, technology development and diffusion, and technological co-operation. Within the EU, policy integration occurs across sectors: transport, agriculture and energy. In particular, the integration of transport policies with other sectoral policies is promoted, for instance making land-use planning an important element of transport policies.
- Policies also provide further incentives for cross-sectoral energy efficiency investments; in this regard there is a strong need for a more innovative policy approach, which contemplates investment in new technologies.
- Integrated climate policies evaluate and investigate the implications of liberalization in the energy market for national policies, addressing its environmental and socio-economic impact. This view was strongly supported by Delia Villagrasa, who expressed her concern about the environmental impact of liberalization.
- With regard to the socio-economic impact of climate policies, Willy Buschak, representative of the European Trade Union Confederation (ETUC), Brussels, proposed that, since jobs will be lost in the energy sector, climate-change policies should be used to generate employment. He stated that 'climate-change mitigation is an option provided job opportunities are generated'.
- Policy integration implies action across groups and policy targets: policies are directed to mould both producers' and consumers' behaviour, on both the supply and the demand sides.
- Consider, as an example of policy integration, the many suggestions that were provided in the debate on transport and climate policies. Participants in the panel and in the subsequent discussions outlined several improvements of transport policies in the EU, e.g.:

    1. The need to control air transport emissions.
    2. The need to shift the automobile car industry to a 'mobility service' industry.
    3. The need to make the transport pricing reform into a package. In fact there is a pressing need for a railway reform, and for moving away from a modal transport system towards an inter-modal system.
    4. The need to correct the biases and inefficiencies of transport policies. Policies in fact always tend to hit car producers at the margin, and they

are never aimed at hitting the stock of cars, thus affecting users. Policies will generate inefficiencies as long as they only hit small groups. In this regard, it was recommended that city mayors should sit at the round tables of policy makers.

5. The need to avoid contradictory targets that generate inefficiencies. For instance transport policies tend to fulfil safety criteria as opposed to environmental friendliness criteria (heavier cars are more polluting).

Research priorities on transport were also identified:

1. Good transport forecasts for the EU.
2. A study of the potential for emissions reductions in air transport: air transport fuel is not taxed and there is no reason to maintain this inefficiency.
3. Evaluation of the costs of fuel efficiency standards.
4. Study the potential for shifts between different types of trucks (small and large), and between road, rail and shipping.

5.3. Strategic issues

- *An effective climate policy cannot neglect the fact that its outcome crucially depends on the actions and interactions of several players.*

For example, the abatement decision of a group of countries, e.g. Annex I, can be costly and ineffective if offset by emissions expansions in other countries. This is the well-known carbon leakage issue, which arises because of the free-riding strategic incentives that characterize the climate problem.

- *Carbon leakage is not adequately addressed in the Kyoto Protocol, and future policies should make an effort to develop measures to counteract it.*

Frank Convery stressed that 'carbon leakage occurs whereby firstly carbon-intensive products become more expensive in signatory countries and imports increase from non-signatory countries, and secondly firms using carbon-emitting technologies in carbon signatory countries move to non-signatory countries. The Kyoto Protocol does not contain any policies or measures to counteract carbon leakage'. Claude Culem, from the Energy Policy European Chemical Industry Council (CEFIC), Brussels, expressed serious concern from the chemical industry side towards carbon leakage in that developing countries have no commitments to limit their emissions.

M. Hoel (1999) made the general point that actions taken by a particular country (or group of countries) will in general affect equilibrium prices of internationally traded goods. This in turn may affect the production and consumption decisions of other countries, and thus emissions from these countries. 'Given the Kyoto agreement, it is only leakage to developing (non-Annex B) countries that is of importance. Moreover, it is shown that differentiation of a carbon tax is not justified by a concern for CO_2 emissions in developing countries. It is more cost-effective to induce these countries to limit their emissions through appropriate transfers. Ignoring the optimal tariff argument, an approximation of the optimal policy is thus to have a uniform carbon tax and no tariffs. Carbon leakage should be taken care of by industrialized countries giving the developing countries transfers conditional on the developing countries implementing climate policies'.

- *Interactions between producers and between producers and consumers at the domestic level should be taken into account by climate policies.*

The RFF proposal for the US could be adopted as an interesting example of a domestic trading permit scheme. The RFF proposal does provide for the auctioning of carbon quotas, with carbon fuel energy producers and importers required to buy the carbon quota equivalent of their output or imports respectively, with a ceiling price of US$25 per tonne of carbon, the permits to last two years, and the revenues to be returned to households in compensation for higher fuel prices.

- *Strategic interactions and preferences for a better environment can be exploited by firms for business purposes.*

BP-Amoco indeed represents a very interesting example of 'far-seeing' climate policy from the industry side; BP-Amoco has in fact adopted a very innovative policy, setting its own reduction target: 22 million tonnes reduction by the year 2010, which is equivalent to a 10% reduction with respect to its 1990 baseline. In order to meet this challenge, BP-Amoco intends to develop a strategy that will result in the greatest GHG emissions reduction at the lowest cost over a specific timeframe.

The implementation of this strategy will require a process of identification of the emissions, by source and type, of measurement and verification of the emissions, of identification of GHG reduction opportunities across the business. Reduction policies aim at enhancing efficiency, promoting new technology, and preventing flaring and venting. Part of this process is the implementation of a pilot trading system, which was launched in late 1998. Several business units are involved in the scheme; the pilot group has a reduction target of 3% below the 1995 baseline by the year 2003; basic emissions rights have been allocated for five annual periods from 1999, with requirements to be met each year; a bank acts as a broker and registers trade across business units; banking, forward sales and purchases are allowed, but no borrowing can take place; trading occurs in CO_2 equivalent units.

The BP-Amoco experiment could serve as a pivotal scheme for industry, with the aim of understanding how to deliver commitments at minimum cost. The identification of opportunities across business could also help industry and governments to develop integrated and more effective climate policies.

- *Industry has important and serious contributions to make to mitigate climate change.*

As stressed by Manfred Bergmann, Paul Metz, Stefano Micossi, and other participants, voluntary commitments and voluntary agreements seem to be the best option by which industry can improve energy efficiency and identify business opportunities from climate change mitigation options. There are very encouraging perspectives particularly from the chemical and oil industry. J. Rogier, from the European Solid Fuel Association, expressed confidence in the instruments proposed at Kyoto, and particularly in the Clean Development Mechanism, which would help China and India to reduce their emissions. However, Stefano Micossi, former General Director of DGIII, European Commission, expressed doubts on the capability of European and National public administrations to efficiently manage voluntary agreements on climate change and the CDM in particular.

5.4. *Some final remarks on climate policy*

- *The likelihood of the ratification of the Kyoto Protocol is still uncertain. Policies should be directed at inducing the US to ratify the Protocol.*

Among several other participants, Nigel Haigh, from the Institute for European Environmental Policies, UK, expressed the growing concern about the likelihood of ratification. If the US drops out, the climate-change process initiated in Kyoto will fail. It was recommended that future policies should be more realistic than they have been so far.

- *The political dimension of climate policies is as important as their environmental and economic dimensions.*

Stefano Micossi explicitly expressed the need for political realism: the Kyoto targets, albeit insufficient from an environmental viewpoint, are likely to be too costly to be achieved, political negotiation procedures at the EU level are too complex, and the EU commitment is too ambitious. Richard Tol claimed that, talking about EU climate change policies, no reference has been made to the EMU. Climate change mitigation options may induce some inflationary pressure within the EMU and this issue cannot be disregarded by policy makers. Albeit controversial, this is an example of the degree of integration of climate policy with all other economic policies.

- *Future policy proposals will have to be concrete with regard to costs, definition of the flexibility mechanisms, and technology advances.*
- *The institutional development of the Kyoto Protocol requires that more attention be devoted to its environmental effectiveness: there is a strong need for clear and objective environmental targets.*

Notes

* This report summarizes the main ideas, proposals, scientific achievements, consensus and conflicting issues that emerged at the Second EFIEA Policy Workshop, held at Fondazione Eni Enrico Mattei, Palazzo delle Stelline, Corso Magenta 63, Milan, Italy, March 4–6, 1999. EFIEA, the European Forum for Integrated Environmental Assessment, is a concerted action funded *by the Environment and Climate Programme of the European Commission, Directorate-General XII (contract no. ENV4-CT97-0450)*. The two main objectives of the EFIEA are: (i) to improve the scientific quality of integrated environmental assessment; (ii) to strengthen the interaction between environmental science and policy making. In addition, the EFIEA fosters co-operation between scientists and decision makers inside the European Union, communication and co-operation outside the EU, and training of IEA techniques. The Second EFIEA Policy Workshop was devoted to improving communication and stimulate co-operation between different stakeholders, policy makers and scientists on the most relevant climate policy issues. The Authors of this report are grateful to Pier Vellinga and Richard Tol for many helpful suggestions and remarks. They have, however, the sole responsibility for the opinions expressed in the report and for any misrepresentation of some participants' opinion.

1. The Second EFIEA Policy Workshop was held at Fondazione Eni Enrico Mattei, Palazzo delle Stelline, Corso Magenta 63, Milan, Italy, March 4–6, 1999.
2. Let us recall that during the EFIEA Conference only the six surveys of the about 80 papers that were circulated and discussed have actually been presented. These surveys have been published in a special issue of *Integrated Assessment*.
3. A call for papers was sent to more than 1000 scientists asking them for recent or new papers which could provide an answer to the questions identified by the organising committee in co-operation with EU policy makers and advisors.
4. Coase (1960): The Coase theorem states that if trade of the externality can occur, then bargaining will lead to an efficient outcome no matter how property rights are allocated.

References

Albrecht, Johan (1998). Environmental consumer subsidies and potential reduction of CO_2 emissions. Centre for Environmental Economics and Environmental Management, University of Gent. Working Paper 98/59, October 1988.

Albrecht, Johan (1999). Making CO_2 emission trading more effective: integrating tradable certificates, Centre for Environmental Economics and Environmental Management, University of Gent. Paper circulated at the Second EFIEA policy workshop, March 1999. Milan: FEEM.

Banuri, T. *et al.* (1996). Equity and social considerations. In: Bruce *et al.* eds. Climate change 1995: economic and social dimension of climate change. Cambridge: Cambridge University Press. 79–124.

Berk, Marcel, den Elzen, M. and Metz, B. (1999). Global climate protection and equitable burden sharing – an exploration of some options. Netherlands National Institute for Public Health and the Environment. Paper circulated at the Second EFIEA policy workshop, March 1999. Milan: FEEM.

Bohm, Peter (1999). International greenhouse gas emission trading – with special reference to the Kyoto Protocol. Department of Economics, Stockholm University. Paper circulated at the Second EFIEA policy workshop, March 1999. Milan: FEEM.

Bosello, Francesco and Roson, Roberto (1999). Distributional consequences of alternative emissions trading schemes. FEEM and University Ca' Foscari of Venice. Paper circulated at the Second EFIEA policy workshop, March 1999. Milan: FEEM.

Acutt, Melinda and Elliott, Caroline (1999). Regulatory conflict? Environmental and economic regulation of electricity generation. Department of Economics and Accounting, University of Liverpool, and Department of Economics, University of Lancaster. Paper circulated at the Second EFIEA policy workshop, March 1999. Milan: FEEM.

Gerlagh, Reyer (1999). Discounting and sustainability in applied IAMs. Institute for Environmental Studies, Vrije Universiteit, Amsterdam. Paper circulated at the Second EFIEA policy workshop, March 1999. Milan: FEEM.

Gottinger, Hans. Economic damage control for greenhouse gas emissions. In: *Keiei to Keizai*, 78(2) September 1998, Economics Society of Nagasaki University.

IPCC (Intergovernmental Panel on Climate Change). Technical Assessment Reports. 1996, 1997.

Janssen, Josef (1998). Joint implementation and clean development mechanism in the natural gas sector. Institute for Economy and the Environment at the University of St. Gallen and FEEM. Paper prepared for the Fourth International Conference on Greenhouse Gas Control Technologies, organized by the International Energy Agency, ABB and the Paul Scherrer Institute, 30 August–2 September, 1998, Interlaken, Switzerland.

Janssen, Josef (1998). Strategies for risk management of joint implementation investments. In: Riemer, P., Smith, A. and Thambimuthu, K., eds. *Greenhouse gas mitigation: technologies for activities implemented jointly*. Elsevier, 357–65.

Janssen, Josef (1999). (Self-) enforcement of joint implementation and Clean Development Mechanism contracts. Institute for Economy and the Environment at the University of St. Gallen and FEEM. Paper circulated at the Second EFIEA policy workshop, March 1999. Milan: FEEM.

Hoel, Michael (1999). International trade and the environment: how to handle carbon leakage. Department of Economics, University of Oslo. Paper circulated at the Second EFIEA policy workshop, March 1999. Milan: FEEM.

Hope, Chris (1999). Stabilisation and delayed cutbacks: incorporating sulphates into the calculation of impacts of two almost identical climate change scenarios. Judge Institute of Management Studies, University of Cambridge. Paper circulated at the Second EFIEA policy workshop, March 1999. Milan: FEEM.

Manne, Alan S. and Richels, Richard G. (1999). The Kyoto Protocol: a cost-effective strategy for meeting environmental objectives? Stanford University and EPRI. Paper circulated at the Second EFIEA policy workshop, March 1999. Milan: FEEM.

Marsiliani, Laura (1999). Imperfect competition and environmental tax reforms: theory and evidence from Italian data. Tilburg University and Center for Economic Research. Paper circulated at the Second EFIEA policy workshop, March 1999. Milan: FEEM.

Marsiliani, Laura and Renström, Thomas (1999a). Inequality and environment protection: a politico-economy approach. Tilburg University and Center for Economic Research. Paper circulated at the Second EFIEA policy workshop, March 1999. Milan: FEEM.

Marsiliani, Laura and Renström, Thomas (1999b). Time inconsistency in environmental policy: tax earmarking as precommitment solution. Tilburg University and Center for Economic Research. Paper circulated at the Second EFIEA policy workshop, March 1999. Milan: FEEM.

Martin, Philippe (1997). Institute for systems, informatics and safety – EC joint research centre 'If you don't know how to fix it, please stop breaking it!' The precautionary principle and climate change. In: *Foundations of Science*. 2:263–92.

Martin, Philippe (1998). Institute for systems, informatics and safety – EC joint research centre 'New estimate of the carbon sink strength of EU forests integrating flux measurements, field surveys, and space observations: 0.17–0.35 Gt(C). *Ambio*. 27(7):1998.

Martin, Philippe (1999a). From communication to design in the European climate policy context. Institute for systems, informatics and safety – EC joint research centre. Paper circulated at the Second EFIEA policy workshop, March 1999. Milan: FEEM.

Martin, Philippe (1999b). Care, clarity and completeness: keys to highlighting costs and identifying opportunities in the context of European climate policy design. Institute for systems, informatics and safety – EC joint research centre. Paper circulated at the Second EFIEA policy workshop, March 1999. Milan: FEEM.

Mason, Pamela, Markyanda, Anil and Hunt, Alistair (1999). Measuring the indirect costs and benefits of greenhouse gas mitigation options: methodology and a case study from Hungary. University of Bath. Paper circulated at the Second EFIEA policy workshop, March 1999. Milan: FEEM.

Michaelowa, Axel and Stronzik, Marcus (1999). Early crediting of emission reduction – a panacea or Pandora's box? Center for European Economic Research (ZEW), Mannheim. Paper circulated at the Second EFIEA policy workshop, March 1999. Milan: FEEM.

Nordhaus, W.D. (1997). Discounting in economics and climate change: an editorial comment. *Climatic Change*. 37:315–28.

Onigkeit, J. and Alcamo, J. A different perspective for global climate policy: combining burden sharing and climate protection. Centre for Environmental Systems Research, University of Kassel, Germany. Paper circulated at the Second EFIEA policy workshop, March 1999. Milan: FEEM.

Proost, S. (1999). Transport, urban air policies and climate change. Center for Economic Studies, Katholieke Universiteit of Leuven. Paper circulated at the Second EFIEA policy workshop, March 1999. Milan: FEEM.

Ringius *et al.* (1997). Differentiation, leaders and fairness: negotiating climate commitments in the European Community. Center for International Climate and Environmental Research, Oslo, Norway. Report 1997–98.

Rose, Adam Z. and Casler, Stephen D. (1998). Carbon dioxide emissions in the U.S. economy – a structural decomposition analysis. *Environmental and Resource Economics*. 11(3–4):1998.

Rose, Adam Z. and Edmonds, Jae (1998). International equity and differentiation in global warming policy. *Environmental and Resource Economics*. 12:25–51.

Rose, Adam Z. (1999). Burden-sharing and climate change policy beyond Kyoto: implications for developing countries. *Environmental and Development Economics*. 3:392–98.

Rose, Adam Z., Bulte, Erwin and Folmer, Henk (1999). Long-run implications for developing countries of joint implementation of greenhouse gas mitigation. Department of Energy, Environmental, and Mineral Economics, Pennsylvania State University. Paper circulated at the Second EFIEA policy workshop, March 1999. Milan: FEEM.

Rose, Adam Z. and Collamer, Nathan (1999). The changing role of transaction costs in the evolution of joint implementation. *International Environmental Affairs*.

Rose, Adam Z. and Stevens, Brandt (1999). A dynamic analysis of fairness in global warming policy: Kyoto, Buenos Aires and beyond. Department of Energy, Environmental, and Mineral Economics, Pennsylvania State University. Paper circulated at the Second EFIEA policy workshop, March 1999. Milan: FEEM.

Shogren, Jason (1999). Benefits and costs of Kyoto. University of Wyoming. Paper circulated at the Second EFIEA policy workshop, March 1999. Milan: FEEM.

Shukla, P.R. (1999). Fairness in climate change: bridging the north–south divide. Indian Institute of Management. Paper circulated at the Second EFIEA policy workshop, March 1999. Milan: FEEM.

Tol, Richard (1998). Time discounting and optimal emission reduction: an application of *FUND*. In: *Climatic Change*. Dordrecht: Kluwer Academic Publishers.

Tol, Richard (1999). Equitable cost–benefit analysis of climate change. Institute for Environmental Studies, Vrije Universiteit, Amsterdam. Paper circulated at the Second EFIEA policy workshop, March 1999. Milan: FEEM.

Tol, Richard (1999). The optimal timing of greenhouse gas emission abatement, individual rationality and intergenerational equity. Nota di Lavoro 3.98. Milan: FEEM.

Tol, Richard (1999). Welfare specifications and optimal control of climate change: an application of *FUND*. Institute for Environmental Studies, Vrije Universiteit, Amsterdam. Paper circulated at the Second EFIEA policy workshop, March 1999. Milan: FEEM.

Tulkens *et al.* (1999). The Kyoto Protocol: an economic and game-theoretic interpretation. CLIMNEG Working Paper circulated at the Second EFIEA policy workshop, March 1999. Milan: FEEM.

List of contributors

J. Albrecht, CEEM, University of Ghent, Ghent, Belgium

J. Alcamo, University of Kassel, Germany

P. Bohm, Department of Economics, Stockholm University, Stockholm, Sweden

C. Böhringer, University of Mannheim, Germany

F. Bosello, Fondazione Eni Enrico Mattei, Venice, Italy

P. Buonanno, Fondazione Eni Enrico Mattei, Venice, Italy

C. Carraro, University of Venice and Fondazione Eni Enrico Mattei, Milan, Italy

E. Castelnuovo, Fondazione Eni Enrico Mattei, Milan, Italy

M. Galeotti, University of Bergamo, Bergamo, Italy

R. Gerlagh, IVM/Vrije Universiteit, Amsterdam, The Netherlands

A. Goria, Fondazione Eni Enrico Mattei, Milan, Italy

J.-C. Hourcade, CIRED-CNRS-EHESS, Nogent-sur-Marne, France

J. Jensen, Ministry of Business and Industry, Denmark

T. Jeppesen, Odense University, Odense, Denmark

A.S. Manne, Department of Engineering Economic Systems & Operations Research, Terman Engineering Center, Stanford University, California, USA

T. Le Pesant, CIRED, France

A. Michaelowa, Hambourg Institute for Economic Research (HWWA), Hamburg, Germany

L. Nielsen, Department of Economics, University of Southern Denmark, Odense, Denmark

J. Onigkeit, Center for Environmental Systems Research, University of Kassel, Kassel, Germany

R. Richels, Electric Power Research Institute, Palo Alto, USA

A. Rose, Department of Energy, Environmental, and Mineral Economics, The Pennsylvania State University, Pennsylvania, USA

R. Roson, University of Venice, Venice, Italy

T. Rutherford, University of Colorado, Boulder, USA

J. Shogren, Department of Economics and Finance, University of Wyoming, Wyoming, USA

B. Stevens, California Energy Commission, USA

M. Stronzik, ZEW, Mannheim, Germany

R.S.J. Tol, Hamburg University, Germany; Vrije Universiteit, The Netherlands; Carnegie Mellon Universities, USA

P. Vellinga, Institute for Environmental Studies, Vrije Universiteit, Amsterdam, The Netherlands

Index

FONDAZIONE ENI ENRICO MATTEI (FEEM) SERIES ON ECONOMICS, ENERGY AND ENVIRONMENT

This series serves as an outlet for the main results of FEEM's research programmes in the areas of economics, energy and environment.

---

1. C. Carraro and D. Siniscalco (eds.), *The European Carbon Tax: An Economic Assessment.* 1993 ISBN 0-7923-2520-6
2. C. Carraro (ed.), *Trade, Innovation, Environment.* 1994 ISBN 0-7923-3033-1
3. C. Dosi and T. Tomasi (eds.), *Nonpoint Source Pollution Regulation: Issues and Analysis.* 1994 ISBN 0-7923-3121-4
4. C. Carraro, Y. Katsoulacos and A. Xepapadeas (eds.), *Environmental Policy and Market Structure.* 1996 ISBN 0-7923-3656-9
5. C. Carraro and A. Haurie (eds.), *Operations Research and Environmental Management.* 1996 ISBN 0-7923-3767-7
6. I. Musu and D. Siniscalco (eds.), *National Accounts and the Environment.* 1996 ISBN 0-7923-3741-7
7. C. Carraro and D. Siniscalco (eds.), *Environmental Fiscal Reform and Unemployment.* 1996 ISBN 0-7923-3750-6
8. A. Beltratti, *Models of Economic Growth with Environmental Assets.* 1996 ISBN 0-7923-4032-9
9. G. Chichilnisky, G. Heal and A. Vercelli (eds.), *Sustainability: Dynamics and Uncertainty.* 1998 ISBN 0-7923-4698-X
10. R. Roson and K.A. Small (eds.), *Environment and Transport in Economic Modelling.* 1998 ISBN 0-7923-4913-X
11. A. Markandya and M. Pavan (eds.), *Green Accounting in Europe - Four Case Studies.* 1998 ISBN 0-7923-5470-2
12. A. Lanza (ed.), *Resources Accounting in China.* 1999 ISBN 0-7923-5485-0
13. C. Carraro (ed.), *International Environmental Agreements on Climate Change.* 1999 ISBN 0-7923-5515-6
14. C. Carraro and F. Lévêque (eds.), *Voluntary Approaches in Environmental Policy.* 1999 ISBN 0-7923-5516-4
15. C. Carraro (ed.), *Efficiency and Equity of Climate Change Policy.* 2000 ISBN 0-7923-6262-4

LUWER ACADEMIC PUBLISHERS – DORDRECHT / BOSTON / LONDON